General Systems Theory and Psychiatry

General Systems Theory and Psychiatry

By 28 Authors

Edited by

WILLIAM GRAY, M.D.
Chief, Mental Hygiene Clinics, Veterans Administration Outpatient Clinics, Worcester and Springfield, Massachusetts

FREDERICK J. DUHL, M.D.
Assistant Professor in Psychiatry, Tufts University School of Medicine; Director of Education, Boston State Hospital, Boston, Massachusetts

NICHOLAS D. RIZZO, Ed.D., M.D.
Consultant in Psychiatry, Phillips Academy and Abbot Academy, Andover; Director, Court Clinic, Lawrence District Court, Lawrence, Massachusetts

Little, Brown and Company
BOSTON

LIBRARY OF CONGRESS CATALOG CARD NO. 78-82924

FIRST EDITION

Published in Great Britain
by J. & A. Churchill Ltd., London
British Standard Book No. 7000 0153 0
PRINTED IN THE UNITED STATES OF AMERICA

Contributing Authors

Silvano Arieti, M.D.
Clinical Professor of Psychiatry, New York Medical College, New York

Edgar H. Auerswald, M.D.
Associate Professor of Clinical Psychiatry, Mount Sinai School of Medicine; Chief of Psychiatric Services, Gouverneur Health Services Program of Beth Israel Medical Center, New York

Warren M. Brodey, M.D.
Associate, Environmental Ecology Laboratory, Boston; formerly of Massachusetts Institute of Technology, Cambridge

Edward J. Carroll, M.D.
Clinical Associate Professor of Psychiatry, University of Miami School of Medicine, Miami

E. Joseph Charny, M.D.
Clinical Assistant Professor of Psychiatry, University of Pittsburgh School of Medicine; Staff, Western Psychiatric Institute and Clinic, Pittsburgh

Leonard J. Duhl, M.D.
Professor of Urban Social Policy and Public Health, University of California, Berkeley

Roy R. Grinker, Sr., M.D.
Director, Institute for Psychosomatic and Psychiatric Research and Training, Michael Reese Hospital and Medical Center, Chicago

Norris Hansell, M.D.
Department of Psychiatry, Northwestern University Medical School, Chicago; Department of Psychiatry, Harvard Medical School, Boston; formerly Rockford Zone Director, H. Douglas Singer Zone Center of the Illinois Department of Mental Health, Rockford

Don D. Jackson, M.D.
Late Director, Mental Research Institute, Palo Alto

Julian S. Kaiser, M.D.
Medical Director, Phillips Academy, Andover, Massachusetts

H. Peter Laqueur, M.D.
Assistant Clinical Professor of Psychiatry, Mount Sinai School of Medicine, New York; Supervising Psychiatrist, Creedmoor State Hospital, Queens Village, New York

John MacIver, M.D.
Director, Psychiatric Services, United States Steel Corporation, Pittsburgh

Judd Marmor, M.D.
Clinical Professor of Psychiatry, School of Medicine, University of California at Los Angeles; Director, Divisions of Psychiatry, Cedars-Sinai Medical Center, Los Angeles

Jules H. Masserman, M.D.
Professor and Co-chairman, Department of Psychiatry, Northwestern University Medical School, Chicago

Karl A. Menninger, M.D.
The Menninger Foundation, Topeka; Senior Consultant, Stone-Brandel Center, Chicago

James G. Miller, M.D., Ph.D.
Vice President for Academic Affairs, Cleveland State University, Cleveland

Norman L. Paul, M.D.
Assistant Clinical Professor of Psychiatry, Tufts University School of Medicine, Boston

Eugene Pumpian-Mindlin, M.D.
Professor and Associate Chairman, Department of Psychiatry, University of Oklahoma School of Medicine, Oklahoma City

Howard P. Rome, M.D.
Professor of Psychiatry, Mayo Graduate School of Medicine, University of Minnesota; Senior Consultant in Psychiatry, Mayo Clinic, Rochester

Jurgen Ruesch, M.D.
Professor of Psychiatry, School of Medicine, University of California at San Francisco; Director, Section of Social Psychiatry, The Langley Porter Neuropsychiatric Institute, San Francisco

Albert E. Scheflen, M.D.
Professor of Psychiatry, Albert Einstein College of Medicine; Acting Research Director, Bronx State Hospital, Bronx

John P. Spiegel, M.D.
Professor of Social Psychiatry, Florence Heller Graduate School for Advanced Studies in Social Welfare, Brandeis University, Waltham, Massachusetts

Montague Ullman, M.D.
Professor of Psychiatry, State University of New York Downstate Medical Center; Director, Department of Psychiatry, Maimonides Medical Center, Brooklyn

Ludwig von Bertalanffy, Ph.D.
Faculty Professor, State University of New York at Buffalo, Buffalo

Raymond W. Waggoner, Sr., M.D., Sc.D.
Professor and Chairman, Department of Psychiatry, University of Michigan Medical School; Director, Neuropsychiatric Institute, University of Michigan, Ann Arbor

Preface

DURING THE PAST FIVE YEARS a rapidly increasing interest in general systems theory has grown among psychiatrists, creating an urgent need for basic reference books on this subject. This volume focuses on the history and development of general systems theory and of psychiatric interest in this area. General Theoretical Considerations, the first part of the book, is primarily the editorial responsibility of William Gray. The second, Specific Theoretical Considerations, is under the supervision of Nicholas D. Rizzo, and the third, General Systems Theory in Action, of Frederick J. Duhl. Many of the chapters are by pioneers in the field.

In accordance with the theme of general systems theory and psychiatry, our authors, with one exception, are all psychiatrists; the exception is an Honorary Fellow of the American Psychiatric Association. We leave it to other books, now fortunately beginning to appear, to present the views of general systems theoreticians primarily identified with other fields and sciences.

W. G.
F. J. D.
N. D. R.

Acknowledgment

WE WISH TO ACKNOWLEDGE the extreme helpfulness of our wives, Lucille Gray, Bunny Duhl, and Muriel Rizzo, in preparing this book for publication. Without their steadfast support, their love, and their wisdom, we would not have been able to survive the multiple and variegated problems that have arisen over the past five years in developing the idea of general systems theory and psychiatry, in the planning and organization of programs on this subject for the American Psychiatric Association, and in the preparation of the book itself. Lucille Gray deserves our particular appreciation, for she has served throughout the years as an ever-cheerful, encouraging, and efficient executive secretary and coordinator.

Contents

Introduction

THIS BOOK IS THE FIRST RESPONSE in a conversation that begins "Just what is general systems theory and what does it have to do with psychiatry?" Conversations, as Warren Brodey has pointed out, are meaningful only if the conversationalists draw closer together in their trajectories, homing in toward some future point of contact.

Our first reply cannot help keeping our young in mind. General systems theory is not an apology for "the system," nor is it the mechanistic handmaiden of government or of the establishment, busily and unwittingly organizing various versions of hell in Vietnam, in our cities, and in our society, doing so with the speed and economy of the computer. It is not a group of valueless concepts in search of a zero-defect world. It is, rather, the conceptual product of humanistic scientists whose respect and concern is for man, and as such it is not surprising that it has been nurtured by psychiatrists. Its earliest author taught at the Menninger Clinic, and its firmest proponents did their research at the Institute for Psychosomatic and Psychiatric Research and Training in Chicago and at the Mental Health Research Institute in Ann Arbor, Michigan. So the question "Just what is general systems theory and what does it have to do with psychiatry?" is important and is what we mean to discuss here and in the collected papers in this book.

To the editors this task becomes one of trying to communicate to the reader the excitement generated in them by this new and humanistic science with its central focus on man—his species, his culture, his history and future, his ecology, and his position of continuing transactions with multiple systems larger and smaller. The markedly humanistic trends of general systems theory appear to have their origin in its high degree of generality, in its strongly global, comprehensive, relativistic and perspectivistic characteristics, and in its historical development as an expansion of scientific theory in response to the growing needs of the biological, behavioral, and psychosocial sciences. Such characteristics, while obviously clarifying the importance of general systems theory to psychiatry, make the task of defining it the difficult one of covering a very wide and rich field through a series of approximating statements.

Simple definitions do violence to its richness, while rigid definitions stifle the growing and developing nature of general systems theory. We must avoid both. The multiple defining statements will therefore be complex and flexible, as in the early part of a conversation, in order to open the way to further discussion, testing, and elaboration. Some will deal with structural, functional, or comparative dimensions; some will deal with historical, implicative, unifying, metascientific, metatheoretical, and philosophical dimensions; and others will be in the form of lists of the more essential general systems principles and properties that have evolved. Our aim in this introduction is to be as brief as possible within the limits of clarity, leaving it to the curiosity of our readers to turn to the book as a whole for more detailed, rigorous, and complete definitions. If it seems that we are being open in our approach to open systems, we are being so deliberately. We believe this is the most scientific and honest one can be, if one is to be true to the contents of this book.

Historically, general systems theory can be regarded as having two distinct origins. The first, and earlier, arose from the need to make the behavioral, biological, and psychosocial sciences "scientific," so that their insights and theories would be as valued as are those of physics and mathematics. This required an expansion of general scientific theory, and was first expressed shortly after World War II by Ludwig von Bertalanffy in the form of a demand for a "general theory of systems" and an outlining of its overall form. As this new part of scientific theory was differentiated and formulated in more complex ways, the totality of scientific theory was also changed and science itself redefined. From this point of view, then, general systems theory can be defined as the development of new forms of scientific conceptualizations, models, and laws able to deal with the organismic processes fundamental to the life sciences. With this, the requirements for an adequate scientific methodology of study began to include inquiries about structure, functioning, evolution, and reproduction.

The second and more recent origin of general systems theory resulted from the need of the more specific systems sciences for a broader theory, a theory of interdisciplinary work. The same trend was represented somewhat earlier as an urgent need to find some way of unifying science in an age of explosively increasing and more intensive specialization. Thus the physical and mathematical sciences on the one hand, and the behavioral, biological, and psychosocial sciences on the other, needed to define their relationship to each other; and it is these relationships that are added to the total theory of science. General systems theory evolved not as a compendium of sciences but as a linkage between them. As soon as the sciences are brought together, the interfaces between them gain major significance.

From this point of view general systems theory can be seen either as a development of *theories about generalized systems* or as a development of *generalized theories about systems*. The more specific systems sciences—which include cybernetics, information theory, communication theory, systems engineering, operations research, computer science, management science, game theory, and decision theory among others—have elaborated markedly in the past fifteen years. Their development was powerfully stimulated by Wiener's formulation of cybernetics, by Shannon and Weaver's formulation of information theory, and by the urgent engineering demands created by increasingly complex production processes.

We must underline an important aspect of general systems theory at this point. If the focus of a generalized theory is only upon the interconnections, as it is in the systems sciences, the tendency is to evolve a mechanistic model. Similarly, if the focus is only upon the specific disciplinary sciences, as was the case previously, the models become limited and isolated. We would insist that the systems sciences do not supplant, and should not be thought of as supplanting, the disciplinary sciences, but rather as enriching and enlarging them. *General systems theory, by including both the systems sciences and the disciplinary sciences and by its concern for their integration, provides a new type of science based on organismic and open models in which humanistic values are a necessary part.*

In viewing general systems theory as a development in its own right, we note that it has as its immediate theoretical precursors organismic theory, the theory of open systems, and gestalt theory. The concept of *organism* became generalized to that of *organized systems* with the recognition that being, acting, and becoming are fundamental aspects of all organized systems. Open systems, with their properties of neg-entropy, equifinality, and steady state, are incorporated in generalized form in general systems theory. One may say, further, that general systems theory evolved as a scientific transformation of holistic theory through the addition of the modern concept of system.

In using the term *system,* we mean a set of objects together with the relationships between the objects and between their attributes. The etymology of *system* is clear; the root word simply refers to that which either holds together or falls together. Anatol Rapoport adds the definition of a system as "a whole which functions as a whole by virtue of the interdependence of its parts." He points out that the aim of general systems theory is to discover how this is brought about in the widest variety of systems [2].

General systems theory may be understood from several vantage points. Metascientifically, it offers a new scientific world view, replacing the previous one of mechanism with that of organized complexity. Metatheoretically, general systems theory offers a framework for the more specific system theories, allowing for a clearer definition of the orientation of the observer-scientist. It may be considered to serve as a science of systems, or as a system of systems, thereby providing the framework for relating otherwise overwhelming collections of isolated facts and theories. Thus many see the usefulness of general systems theory in terms of its bridging functions, in which its cross-disciplinary language, its strongly convergent trends, and its emphasis on isomorphism play large roles. In these ways it offers potential for lessening the gap between the biological and the mental, the social and the cultural, the specialist and the generalist, and between theory and practice. In its open, organismic, and evolutionary qualities it provides a most congenial framework within which to consider problems of growth and change.

We must underline the value of general systems theory in terms of its perspectivistic qualities—that is, its insistence that no world view, its own included, is ultimate truth or ultimate reality. In terms of its concepts of growth and development, it foresees its own eventual replacement by more adequate and relevant theories as they develop.

Many recurrent themes in general systems theory will be found in the various chapters in this volume, and are referred to as general systems principles. A partial list includes open system, steady state, isomorphism, negentropy, anamorphosis, hierarchical organization, equifinality, growth, negative and positive feedback, process, information, matter, information processing, matter-energy processing, transaction, cross-level transaction, regulation, maintenance in change of component elements, goal-directedness, steering and trigger causality, multivariable dynamic interaction, progressive differentiation, progressive centralization, progressive mechanization, and progressive deanthropomorphization.

This quick summary cannot provide a sufficient understanding of the value of general systems theory, which goes beyond the mechanistic, utilitarian, and factory concepts of systems to a humanistic one, beyond the craft state of knowledge of systems to a scientific one, beyond isolated views to unified ones, and beyond the consideration of parts to the consideration of wholes. The static is surpassed in the theory's recognition of the temporal quality of life and of the omnipresence and inevitability of change. The analytical method of earlier science gives way to the development of a science and mathematics both analytical and synthetic, and applicable to wholes. The interdisciplinary approach is enriched by a focus on interface, relationship, and process, while concepts of interaction develop into those of transaction. Homeostasis is expanded to a consideration of steady states and heterostasis. Most valuable is its human orientation, focusing on man as a system ecologically suspended in multiple systems. Finally, general systems theory goes beyond previous theories and models in its high degree of relevance for our modern world with its continually increasing complexity and change. It is by its relevance to our present world that general systems theory must stand or fall—as must this volume.

By serving as a general theory for the more specific system theories of psychiatry, general systems theory brings the highly developed biological and intrapsychic system theories of psychiatry into transaction with their relevant ecological systems. In going beyond the mechanistic and stimulus-response models of man to that of man as an active personality system, it contributes immensely in preserving the humanization of psychiatry. By its inclusion of concepts of subsystem integration and of the hierarchical organization of systems, it makes possible the relevance to psychiatry of theories of groups, families, societies, and cultures, and brings them into transaction in the form of *general systems behavior theory*, which is gaining increasing recognition as *the* theory of community psychiatry. Since its terms are psychophysically neutral, it has great promise for bridging the basic mind-body dichotomy of psychiatry. Equally important, it promises to span the gap between the humanities and sciences, thus making it possible to continue to build psychiatric theories that are both humanistic and scientific.

This, then, is the meaningfulness of general systems theory to the psychiatrist and the reason for this book. The changing psychiatrist of our day, the Absurd Healer of Matthew Dumont [1], must deal with his puzzlement by reaching out for a theory that does not supplant his understandings but enlarges them. He, too, like all of life, must grow, change, differentiate, and reorganize to adapt to his ecological world. This, we propose, is why he needs to understand general systems theory. The crisis of our generation is that what we, the old, must yet learn the young already understand but cannot make operational. The crisis of psychiatry is that it needs to find new concepts—concepts the young already know—to survive as a competent, scientific, and humanistic servant of man. We of the older generation must struggle to make these concepts clearer to ourselves so that we do more than survive, so that we may act to serve man and to serve him well.

Now let our conversation continue.

References

1. Dumont, M. *The Absurd Healer.* New York: Science House, 1968.
2. Rapoport, A. Foreword. In Buckley, W. (Ed.), *Modern Systems Research for the Behavioral Scientist*. Chicago: Aldine, 1968.

General Systems Theory and Psychiatry

PART I

GENERAL THEORETICAL CONSIDERATIONS

IN PART I our focus is on the changes in general theory that occur both in general systems theory and in psychiatric theory as they enter into a transactional relationship with each other. Our authors are particularly well suited to this task, having gone beyond their actual occupations—as teachers, researchers, practitioners, editors of basic textbooks and periodicals, and developers and directors of important institutions—to positions of eminence as innovative and creative general theoreticians. They have infused a new vitality and relevance into their fields, enlarged and enriched basic theory in revolutionary ways, and brought these new generalized theories into effective transaction with their fellow inhabitants of the highly populated universe of scientific theories.

With one exception our authors are all psychiatrists. However, this exception, Ludwig von Bertalanffy, to whose genius and courage we are indebted for the original conception of general systems theory and for much of its subsequent development, has also for many years been a psychiatrist by identification in terms of the direction of his work and his concerns, and he is now a psychiatrist by assimilation, having been elected an Honorary Fellow of the American Psychiatric Association at its 1967 annual meeting.

What we present in this book is therefore biased, for it is the view of psychiatric general theoreticians on the changes taking place in general systems theory and in psychiatric theory and on the implications of these changes. We leave it to other books, now fortunately beginning to appear, to present the views of general systems theoreticians from other fields and sciences. Our selectivity, however, has not led to a book valuable only to specialists. This is because of the particular character of our authors, whose work has carried them beyond the boundaries of their own fields and whose views are profoundly interdisciplinary, multidisciplinary, and cross-disciplinary, and because it is the special attribute of general systems theory that it makes possible a new and vibrant unification of science, so that the views of general systems theorists, whatever their home disciplines, have widespread implications and applications.

The first paper aims at providing a historical orientation for the reader, a view of the changing cultural ecology in which general systems theory arose, in which it has developed, and in which its relevance for psychiatry has emerged. The fundamental scientific, humanistic, and philosophical issues with which it deals are outlined. As far as we know, it is the first historical paper on general systems theory to be published. Its authors, who are also two of the three editors of this volume, share with the other contributing authors a wide variety of interests and activities, and in particular have had a long background in the relationship of general systems theory and holistic psychology to psychiatry. William Gray is one of the early members of the Society for General Systems Research and a past president of the Boston Systems Group. He has been primarily responsible for the organization of meetings on general systems theory and psychiatry for the American Psychiatric Association. Nicholas D. Rizzo, who holds doctorates in both education and medicine, began his interest in general systems theory and holistic psychology in his undergraduate days at the University of Kansas, where he worked closely with Raymond W. Wheeler, the American pioneer in holistic psychology. Much of the credit for the organization of meetings within the American Psychiatric Association belongs to Dr. Rizzo, to our third editor, Frederick J. Duhl, and to many of the contributors to this volume.

The second paper, which Dr. von Bertalanffy presented at the 1967 annual meeting of the American Psychiatric Association when he was elected an Honorary Fellow, is a precipitous plunge into the real issues and challenges to which this book is addressed. It warns us that our time as effective human beings and behavioral scientists is limited, as is the time of the world as an effectively human ecology, unless we seriously and creatively face the basic thesis of this book, that the theory of all the behavioral sciences has been and continues to be mechanistic, and therefore capable only of producing more excellent or efficient robots. The urgent challenge is that of changing our world view from one of mechanism to one of organized complexity and of employing models that are open, organismic, and humanistic.

From the time of his first public presentation of the concept of general systems theory, in 1945, Dr. von Bertalanffy's genius has continued to enrich, enlarge, and further delineate this rapidly growing new field. In the breadth and variety of his interests and in the freshness of his views, Dr. von Bertalanffy is truly a modern Renaissance man. After earning his Ph.D. at the University of Vienna, his teaching, research, and productivity have carried him to institutions in many countries and have made him a world-renowned figure.

Because of the importance of Dr. von Bertalanffy's paper to psychiatry, and because of the importance of their own contributions to general systems theory and psychiatry, the discussions of von Bertalanffy's paper by Silvano E. Arieti and Karl A. Menninger are included. Both men are too well known in psychiatry to require details on their massive and wide-ranging contributions. A paper by Dr. Arieti will be found in Part II. We deeply regret, however, that technical considerations have made it impossible to include sections from Dr. Menninger's *The Vital Balance* [2], one of the earliest and most important textbooks on general systems theory and psychiatry.

James G. Miller, the author of our third paper, is one of the earliest and firmest proponents of general systems theory and psychiatry. As Director for over ten years of the Mental Health Research Institute at Ann Arbor, Michigan, he has been in charge of the most complete program of research on general systems theory in the area of living systems that has yet been carried out. He is also responsible for introducing the term *behavioral science* in his earlier work at the University of Chicago. Here we present a remarkably brilliant account of the basic general theoretical considerations of general systems theory as they relate to the living subset of all systems, what Miller calls "general systems behavior theory." Dr. Miller's work is appropriately gaining rapid recognition as the basic theory of community and social psychiatry.

Our fourth author, Roy R. Grinker, Sr., is widely recognized as the most prominent American author on general systems theory and psychiatry and shares with Miller the position of earliest and firmest proponent of the field. Grinker's position as one of the great general theoreticians of psychiatry is illustrated by his recognition in the 1940's that psychiatry stood in danger of running out of intellectual capital because of the narrowness of its basic theory. Coupling this with his exceptional skills as leader, teacher, and organizer, he was able to arrange a series of four multidisciplinary conferences, held in Chicago in the early 1950's, to investigate the possibilities of developing a unified theory of human behavior. These conferences, whose participants included Jurgen Ruesch, John P. Spiegel, and Lawrence K. Frank among others, represent one of the basic turning points in American psychiatry. They are available in *Toward a Unified Theory of Human Behavior*, edited by Dr. Grinker [1]. The new Introduction by Grinker and Epilogue by Jurgen Ruesch, appearing in the second edition, are classic contributions to the general theoretical implications of the increasing transaction between general systems theory and psychiatry. We deeply regret that technical considerations have prevented us from including them in this volume. However, the paper of Dr. Grinker's that appears here will give the reader a taste of the richness and importance of his thinking, and should encourage the study of *Toward a Unified Theory of Human Behavior* as essential background.

Jurgen Ruesch, our fifth author, is the most prominent proponent of the importance of information and communication theory for psychiatry, and is recognized as the leading American writer in this area. He is primarily a general theoretician, focusing on the revolutionary changes in psychiatric theory that occur as we enlarge our horizons. His many papers, required reading for the modern psychiatrist, include the classic paper on the "double-bind," one of his earliest contributions, coauthored with Gregory Bateson and Don D. Jackson. In this volume we include one of Dr. Ruesch's most recent papers, "A General Systems Theory Based on Human Communication," presented at the 1966 annual meeting of the American Psychiatric Association as part of the initial program on general systems theory and psychiatry. In an editorial entitled "The Old World and the New" [3] Dr. Ruesch describes with awesome clarity the transition of our world view from one of "person-orientation" to one of "system-orientation." While we agree with Dr. Ruesch that this is a possible outcome, we do not agree that it is necessarily so. It is our hope that general systems theory, as distinguished from systems theory, will be one of the main forces preventing such an outcome.

Our sixth contributor, Albert E. Scheflen, is another early worker in general systems theory and psychiatry whose contributions to the field are of both general and specific theoretical nature. We consider his paper "Systems and Psychosomatics" one of the clearest expositions yet written on the widespread and fundamental changes in general theory that result from the application of the general systems point of view. Dr. Scheflen's contributions will be discussed more fully in Part II, where another paper of his appears.

Our seventh and final contributor to Part I, Jules H. Masserman, shares with the others a position of greatness in American psychiatry as teacher, researcher, and innovator, but primarily as a general theoretician responsible for fundamental advances in psychiatric theory. He is singular in the richness of his activities, interests, and concerns and in his ability to transact effectively with a great many areas for the benefit of psychiatry. His recent activities in organizing programs on the arts and humanities and on transcultural psychiatry for the American Psychiatric Association have been outstanding; his many papers on enlarging and expanding the basic general theory of psychiatry are classics. His contributions enhance, complement, and parallel those more directly derived from general systems theory, as will be seen in his paper appearing in this volume.

W. G.

References

1. Grinker, R. R., Sr. (Ed.) *Toward a Unified Theory of Human Behavior* (2d ed.). New York: Basic Books, 1967.
2. Menninger, K. A., with Mayman, M., and Pruyser, P. *The Vital Balance.* New York: Viking, 1963.
3. Ruesch, J. The old world and the new. *Amer. J. Psychiat.* 124: 225, 1967.

1

History and Development of General Systems Theory

WILLIAM GRAY, M.D., AND NICHOLAS D. RIZZO, ED.D., M.D.

GENERAL SYSTEMS THEORY is a bold new development in human thought which has brought into the range of the scientific approach a great number of complex fields previously unapproachable through classic science. It is an answer to Alfred North Whitehead's warning, made in 1920, that the intellectual capital of science was running out so long as it remained exclusively mechanistic in its approach [81]. General systems theory is a continually evolving body of ideas, increasingly useful to large numbers of persons in diverse scientific disciplines, and located in widespread geographic areas. In the words of its primary author, Ludwig von Bertalanffy, a leading contemporary biologist with wide-ranging scientific and cultural interests, "General Systems Theory is a new discipline whose subject matter is the formulation and derivation of those principles which are valid for systems in general [63]." A system is defined by von Bertalanffy as sets of elements standing in interaction. General systems theory is a logical-mathematical field which deals with the new scientific doctrines of wholeness, dynamic interaction, and organization. It is a new approach to the unity-of-science problem which sees organization rather than reduction as the unifying principle, and which therefore searches for general structural isomorphisms in systems.

General systems theory is thought by Kenneth Boulding to consist of theoretical model building which lies at a level somewhere between the highly generalized constructions of pure mathematics and the specific theories of the specialized disciplines. One of the main objectives, then, of general systems theory is to deal with the critical problem of loss of relevant communication between scientific specialists, and as such to become a theory of interdisciplinary scope. Boulding considers the quest of general systems theory to be that of building a body of theoretical constructs which will discuss and describe the general relationships of the empirical world, and which, in so doing, will develop something like a spectrum of theories, or a system of systems, or a metatheoretical framework which may perform the function of a gestalt in theoretical construction [10].

General systems theory is considered by Anatol Rapoport to be a methodology which emphasizes those aspects of objects or events derived from the general properties of systems. He views general systems theory as having received its historical impetus from two sources: the first, the realization of the inadequacy of mechanism as a universal model in science; the second, a tendency to counteract the fragmentation of science into isolated specialties. To Rapoport, general systems theory is an effort to fuse the mechanistic and organismic approaches and so to reintroduce into the theory of physical processes such teleological notions as purpose and goal-seeking behavior, which become then no longer truly teleological. He considers the development of general systems theory to be based on recognition of the fundamental distinction between isolated and nonisolated systems formulated by von Bertalanffy in his theory of open systems [49].

In the view of W. Ross Ashby, the emergence of general systems theory is symptomatic of a new movement that has been developing in science during the past decade, in which science is at last giving attention to systems that are intrinsically complex. Ashby distinguishes between two main methods of study that general systems theory can follow in intact, highly complex systems. The first, an empirical intuitive approach which has been highly developed by von Bertalanffy and his co-workers, examines the various systems that occur in the world as we find it; the second is a deductive systems theory, followed by Ashby and his co-workers, in which the set of all conceivable systems is initially considered and then reduced to sets of more reasonable size [6].

It is most important to gain an understanding of the history and development of a concept as seminal as that of general systems theory. Although it is an evolutionary development in the history of human thought and therefore the product of many minds, its historical development can best be portrayed and understood in terms of the thinking of Ludwig von Bertalanffy,* who first gave the theory its name.

Early Phases

General systems theory was not elaborated by von Bertalanffy all at once, but rather after a long and interesting evolution of views which included two concepts in particular as the chief precursors of general systems theory. These were the theory of organismic biology, about which he wrote in 1928, and the theory of open systems, presented in 1932. Von Bertalanffy did not publish his concept of general systems theory until 1945, at the conclusion of World War II, although he had developed the idea in the mid-thirties and had presented it at various lectures, most notably in a Charles Morris Philosophy Lecture at the University of Chicago in 1937. He explains the delay in publication as being the result of fear of overwhelming criticism by the classical and orthodox biologists of the day. On publishing, he discovered that an interesting and surprising change in intellectual climate had taken place, making model building and abstract generalizations fashionable, and that a number of scientists were following similar lines of thought [62].

The climate of the times in which von Bertalanffy began his work was one in which physics reigned as the queen of science, and in fact was considered to be the only science. The theory of mechanism was looked upon as the only scientific approach. It was the Laplacean ideal which resolved the world into an aimless play of atoms governed by the laws of chance, with the future entirely determined by initial conditions. But the climate of the times also included an air of ten-

* Born in Austria in 1901, Ludwig von Bertalanffy received a Ph.D. from the University of Vienna and taught there until 1949 when he migrated to Canada to become Professor of Biology at the University of Ottawa. He was a Fellow at the Center for Advanced Study in the Behavioral Sciences in 1954–1955 and then became Director of Biological Research at Mt. Sinai Hospital, Los Angeles, California, and Visiting Professor at the University of Southern California from 1955 to 1958. In 1958–1959, he was Sloan Visiting Professor at the Menninger Foundation. Since then he has been at the University of Alberta, where he has occupied the chair of Theoretical Biology, has been a founding member of the Center for Advanced Study in Theoretical Psychology, and has, since 1968, been awarded the position of University Professor. At the 1967 Annual Convention of the American Psychiatric Association, von Bertalanffy was voted into Honorary Fellowship of the American Psychiatric Association in view of his many distinguished contributions.

sion in the scientific world as, more and more, the study of systems of greater complexity was undertaken—that is, the systems of biology and the behavioral sciences—and it was found that these could not be understood through use of the mechanistic approach. It became increasingly apparent that such systems were not additive, and that understanding the whole could not be accomplished simply by understanding the various parts and then adding them together.

The fervor of the adherents to the mechanistic approach can be understood in terms of the previous history of science, in which Aristotelian teleology had blocked scientific advance until it was replaced by the Galilean concept of mechanism. The concept of mechanism made possible the development of science and led to its flowering in the nineteenth and early part of the twentieth centuries. But biological and behavioral explanations, in which the concepts of goal or purpose seemed necessary, were viewed with great fear, as if the danger of a return to prescientific Aristotelian teleology might result. This led to a conflict in intellectual and scientific circles, which von Bertalanffy has described as a "mechanistic-vitalistic conflict" [65].

MECHANISTIC-VITALISTIC CONFLICT

The mechanistic-vitalistic conflict was typified by the argument between those who insisted that the teleological type of systems found in biology and the behavioral sciences could somehow be explained in a mechanistic way, thus preserving the validity of the reductionist scientific approach, and those who insisted that the teleological, goal-directed systems of biology and behavioral science could be understood only by the assumption that a vitalistic element was present. Vitalism depended upon the action of a higher intelligence, and therefore was beyond scientific understanding. General systems theory resolves this controversy by expanding the mechanistic approach so that teleological systems could be dealt with and understood in terms of physical properties. Thus, the validity of the scientific approach was maintained, and vitalism became an unnecessary assumption.

It is possible to capture some of the spirit of the times when this controversy occurred by describing Hans Driesch's famous experiments of the 1890's [17]. Driesch's work and the conclusions he drew from it played a central role in stimulating von Bertalanffy's thinking in the direction of solving the apparently insoluble paradox of a scientific approach to teleological systems. In his experiment, Driesch noted that a sea urchin embryo divided in half developed into a whole sea urchin larva, smaller than usual, but normal and complete, instead of developing into the half sea urchin that one might expect. The strange result of this experiment was explained by Driesch by the principle of equifinality, an equifinal event being one in which the same goal is reached from different starting points and in different ways. Since, according to the classical mechanistic science of the day, a final state is unequivocally determined by the initial conditions, Driesch concluded that equifinality contradicts the laws of physics and could be accomplished only by a soullike vitalistic factor which governed the processes in foresight of the goal, that is, the normal organism to be established. This conclusion was nothing less than a declaration of the bankruptcy of traditional science. Driesch's choice of a nonscientific explanation, in which he assumed the existence of entelechy, was logical in many ways and made the paradox facing the scientific paradigm of the day extremely clear. In this sense Driesch made an important contribution to the evolution of modern scientific thought. It is the resolution of the Driesch paradox and its challenge to scientific thinking that constitutes one leading aspect of the history and development of general systems theory.

ORGANISMIC THEORIES

A second factor in the climate of the times influencing the development of general systems theory consisted of the parallel development of organismic theories in a number of fields. These include H. Werner's studies of the comparative psychology of mental development in which he formulated an organismic psychology, von Bertalanffy's studies of modern theories of development in which he proposed an organismic biology, Cassirer's work on the philosophy of symbolic forms, and Piaget's studies on language and learning.

The organismic view insists on studying not isolated parts of processes, but the organizing relationships themselves that result from dynamic interaction and make the behavior of parts different, when studied in isolation, than when studied within the whole. These parallel developments in various fields toward an organismic approach are even more dramatic when one considers that they developed independently.

The meaning of organismic biology was clearly that living organisms are organized entities and that biologists must deal with them as such. Organismic theory emphasized the idea that notions which did not appear in conventional physics were characteristic of organization, whether dealing with a living organism or a society, and that such notions included those of wholeness, growth, differentiation, integration, hierarchical order, dominance, control, competition, centralization, leading part, finality, equifinality, and others.

DEGRADATION-EVOLUTION PARADOX

A third aspect of the intellectual climate of the times in which von Bertalanffy began his work was the increasing awareness of the contrast between inanimate and animate nature. It could be seen in the violent contradiction between Lord Kelvin's degradation and Darwin's evolution, that is, the contrast between the laws of dissipation in physics and the law of evolution in biology. According to the second principle of thermodynamics, the general trend of events in physical nature is toward states of maximum disorder and leveling down of differences, with the so-called heat death of the universe as a final outcome. In marked contrast to this, the living world shows in embryonic development and in evolution a transition toward higher order, heterogeneity, and organization. This contradiction was a second source of support for the vitalistic arguments.

Von Bertalanffy's theory of open systems and steady states gave an answer to the degradation-evolution paradox, once again undercutting the vitalistic argument. Von Bertalanffy had pointed out the extremely important observation that no physical theory of open systems existed at the time and that the principles of thermodynamics, particularly the second law, would require expansion as well as modification in order to be applicable to the operation of living systems. He emphasized that living organisms are essentially open systems, maintaining themselves in a continuous state of inflow and outflow, building up and breaking down components, never being, so long as they are alive, in a state of chemical or thermodynamic equilibrium, but rather in a steady state of balanced tension.

With the theory of open systems, the apparent contradiction between entropy and evolution disappeared. In open systems there is not only the production of entropy as in closed systems, but also the import of entropy which may be negative. In the case of the living organism, complex molecules high in free energy, and so negentropic, are imported. Living systems, therefore, can avoid an increase in entropy, and even more may develop toward higher states of order and organization [63]. Ernst Schroedinger made this essential point effectively and picturesquely in his famous statement that life feeds on negative entropy [53].

UNIFICATION OF SCIENCE

A fourth factor in the intellectual climate which led to the formulation of general systems theory was concern for the increasing lack of unity among the sciences. Boulding notes that the crisis in science arose because of the increasing difficulty of profitable talk among scientists as a whole, in view of the high order of specialization and in view of the specialized deafness that resulted therefrom. One of the main objectives of general systems theory has been to develop "generalized ears," that is, to develop a framework of general theory which would enable one specialist to understand relevant communication from others [13]. The integrative potential of general systems theory is one of its most important characteristics. It contributes to the unity of science in a more realistic fashion than did previous reductionist attempts. It constitutes a modern approach to the unification of science and is based on the isomorphism of laws in different fields. Von Bertalanffy calls this view "perspectivism" and comments that the unifying principle is that of organization. What is known, with respect to national economy, as Pareto's law of distribution of income within a nation is represented in biology by the law of allometric growth, describing the relative increase of organs, chemical compounds, or physiological activities with respect to body size [61].

Similar isomorphism can be found in relation to competiton and related phenomena. Thus the conceptual models for phenomena such as the struggle for existence (developed in the work on population dynamics by Volterra, Lotka, Gause, and others) not only have been of great practical use to the fishing industry, leading to their adoption by the Food and Agriculture Organization of the United Nations, but can be applied also in areas as diverse as econometrics, the study of armament races, and laws governing business cycles [65].

In several publications on problems of mass diffusion, Rapoport discusses differential equations that govern diffusion in chemical reactions and emphasizes that such equations may well serve as mathematical models of such apparently diverse systems as the spread of information, the spread of rumor, divorce rates, accident rates, and others. He also points out that the variables in such equations might represent populations of several species of organisms, and that they may therefore represent a rough model of ecological systems as well as chemical ones [48, 49]. Boulding has made the theme of isomorphism of laws of growth a particular area of emphasis, and comments that "the remarkable universality of principles enunciated here with regard to a general theory of growth indicate that perhaps there is emerging from the welter of the sciences something like a general theory of growth." He adds that it is helpful to the economist to know that the equations he uses are isomorphic with the field equations of modern physics [13, 14].

Recent Phases

The recent contributions of Ludwig von Bertalanffy to general systems theory include principles not previously mentioned and tending to be of particular importance in the biological and behavioral sciences. The first, *progressive differentiation,* is the primary process through which biological wholes achieve higher states of organization, in contrast to physical wholes in which higher organization appears to result from the union of preexisting elements. Progressive differentiation implies another principle, that of *progressive segregation,* exemplified by the living organism as it passes from a state of undifferentiated wholeness to one of differentiated parts. Since this implies that parts become fixed with respect to certain actions, a third principle, that of *progressive mechanization,* must be added to our list. The principle of progressive mechanization leads to yet another important development, the creation of *leading parts,* in which, because the system becomes centered around a particular part, a small change in this part will be amplified throughout the total system. The small part has then become a zone of *trigger action.* In such cases, instead of *conservation causality,* in which the effect is equal to the cause, one finds *instigation causality,* in which an energetically insignificant change in one part of a system causes a considerable change in the total system. The time-dependent evolution of leading parts results in *progressive centralization,* and since an individual can be defined as a centralized system, *progressive individualization* also follows. Thus, leading parts, trigger action, instigation causality, progressive centralization, and progressive individualization are the recently developed principles of general systems theory. *Progressive stratification* and *hierarchization* must also be added, since they are implied by progressive centralization. In most living systems centralization and hierarchical order are achieved by stratification, that is, the superimposition of higher layers which take the role of leading parts. Further, *progressive organization* must be added to this newer list of general systems concepts, since it is a necessary consequence of progressive differentiation if integration is to be maintained [64, 68, 70, 71].

Sufficient further clarification of previously developed general systems principles has occurred to warrant their inclusion in the present list. Thus, *anamorphosis*, or the spontaneous transition toward higher order, is now recognized as a principle applicable to all living organisms and to certain other open systems not in the biological area. General systems theory views the living organism as an open system with autonomous activity and anamorphosis. This type of model makes more clear the intrinsic potential of living systems for growth and development and for creativity [64]. The origin of anamorphosis remains one of the major problems facing general systems theory.

Further developments in regard to the concept of *equifinality* have led to the concept of *equifinal steps* or *equifinal levels*, which should be included in this newer list of general systems principles. Equifinality is characteristic of open systems tending toward a steady state. Equifinal phases are reached from different starting points and are maintained for an interval until anamorphic pressures evoke new development, and with it a new equifinal phase. Characteristically such equifinal levels are relatively stable systems. These concepts are isomorphic with those of *step function*, introduced and developed by Ashby.

The general systems principles limited to and defining human systems are those of *symbolism* and *progressive deanthropomorphization*. Von Bertalanffy has commented that man lives in a universe not of things but of symbols—a symbolic world made up of language, thought, social entities, money, science, religion, art forms, and the world of objects around him. Increasingly these objects represent the materialization of his own symbolic activities.

In his most recent book von Bertalanffy proposes a new image of man, seeing him as more than a machine or a robot, and stressing the human character of man as contained in his concept of man—a psychophysical organism with a capacity for spontaneous activity and the power to change goals and to build symbolic structures. Equally momentous is von Bertalanffy's insistence that science itself must become an open system so that a new natural philosophy relevant to the new image of man will develop. General systems theory in its metascientific aspects promises to create a revolution in the conceptual orientation of science as did Copernicus, Darwin, Freud, and Einstein.

Further fundamental contributions to general systems theory have come through the work of the Mental Health Research Institute at the University of Michigan under the leadership of its former director, James G. Miller. It was there that the first interdisciplinary group was formed to undertake a systematic general systems approach to the study of biological and behavioral phenomena of relevance in the areas of mental health and mental illness. Its research has systematically been conducted on normal and abnormal behavior at various levels: on molecules and cells, organ systems, individuals, groups, and social institutions, all in an effort to understand not only what makes a particular living system behave as it does, but also to find general principles or isomorphisms that apply to the different levels of behavior. The Institute's members have coordinated their research projects into five areas: biological sciences, psychological sciences, systems sciences, societal sciences, and interdisciplinary sciences, and again here have pioneered in giving as much weight to the systems sciences as to the more traditional disciplines. The work of Miller and his collaborators on *information input overload* at each level is a classical contribution to a fundamental aspect of systems theory that had previously been neglected. More recently, William Bolman has proposed the application of Miller's general systems behavior theory as a model for community mental health activities [11]. One of Miller's papers describing his general systems theory of behavior is reprinted in this volume [47].

A rich source of fundamental contributions to general systems theory has been the work of Roy R. Grinker, Sr., of Chicago, Jurgen Ruesch, John P. Spiegel, Lawrence K. Frank, and their associates, drawn from nearly all disciplines whose chief concern is understanding human behavior. Four conferences were held under Grinker's leadership, beginning in 1951, with the purpose of developing leads toward the unification of behavior theory. The report of the symposia was published in 1956 under the title *Toward a Unified Theory of Human Behavior* [34], and republished in a second edition in 1967 with a new introduction by Roy R. Grinker, Sr., and a new epilogue by Jurgen Ruesch [35]. Without the prior knowledge of Grinker and his colleagues their work had won recognition from those working in general systems theory as a classical contribution to the field. Thus, in a recent survey, the report of the Chicago symposia was ranked second only to the various volumes of *General Systems* as a source book for general systems concepts [84]. Perhaps the most important contribution of the Grinker symposia is that their publication made available for the first time to those interested in or working in the area of general systems the thinking, the frames of reference, and the concepts of a group of very able men from a number of disciplines, all interested in the possibility of developing a unified theory, and all willing to work toward this goal. That it was possible to bring together such a large group of highly competent individuals to transact effectively with one another is a high tribute to Grinker's organizational and leadership qualities.

Grinker has stated that three unifying principles evolved from the conferences. These were (1) a principle of stability designated as homeostasis, (2) a principle of reciprocating relationships designated as transactional, and (3) a principle of the universality in all systems of processes of communication and information. Grinker's introduction to the second edition of *Toward a Unified Theory of Human Behavior* is enlightening for the student of the historical development, significance, and implications for the future of general systems theory. The indications of ferment and change during the past twenty years are considered to be a third revolution in the field of psychiatry. Depending upon their special interests, some have designated the development of group therapy, others the development of community psychiatry, as the third revolution. Grinker, however, considers the slow shift in theory toward a tendency to develop more general, all-encompassing, or unified theories of human behavior as the momentous fact of the third revolution. Although social scientists other than psychiatrists had studied, understood, and applied systems theory, it was not until 1966 that suddenly, under the leadership of William Gray of Boston, a threshold was reached, and at the

122nd annual meeting of the American Psychiatric Association two sessions were held at which this theory was presented, with the result that psychiatrists were introduced to general systems theory and could participate in its further development.

Grinker has distinguished clearly the necessary interdependence of global and particularistic theories. In and of themselves global theories are not operational, but they may become extraordinarily fruitful when they act as umbrellas encompassing subordinate operational theories. Grinker, as well as many others, has emphasized that it was von Bertalanffy's genius to bring together such particularistic theories into a harmonious global theory which spoke out against the physicochemical model for biology, psychology, and sociology.

Jurgen Ruesch has stressed that the speed with which the former person-oriented world was converted into a system-oriented world was not anticipated. In view of this change, task orientation has become dominant. The modern approach to man and behavior deals with total systems, and is accompanied by profound alterations in scientific thinking and human ethics [35]. Ruesch has warned of the danger in certain modern, computerized approaches that deal with man as an impersonal, replaceable, and expendable unit in their attempt to adapt the world to system design. In contrast, one of the strongest characteristics of general systems theory is its insistence on going beyond mechanism and focusing on the living, human aspects of systems.

Further contributions of pervading importance are John Spiegel's concept of foci in a transactional field and those of Lawrence K. Frank on organized complexity. Frank has insisted that our current political theory, which was formulated in the late eighteenth century, is in urgent need of replacement by a general systems theory encompassing organized complexities, open systems, cybernetics, and self-organizing systems. Another fruitful approach has been provided by cybernetics as formulated by the late Norbert Wiener.

Cybernetics has frequently been misused by limiting its application to static entities and inertial systems. Also, feedbacks have frequently been misinterpreted as externally applied forces. Frank, however, emphasizes their role as probes, to evoke from the environment data necessary for correction and direction of coordination and growth. Ashby's concepts of step functions support this view [18].

Important precursors to the Grinker symposia were the interdisciplinary symposia sponsored by the Josiah Macy, Jr. Foundation in the 1940's and early 1950's [72–76]. The early conferences were devoted to the study of teleological mechanisms and circular causal systems, with the topic being shifted to cybernetics following the publication of Norbert Wiener's *Cybernetics* in 1948 [82].

The first general systems theory–oriented textbook of psychiatry, *The Vital Balance* by Karl A. Menninger, published in 1963 [44], skillfully blends psychiatric theory with general systems theory. Psychopathology is presented in systems-theoretical language, with mental illness being the result of various degrees of dedifferentiation or disorganization of the personality. In order to emphasize the reversibility of certain states, Menninger makes free use of the prefix *dys-*. He has classified emotional disorders into five categories of increasing dysfunction, increasing dyscontrol, and increasing dysorganization. Menninger goes beyond the principle of homeostasis and feels that the phenomena of growth and development demand an entirely different principle, namely, heterostasis, the progressive moving away from the status quo. In human organisms the ego may be considered as an advanced subsystem, a product of progressive differentiation, which in effect is the controlling agency of the vital balance.

Scheflen reminds us that there is nothing new in principle about a systems view, stating that it is the newest form of a classical approach to the study of science, one that evolved from an ancient approach to the naturalistic sciences, appearing as the holistic view in psychiatry. In psychiatry, even though the naturalistic view has remained the backbone of *clinical* practice and observation, since World War II the isolation of variables, or an American experimental methodology, has become ascendant in *research*. However, the naturalistic view was consistently advocated in psychology by such men as Wertheimer and the other gestalt theorists. Scheflen has commented that it is a strange paradox to find that psychiatric research should shift to the Newtonian method while other sciences are experiencing a strong resurgence of naturalistic orientation.

Parallel Developments

Attention is now drawn to a number of modern theories that constitute parallel developments to general systems theory and that can be considered as part of the modern systems movement. Specific comments will be made about operations research, cybernetics, mathematical general systems theory, and gestalt psychology. Additional parallel developments include information theory, communication theory, game theory, decision theory, topology, factor analysis, systems engineering, and human engineering.

OPERATIONS RESEARCH

Russell Ackoff, a leading figure in the development of operations research, has described and summarized the spread of the systems concept. Systems have been studied for centuries, but in the past two decades something new has been added—namely, the tendency to study systems as a whole rather than as a conglomeration of parts. Leading contemporary scientists are no longer content to isolate phenomena into narrowly confined contexts, but choose to open all types of interactions for examination and to study larger and larger slices of nature. Ackoff believes that the difference between general systems theory and systems or operations research primarily involves methodology. Whereas von Bertalanffy tried to unify science by reassembling aspects of nature which science had already disassembled, Ackoff has argued that science can be unified without going through an initial disunification and that this is the pathway that has been followed by operations research.

Systems are taken as they are found—"holistically and in all their multidisciplinary glory"—with unification of the disciplines resulting from the way research is conducted, and producing facts, laws, and theories which are multidisciplinary in character. Inventory theory, for example, is applicable to all open systems in which the exchange of material energy and information with the environment is at least partially controllable. If increases in the rate of input are accompanied by an increase in at least one well-defined benefit and one well-defined loss, the conditions necessary for the use of inventory theory are present. This theory is most familiarly used in the inventory processes of business and industry where it is applied to such diverse phenomena as the acquisition and use of operating capital, the hiring and training of personnel, and the determination of how much and how frequently productive capacity is to be acquired. But it is also applicable to the study of the metabolic processes of living organisms, the natural water system of a geographical region, and the operation of heating systems, computer centers, and documentation centers. Operations research has produced a number of other theories with similar characteristics, such as allocation, queuing, sequencing, routing, replacement, competition, and search theories.

CYBERNETICS

The need to counteract excessive specialization in science, which had been threatening to sever all communication among scientists for want of a common technical language, was forcefully voiced by Norbert Wiener in his book *Cybernetics*, published in 1948 [82]. Cybernetics, the science of communication and control, is an example of a newer discipline which cuts across several established and older disciplines, thereby providing opportunity for communication among different disciplines. Cybernetics is a concrete example of how systems concepts can be developed without departing from the standards of rigor demanded in physical science, since cybernetics is a mathematical method specifically developed to describe organized complexity [49]. The subsequent development of information theory and its fundamental importance to cybernetics is described most clearly by Rapoport. The contributions of Wiener in cybernetics and Shannon in mathematical theory of communication [57] recognized the cardinal principle basic to these formulations, that of the amount of information. The concept of information is as central in cybernetics and communication engineering as the concept of energy is in classical physics. By means of these newer theories, organismic and teleological notions of goal-seeking behavior have been reintroduced into the theory of the physical processes. In the modern version, these ideas have been derived not from metaphysical speculations on the nature of the behaving entities, but from the mathematical structures of systems characterized as highly organized complexities. Cybernetics, information theory, and general systems theory most closely approach each other in the area of considering the most fundamental property of the living organism as its ability to maintain and further its organized state against a constant tendency toward disorganization.

Ashby's [5] work on the concept of negative feedback has clarified some of the confusion remaining in the mechanistic-vitalistic controversy in regard to adaptation of living systems to the future.

MATHEMATICAL GENERAL SYSTEMS THEORY

Mathematical general systems theory is a field in its own right and therefore can be considered a third parallel development. It is closely related to the contributions of mathematical biology, set theory, mathematical modeling, and others. Rapoport has stated that the mathematical aspects of general systems theory can be viewed as an effort to fuse the mechanistic and organismic approaches so as to utilize the advantages of each. Mathematical language is best qualified to serve as the language of general systems theory precisely because it is devoid of content and expresses only the relational features of a situation. The limitations of the mathematical approach are minimized when the corresponding mathematical models are sufficiently faithful representations of the systems involved. Mathematical general systems theory is an important addition to the conceptual repertoire of the scientist rather than a method destined to drive all the older methods, such as purely organismic approaches, into obscurity [49].

The work of Ashby [8] on set theory and others [28, 32, 33, 46] on game theory, decision theory, organization theory, and systems science (which includes systems analysis, systems design, and systems engineering) covers vast and far-reaching fields whose development parallels the growth of general systems theory. For the general reader it is necessary to state that these additional areas, such as computer science, programming, modeling, and simulation, have been dealt with by the yearbooks published by the Society for General Systems Research under the title of *General Systems*.

GESTALT PSYCHOLOGY

Another parallel development was gestalt psychology, developed after World War I by Wolfgang Kohler, Max Wertheimer, and Kurt Kaffka. The similarity between the findings of gestalt psychology and general systems theory can be stated in relatively simple terms. Certain anatomical or structural arrangements force processes to take a course which helps the organism to survive. But basically, what are such "arrangements"? Are they similar to the rigid constraints found in inanimate mechanisms? In a fundamental way, they are not. The manner in which they exist differs widely from the constraints in machines where, for the most part, constraints consist of solid objects. They are permanent materials of one shape or another, depending upon their particular purpose. No part of the anatomy of a living organism is a permanent object. Rather, any part is regarded as a steady state, only the shape of which persists while its substance is perpetually being altered by metabolic events [40]. The key concept of dynamic regulation is essentially similar to von Bertalanffy's concepts of steady state and equifinality, and stresses perhaps even more strongly that regulation is a property of the system as a whole, rather than a function of cybernetic controls which are considered secondary. It is not possible to conceive a gestalt except in the presence of dynamic regulation, in the same way that it is not possible to conceive of an integrated system without the presence of dynamic regulation in such forms as steady states and equifinality.

Similar views were propounded in the United States by Raymond H. Wheeler, Chairman of the Department of Psychology at the University of Kansas. Under the titles *The Science of Psychology*, *Principles of Mental Development*, and *The Laws of Human Nature*, he published a series of books on gestalt psychology (which he referred to as organismic psychology) which elaborate and clarify the various dynamic principles of psychology as conceived by Kohler and his associates [78–80]. Kohler, Kaffka, and Wertheimer concluded their distinguished careers as professors at American universities. The work of these three German psychologists and of R. H. Wheeler in America was in sharp contrast to atomistic or mechanistic psychology, which was taught in America during the half century before 1930.

Society for General Systems Research

In 1954 the feeling that a scientific society centering around general systems theory would fill an eminent need crystallized at the Center for Advanced Study in the Behavioral Sciences at Stanford, California. A manifesto to this effect issued in that year received an extremely encouraging response. As a result the Society for the Advancement of General Systems Theory was launched at the 1954 meeting of the American Association for the Advancement of Science in Berkeley, California, with the appointment of an organizing committee consisting of Ludwig von Bertalanffy, Kenneth E. Boulding, Ralph W. Gerard, and Anatol Rapoport. Since then, annual meetings have been held under the auspices of the American Association for the Advancement of Science. The first *General Systems* yearbook appeared in 1956, and since that date new issues have appeared annually. In 1957 the name of the society was changed to the Society for General Systems Research. The Society's membership includes representatives of twenty-five nations; these include the United States, England, Canada, New Zealand, Australia, France, Mexico, Italy, the Netherlands, Switzerland, Austria, Germany, South Africa, the Sudan, Sweden, Japan, Belgium, India, Israel, Argentina, Chile, Colombia, Brazil, Czechoslovakia, and the U.S.S.R. Further, papers appearing in *General Systems* have been contributed by authors from the United States, England, Ghana, Spain, Canada, Germany, the U.S.S.R., Roumania, and Sweden.

The twelve annual volumes of *General Systems* constitute the richest basic source material available for students of general systems theory. Each volume to date has covered large, significant areas of modern scientific inquiry, including exposition of general systems theory; mathematical models; unification of science; principles of mass behavior; biological concepts; population dynamics; social science, education, and psychology; organization theory; information processing and cybernetics; and game theory and linguistics. Certain special topics have included automata and artificial intelligence; theories of stress; systems theory in the U.S.S.R.; political science and economics; philosophy of knowledge; and evolution and the theory of history.

Significance for Present and Future

GENERAL SYSTEMS THEORY AS A NEW WORLD VIEW

The most important issue of our time is that of finding concepts, languages, and frames of reference that are relevant and useful for dealing with the particular complexities of our particular historic era. It is a revolutionary age of the type described by Ortega y Gasset, in which a new civilization emerges to replace the older classical civilization. Kenneth Boulding has this in mind when he speaks of "our entry into the post-civilized stage," meaning that the city will no longer be the main unit of civilization. It is abundantly clear that inexorable forces from many sides push us toward a model of world civilization in which regarding parts of the world as isolated universes will no longer be relevant. It will deal with the whole human species living in the whole world. The new world view which will emerge will be an organismic one. It will have a new vocabulary in the form of key words representing the new relevant concepts. These will include organization, ecology, biocenosis, dynamic regulation, negative feedback, anamorphosis, open systems, negative entropy, complexity, hierarchization, isomorphism, centralization, progressive mechanization, progressive differentiation, progressive deanthropomorphization, positive feedback, self-amplifying mechanisms, leading parts, trigger causality, spontaneous activity of the psychophysical organism, modeling, programming, symbol, and system. Of course, more could be added, including what are perhaps the greatest omissions: technology and computer science. One might well add the rapidly growing interest in psycholinguistics, in language development, and in the creation of artificial languages, as well as current developments in molecular biology. Ideas of organization, of general systems, of symbol, of growth and development, of steady states, and our expanding scientific understanding of open systems would be found an integral part of the new world view. This list of words representing key concepts essential to our understanding of our modern, ongoing world can readily be seen as the same list of words and concepts that have been found necessary in the elaboration of the new world view that is called general systems theory.

One can say that general systems theory as a scientific theory has a measurably greater isomorphism with the modern, ongoing world than our previous, rather mechanistic scientific concepts possessed, and this is true whether such concepts were in the so-called hard sciences or in the behavioral or social sciences. It should be noted that general systems theory does not invalidate previous scientific work, but places it in a new frame of reference, a new way of viewing the world. This will permit the very large amount of knowledge already accumulated, although often in a fragmented and isolated way, to become integrated and be placed in perspective in terms of the growth of knowledge itself, thus giving guidelines as to what further needs to be done. It will be instrumental in bringing the findings and knowledge of the various specialties into effective transaction with one another and will lead hopefully to greater unity in science.

RELEVANCE FOR THE MODERN WORLD

But these are not the most important effects of general systems theory, although they are extremely important. The most important effect will be that of bringing scientific work in all areas into a relevant and transacting relationship to our modern world as it is in this latter part of the twentieth century and as it promises to be in at least the next fifty to one hundred years. At that time it is probable that general systems theory will have become over-institutionalized and over-rigidified as part of the growth cycle of all world views. It is most important to recognize that general systems theory, as all world views, should not be accepted as the only possible way of regarding our present world, and it certainly should not be accepted as one that will escape the institutionalization and rigidification which in time will mean its replacement by some wider and more relevant world view.

REDEFINITION OF SCIENCE

For the present, though, general systems theory represents a widening of previous world views and definitions of what constitutes science and a change in basic categories of thought. It offers a working frame of reference and key concepts that have to some degree rescued modern science from the mechanistic model to which it still clings in many areas. There is in some quarters increasing realization that modern science has already become scientism, and the danger of its becoming more so is an ominous and shuddering possibility. Of hopeful importance is the potential of general systems theory in allowing the development of a truly humanistic science by ending the artificial barriers that separate science and the humanities. It seems likely that the fragmentation and separation of these two large areas of human knowledge was a necessary stage of development, but in the present state of our world it is no longer necessary. It should be understood that general systems theory is now barely twenty-five years old and that problems outweigh answers. As a guideline for the investigator of problems from a new conceptual framework, it is proving invaluable. What might be called the number one problem is the almost complete mystery that remains as to the origin of anamorphosis.

RELEVANCE FOR PSYCHIATRY

What does this mean to our field of psychiatry, and what can we look forward to as the influence of general systems theory on our own field of endeavor continues to increase? It will not eliminate the need for psychiatry, nor will it make useless the fund of special knowledge of technical approaches, of special points of view that constitute our field of expertise; instead, it will give to psychiatrists a wider frame of reference, a general point of view which will permit a more rewarding and satisfying way of placing what we already know in a general frame of reference which will aid us greatly in seeing the relevance of our knowledge to that of other thinkers, learners, and practitioners. It will do much to overcome the problem we have faced in trying to understand the relevance of psychiatry to the problems of our modern and rapidly changing world. One sees this in our community mental health programs in which, essentially, we operate without a comprehensive theory, a situation which is temporarily tolerable but which must give way to the development of theory so that theory and experience can supplement one another and produce growth and development. It seems likely that our new theories will be of general systems theory form, that they will be multiple and will exist as families of general systems theory, and that they will help our psychiatric theories and practice become more relevant in a world that will hopefully become more unified and humanistic as well as more scientifically advanced.

References

1. Ackoff, R. L. Games, decisions and organizations. *Gen. Syst.* 4:145, 1959.
2. Ackoff, R. L. General system theory and systems research: Contrasting conceptions of systems science. *Gen. Syst.* 8:117, 1963.
3. Akhmanova, O. S. On psycholinguistics. *Gen. Syst.* 5:181, 1960.
4. Allport, F. H. *Theories of Perception and the Concept of Structure.* New York: Wiley, 1955.
5. Ashby, W. R. Adaptiveness and equilibrium. *J. Ment. Sci.* 86: 478, 1940.
6. Ashby, W. R. General systems theory as a new discipline. *Gen. Syst.* 3:1, 1958.
7. Ashby, W. R. Cybernetics today and its future contribution to the engineering sciences. *Gen. Syst.* 8:207, 1963.
8. Ashby, W. R. The set theory of mechanism and homeostasis. *Gen. Syst.* 9:83, 1964.
9. Auerswald, E. H. Interdisciplinary versus Ecological Approach. Chapter 18, this book.

10. Boguslaw, R. *The New Utopians*. Englewood Cliffs, N.J.: Prentice-Hall, 1965.
11. Bolman, W. M. Theoretical and empirical bases of community mental health. *Amer. J. Psychiat.* 124:4, 1967.
12. Botnariuc, N. The wholeness of living systems and some basic biological problems. *Gen. Syst.* 11:93, 1966.
13. Boulding, K. E. General system theory—the skeleton of science. *Gen. Syst.* 1:11, 1956.
14. Boulding, K. E. Toward a general theory of growth. *Gen. Syst.* 1:66, 1956.
15. Boulding, K. E. Political implications of general systems research. *Gen. Syst.* 6:1, 1961.
16. Cannon, W. B. *The Wisdom of the Body*. New York: Norton, 1939.
17. Driesch, H. A. E. *The Science and Philosophy of the Organism* (2d ed.). Gifford Lectures delivered at Aberdeen University, Scotland, 1907–1908. London: Black, 1929.
18. Frank, L. K. The need for a new political theory. Proceedings of the American Academy of Arts and Sciences. *Daedalus* 96: No. 3, (Summer) 1967.
19. Freud, S. Beyond the pleasure principle. In *Complete Psychological Works of Sigmund Freud*. London: Hogarth, 1955. Vol. 18, pp. 1–64.
20. *Gen. Syst.* Vol. 1, 1956.
21. *Gen. Syst.* Vol. 2, 1957.
22. *Gen. Syst.* Vol. 3, 1958.
23. *Gen. Syst.* Vol. 4, 1959.
24. *Gen. Syst.* Vol. 5, 1960.
25. *Gen. Syst.* Vol. 6, 1961.
26. *Gen. Syst.* Vol. 7, 1962.
27. *Gen. Syst.* Vol. 8, 1963.
28. *Gen. Syst.* Vol. 9, 1964.
29. *Gen. Syst.* Vol. 10, 1965.
30. *Gen. Syst.* Vol. 11, 1966.
31. *Gen. Syst.* Vol. 12, 1967.
32. Gerard, R. W. A biologist's view of society. *Gen. Syst.* 1:55, 1956.
33. Gerard, R. W. The rights of man: A biological approach. *Gen. Syst.* 1:161, 1956.
34. Grinker, R. R. (Ed.) *Toward a Unified Theory of Human Behavior*. New York: Basic Books, 1956.
35. Grinker, R. R., Sr. (Ed.) *Toward a Unified Theory of Human Behavior* (2d ed.). New York: Basic Books, 1967.
36. Kel'zon, A. S. Dynamic properties of cybernetics. *Gen. Syst.* 5:209, 1960.
37. Khailov, K. M. The problem of systemic organization in theoretical biology. *Gen. Syst.* 9:151, 1964.

38. Klír, J. The general system as a methodological tool. *Gen. Syst.* 10:29, 1965.
39. Koestler, A. *The Ghost in the Machine.* New York: Macmillan, 1967.
40. Köhler, W. *Dynamics in Psychology.* New York: Liveright, 1940.
41. Kremyanskiy, V. I. Certain peculiarities of organisms as a "system" from the point of view of physics, cybernetics, and biology. *Gen. Syst.* 5:221, 1960.
42. Lektorsky, V. A., and Sadovsky, V. N. On principles of system research. *Gen. Syst.* 5:171, 1960.
43. Maruyama, M. The second cybernetics: Deviation-amplifying mutual causal processes. *Gen. Syst.* 8:233, 1963.
44. Menninger, K. A. Psychological aspects of the organism under stress. *Gen. Syst.* 2:142, 1957.
45. Menninger, K., Mayman, M., and Pruyser, P. *The Vital Balance.* New York: Viking, 1963.
46. Mesarovic, M. D. General Systems Theory and Its Mathematical Foundation. Record of the IEEE 1967 Systems Science and Cybernetics Conference. New York: Institute of Electrical and Electronics Engineers, 1967.
47. Miller, J. G. Living systems: Basic concepts. *Behav. Sci.* 10:3, 1965.
48. Rapoport, A. The diffusion problem in mass behavior. *Gen. Syst.* 1:48, 1956.
49. Rapoport, A. Mathematical aspects of general systems analysis. *Gen. Syst.* 11:3, 1966.
50. Rosenblueth, A., Wiener, N., and Bigelow, J. Behavior, purpose and teleology. *Phil. Sci.* 10:18, 1943.
51. Ruesch, J. The old world and the new. *Amer. J. Psychiat.* 124:2, 1967.
52. Scheflen, A. E. Systems and psychosomatics. *Psychosom. Med.* 28:4, 1966.
53. Schroedinger, E. *What Is Life?* New York: Macmillan, 1945.
54. Sčur, G. S. Some considerations of the notion of invariant field in linguistics. *Gen. Syst.* 10:97, 1965.
55. Sčur, G. S. On some general categories of linguistics. *Gen. Syst.* 11:149, 1966.
56. Sčur, G. S. On the relations among some categories of linguistics. *Gen. Syst.* 11:157, 1966.
57. Shannon, C. E., and Weaver, W. *The Mathematical Theory of Communication.* Urbana, Ill.: University of Illinois Press, 1949.
58. Shchedrovitzsky, G. P. Methodological problems of system research. *Gen. Syst.* 11:27, 1966.
59. *Tenth Annual Report.* The University of Michigan Mental Health Research Institute, Ann Arbor, Michigan, 1965.
60. Toward the Year 2000: Work in progress. Proceedings of the

American Academy of Arts and Sciences. *Daedalus* 96: No. 3, (Summer) 1967.
61. von Bertalanffy, L. An outline of general system theory. *Brit. J. Phil. Sci.* 1:2, 1950.
62. von Bertalanffy, L. *Problems of Life*. New York: Wiley, 1952.
63. von Bertalanffy, L. General system theory. *Gen. Syst.* 1:1, 1956.
64. von Bertalanffy, L. In J. M. Tanner and B. Inhelder (Eds.), *Discussions on Child Development*. Proceedings of the Fourth Meeting of the World Health Organization Study Group on the Psychobiological Development of the Child, Geneva, 1956. New York: International Universities Press, 1960. Vol. 4, pp. 69–76, 88–89, 155–175.
65. von Bertalanffy, L. General system theory—a critical review. *Gen. Syst.* 7:1, 1962.
66. von Bertalanffy, L. An essay on the relativity of categories. *Gen. Syst.* 7:71, 1962.
67. von Bertalanffy, L. Discussion at Session I of Symposium on Cognitive Processes and Psychopathology. Annual Meeting of the Academy of Psychoanalysis, Montreal, December, 1964.
68. von Bertalanffy, L. General System Theory and Psychiatry. In S. Arieti (Ed.), *American Handbook of Psychiatry*, Vol. III. New York: Basic Books, 1966.
69. von Bertalanffy, L. The Open System of Science. Inaugural Lecture in the Heinz Werner Lecture Series in Developmental Psychology and Related Disciplines, Clark University, Worcester, Massachusetts, January, 1966.
70. von Bertalanffy, L. *Robots, Man and Minds*. New York: Braziller, 1967.
71. von Bertalanffy, L. General Systems Theory and Psychiatry—An Overview. Chapter 2, this book.
72. Von Foerster, H. (Ed.) *Cybernetics*. Transactions of the Sixth Conference. New York: Macy Foundation, 1949.
73. Von Foerster, H., Mead, M., and Teuber, H. L. (Eds.) *Cybernetics*. Transactions of the Seventh Conference. New York: Macy Foundation, 1950.
74. Von Foerster, H., Mead, M., and Teuber, H. L. (Eds.) *Cybernetics*. Transactions of the Eighth Conference. New York: Macy Foundation, 1951.
75. Von Foerster, H., Mead, M., and Teuber, H. L. (Eds.) *Cybernetics*. Transactions of the Ninth Conference. New York: Macy Foundation, 1952.
76. Von Foerster, H., Mead, M., and Teuber, H. L. (Eds.) *Cybernetics*. Transactions of the Tenth Conference. New York: Macy Foundation, 1953.
77. Von Foerster, H., and Zopf, G. W., Jr. (Eds.) *Principles of Self-Organization*. New York: Pergamon, 1962.

78. Wheeler, R. H. *The Laws of Human Nature.* New York: Century, 1932.
79. Wheeler, R. H. *The Science of Psychology.* New York: Crowell, 1940.
80. Wheeler, R. H., and Perkins, F. T. *Principles of Mental Development.* New York: Crowell, 1932.
81. Whitehead, A. N. *Science and the Modern World.* New York: Pelican, 1948.
82. Wiener, N. *Cybernetics.* New York: Wiley, 1948.
83. Young, O. R. The impact of general systems theory on political science. *Gen. Syst.* 9:239, 1964.
84. Young, O. R. A survey of general systems theory. *Gen. Syst.* 9:61, 1964.

2

General Systems Theory and Psychiatry — An Overview*

LUDWIG VON BERTALANFFY, PH.D.

IT GIVES ME great pleasure that the honor was conferred on me to open the present session on "General Systems Theory and Psychiatry." Having been engaged some thirty years in this field, my task is to outline what systems theory means, how it came about, and of what use it may be to psychiatry. Obviously, a brief survey imposes oversimplification of difficult problems and manifold research activities. For this I must apologize, making reference to the extensive literature available [4, 5, 13–16].

So much is spoken these days about systems, systems engineering, a systems approach to all sorts of problems from physics to politics, that the term is in danger of becoming a catchword, seemingly announcing profound insight when, in fact, only a fashionable verbal label was used. Moreover, the dangers of "systems" are apparent. Systems designers, analysts, researchers, behavioral engineers, and members of similar professions—New Utopians, as Boguslaw [3] called them—contribute to or even lord over the industrial-military establishment. Elaborating weapons systems, dominating advertising, mass media, and propaganda, and in general preparing a cybernetic society of the future, they must of necessity tend to exclude or suppress the "human element." For this human element, individualistic, capricious, often unforeseeable, is precisely the unreliable part of the "megamachine" (to use L. Mumford's term) of the present and future; hence, it should be either replaced by computers and other

* Presented at the 123rd annual meeting of the American Psychiatric Association, Detroit, Mich., May 8–12, 1967.

hardware or else made "reliable"—that is, as machinelike, automated, uniform, conformist, and controlled as possible. "Systems" thus appears to be the epitome of the automated wasteland of *Brave New World* and *1984*.

We must therefore look into the origins of the modern "systems approach" and try to evaluate its various expressions and consequences.

One way to explain why systems concepts, research, and theory have become necessary is by the heightened complexity of the modern world. So long as the question was to build a steam engine or automobile, engineers and workmen did well enough. But when it comes to constructing ballistic missiles, with myriads of components originating in mechanical, electronic, and chemical industries, with man-machine adjustments and innumerable economic and political problems thrown in—then it is time to call in systems engineers to make some sense and optimization in an extremely complex situation.

Linear Causality

Such an example stands for countless similar situations. It also epitomizes a far wider problem. Up to recent times, the lodestar of science was the "causal method," relations of "if-then," and explanation, prediction, and control according to this scheme. Using more technical terms, the question was about "linear causality" or relations between two or a few variables. If you put a thermometer into hot water, the mercury will rise. If you run a rat in a Skinner box, a certain conditioning will take place. If you administer an antibiotic, an infection, hopefully, will clear up. But what if you have a large number of interacting and even partly unknown variables? This is what happens in biological, medical, and technological situations. You cannot simply say: If a person is infected with tubercle bacilli, he will contract the disease; it depends on so many factors of constitution, nutrition, and so forth. Nor can you say: If a child comes from a broken home, he will become neurotic or delinquent; he may just as well become a genius, some other conditions, genetic and environmental, being given. But this is precisely the problem of "systems," the problem of interaction of many and partly unknown variables.

It so happens that basic problems in biology, psychology, and the social sciences are of this kind. Suppose we compare a dog—or, for that matter, a human—when alive and healthy, when sick, and when dead. What can you say about the difference from the viewpoint of traditional science? Extremely little, I am afraid. In such comparison, you will analyze the processes going on in the animal and in the corpse; presupposing you have superhuman knowledge and skill, you will eventually come out with neat formulas of innumerable physical and chemical processes. Granted these processes and formulas will be different in the live and dead dog. But there is nothing to tell the difference which is obvious to the naive observer. The laws of physics and chemistry remain the same; they do not care whether dogs are healthy, sick, or dead. It also makes no difference when you introduce the latest refinements of molecular biology. One DNA molecule, protein, enzyme reaction, hormonal process, and so forth is as good as another—each determined by physical and chemical laws, but none better, healthier, or more normal than the other.

But, of course, there is a tremendous difference between a live and a dead dog and even more between a person and a corpse. In the living being, processes are so *ordered* as to maintain the organism. The living being has a marvelous *organization* which is impaired in sickness and decays after death. In apparent contradiction to a well-known principle of physics, this organization—an utmost improbable state—is maintained in a stream of processes which should go toward most probable equilibrium states; improbability even increases, in the dog's ontogenesis and evolution, by way of progressive *differentiation*. In its behavior the dog (not to speak of man) appears to be *goal-directed;* it actively strives toward things it wants, such as food or a mate; and so forth ad infinitum. But all these concepts and terms (order, organization, differentiation, goal-directedness, and many similar ones) do not occur in the vocabulary of physics. Not because, as is unfortunately true, our knowledge is imperfect so that future discoveries will have to fill the gap, but for a different reason. Science, in the way it has developed since Galileo and Newton, was concerned with undirected events, isolable causal trains, one-way causality, relations between an independent and a dependent variable, rejection of any form of teleology, all these and other expressions being aspects of the same viewpoint. But what we are concerned with when envisaging living and dead organisms are questions of organized wholes, of directedness and order of events, of interactions among many variables, of goal-seeking and the like—all these terms, again, being different aspects of one central problem.

In this situation we can take two different attitudes. First, we can decree that such problems do not exist and declare them to be anthropomorphic delusion and metaphysical nonsense. This is, in fact, what so-called mechanistic science did. But then we run counter to everyday observation and to the actual practice of biology and medicine which cannot speak of one single organ or function without considering its significance to the organism as a whole; and we do so under the bias of a mechanistic metaphysics which is not tenable anymore in physics itself. Our second alternative is to admit honestly that our present science—which, after all, is not a mirror of ultimate reality, but a very human, conceptual construct—apparently does not tell us everything, in which case we must decide to do something about it. Since all problems mentioned are in some way aspects of "wholes" or "systems," this probably will amount to the demand of a general systems theory.

This, then, is the aim of the systems approach: looking into those organismic features of life, behavior, society; taking them seriously and not bypassing or denying them; finding conceptual tools to handle them; developing models to represent them in conceptual constructs; making these models work in the scientific ways of logical deduction, of construction of material analogues, computer simulation and so forth; and so come to better understanding, explanation, prediction, control of what makes an organism, a psyche, or a society function.

Obviously, this is an enormous order which, if achieved, would give us a greatly deepened understanding of life, man, and society. Equally obvious, we are in the first beginnings. We have some scraps of theory, some mathematical and nonmathematical models, some instances of successful application of systems theory, and a vast ocean of ignorance hardly touched by the imperfect craft of our present concepts. However fully we realize these shortcomings, we must make a start and embark in the enterprise in one way or another.

As a matter of fact, there are a number of privateering companies, to keep the simile a moment longer, which have set out in the pursuit of such endeavor. They go under flags such as general systems theory, cybernetics, information theory, game theory, and others.

They are a heterogeneous lot, different in their presuppositions, mathematical equipment, and goals, sometimes quarreling, not a unified company. But they are in agreement in their general purpose —namely, the exploration of a continent previously missed.

Mechanismic and Organismic Trends

This, in broad outline, is the idea of general systems. Within this new crop of theories, two main trends can be distinguished. The *mechanistic trend* is grounded in technological, industrial, and social developments such as control techniques, automation, servomechanisms, computerization, and so forth. Cybernetics, operations research, linear programming, theory of automata, and systems engineering may be named as characteristic examples. This trend finds its main expression in the theory and practice of *self-regulating mechanisms* governed by the principles of cybernetics. In the terms mentioned above, linear causality was supplemented by *circular causality*, introducing feedback of output into input and so making the system self-regulating with respect to maintenance of a desired variable or a target to be reached.

The *organismic trend* starts with the trite consideration that organisms are organized things; we have to look for general principles and laws concerning "organization," "wholeness," "order of parts and processes," "multivariable interaction," "growth," "competition," "negentropy," and others. These traits are common to biological, behavioral, psychological, and social phenomena; there are, as it is called, *isomorphisms* between biological, behavioral, and social phenomena and sciences. Therefore, we should try to develop a "general theory of systems." Interaction among many variables and free *dynamic order* may be indicated as central notions.

A clarification is required here. "Systems theory" is frequently identified with cybernetics, control techniques, theory of automata, and the like. It should be apparent by now that this is an error. Self-regulating mechanisms, cybernetic systems (that is, systems regulated by feedback or circular causality), and so forth, are but a rather restricted subclass of "general systems."

It cannot be our aim to review the present status of general systems theory, its structure, applications, hopes, and shortcomings. But we can try to illustrate it by answering our initial question: What can systems-theoretical notions contribute to psychiatric theory and practice?

The idea and term *general systems theory* was developed by this writer from biological and biophysical considerations which cannot be reviewed here. But, of course, there have been many organismic or systems-theoretical developments in psychiatry which can be traced back, in America, to Adolph Meyer. Goldstein, Menninger, Grinker, Carl Rogers, Arieti, and others have followed a similar approach, and comparable emphasis can be found in the phenomenological and existentialist schools. In normal psychology, Gordon Allport, Karl and Charlotte Bühler, Werner, Piaget, Maslow, J. Bruner, and the "new look" in perception are a few representatives of the trend.

The Systems Model

However, general systems theory can, I believe, make essential contributions. It emphasizes the communalities in modern developments of fields which otherwise are different in content. The systems model provides a conceptual framework in which otherwise unconnected currents are integrated, a synthetic view in which many different pieces fall into place. In particular, it seems to offer a reorientation of psychiatric thinking and suggestions for clinical work, signifying a realignment and reorientation.

PERSONALITY AS SYSTEM

The basic change I would venture, first, is the replacement of the image of man as *robot* by that of *active personality system.* We have become conditioned to the image of man as a machine or automaton, reacting to stimuli, motivated by biological drives such as hunger and sex. To be sure, we are self-regulating, goal-seeking, and learning machines; but so are servomechanisms and computers in modern technology. Furthermore, technology, both by the impact of a machine-dominated society and by the techniques of manipulating psychology, ever more *makes* us into robots, machines, or wheels-within-wheels, mass-produced, standardized, replaceable, so as to serve the larger machinery of modern society.

Man as robot means a bundle of nerve circuits, functions, drives, conditioned responses, feedback circuits, complexes, and so on. A different concept of organism and personality is that of *system*—that is, a dynamic order of parts and processes. That this is so is confirmed by psychiatric experience. Mental dysfunction is a *system disturbance* rather than a loss or disturbance of single functions. Even in localized traumas such as cortical lesions the effect is impairment of the total action system, particularly with respect to higher and more demanding functions. Conversely, the system has considerable regulative capacities as a rigid, psychophysiological machine would not. This has been known since the research of Goldstein, Lashley, and others.

INTERNALLY ACTIVE SYSTEM

Secondly, the organism is an *internally active system.* American psychology and also psychiatry were dominated by the stimulus-response or S-R scheme. Behavior and mental life were essentially response to stimuli, or response conditioned by stimuli in the past. Riesman has called this the other-directedness of present humanity. But biological and psychological experience shows otherwise. Internal activity or spontaneity of the psychophysical organism is one of its fundamental characteristics, ultimately based upon its being a so-called open system. This is shown by the fact that, both in individual development and in evolution, autonomous active behavior precedes reaction to external stimuli. Understandably, the spontaneity of the organism was not seen in laboratory rats put into the S-R machines of experimental psychology. Hence, it was denied in classical behaviorism and only grudgingly acknowledged in more recent behavioral theory. Nevertheless, it is basic for animal behavior in the natural habitat. In man it comprises the spectrum from the first exploratory activities of the baby whereby he "creates" the world around him—playing, being curious, creating—up to culture in general, none of which is simply a utilitarian answer to outside demands but rather man's free creation in his specific sphere of activity which we call "symbolic" (cf. Huizinga [6]).

There are obvious consequences for psychiatry. Personality is not a robot; but in mental disease when spontaneity is impaired, the patient may *become* an S-R automaton, as he does in the stereotyped behavior of compulsives, after brain lesions, in the waning of activity in catatonia, and in similar conditions. Meaningless rituals and machinelike responses are thus found in neuroses, psychoses, brain lesions, and so forth. The robotization in modern society actively suppresses spontaneity or creativity; no wonder that either neurosis or socially unacceptable behavior—tantrums, drug addiction, vandalism, and crime as an outlet—is the result.

BEYOND THE HOMEOSTASIS MODEL

This indicates the limits of the often-used *homeostasis model*—that is, the concept that behavior is composed of the gratification of biological needs, satisfaction of drives, relaxation of tensions, adjustment, reestablishment of biological, psychological, and social "equilibrium," and so forth. The drive, equilibrium, and homeostasis schemes cover only a limited aspect of behavior and mental life. Even a rat does not behave according to the S-R model: It runs around anyway, explores its environment without reinforcement and gratification, even looks out for "problems" that have nothing to do with such primary needs as food and sex. The same, of course, applies even more to human behavior, especially at the symbolic level characteristic of man. Hence, cultural and creative activities have little to do with gratification of biological needs, and cannot be reduced to primary drives or utilitarian motives of preservation of the individual or species with which they often are in conflict. Processes of development, evolution, and increasing order are, in general, beyond the homeostasis and equilibrium scheme.

PROGRESSIVE MECHANIZATION

As already implied, a *principle of progressive mechanization* applies to the biological organism as well as to behavior and personality. It is not a machine, but with progressing development functions in many respects like a machine. Mechanization takes place on both the evolutionary and the individual scale, being a prerequisite for higher development of functions. Only by mechanization of relatively lower functions can the organism develop progressively higher ones. Hence, mechanization is found in all respects, from the fixation of physiological functions in bodily mechanisms to learning processes at the involuntary (for example, operant conditioning) or conscious, voluntary level, such as learning of piano playing or calculus. By such processes becoming routine and mechanized (presumably by fixation of neural pathways for particular functions), higher functions become possible. Civilization would be impossible were not innumerable functions mechanized and unconscious, and consciousness thus freed for activity at a higher level. This implies a paradox, however: Specialization and mechanization also imply inhibition and loss of potentialities present, loss of adaptability to changing conditions, and of equipotentiality, that is, functional replacement of parts lost or impaired.

By the same token regulations are progressively mechanized in the way of feedbacks and circular causality, hence, the wide applicability of the cybernetic or feedback model to many well-worked-out examples in physiology, neurophysiology, behavior, psychology, and psychiatry.

In psychopathology, nonhomeostatic functions often decline. Thus, Karl Menninger [9] has described the progress of mental disease as a series of adjustments to ever lower homeostatic levels; Arieti [1] similarly defined schizophrenia as progressive teleological regression.

By the same token there are innumerable human activities beyond the homeostasis scheme, and this helps to answer the great psychosocial paradox of our society. According to homeostatic theory, mental health should prevail under conditions of gratification of needs, satisfaction of drives, and relaxation of tensions. But conditions of extreme stress and need such as those in Europe during World War II did not produce but rather reduced neuroses and psychoses. In contrast, affluent society does provide for needs by economic abundance; it satisfies drives by sexual mores, and takes pains to relieve tensions by permissive education, lowered scholastic standards, and so forth. Contrary to expectation, this is accompanied by an unprecedented increase of mental disorder, juvenile delinquency, and crime not for want but for "fun." This is vivid testimony that the model of reactive or homeostatic personality is wrong, and that it should be complemented by a more adequate one. Retirement neurosis and similar psychopathology belong in the same picture.

SELF-DIFFERENTIATION

Conventional theory sees in the formation of personality essentially a conditioning process. In this respect theories otherwise as different as psychoanalysis and behaviorism agree, the first emphasizing early childhood experience, the second emphasizing conditioning in its classic or operational variants. The role of conditioning needs no emphasis. It is basic in the gamut of learning processes; and in present society, advertising ranging from detergents to presidents is based on the techniques of dog-conditioning of Pavlov and rat-conditioning of Skinner. Unfortunately, this seems necessary to keep our commercial-industrial-military society going.

Nevertheless, there is something more, investigated and conceptualized in different ways by Werner, Piaget, Schachtel, G. Allport, and others. In systems terms, it can be called *self-differentiation*. Psychophysiological development is not exhausted by conditioning, accumulation of traces of past experience and their neurophysiological counterparts. Rather, development—ontogenetic, cultural, microgenic—proceeds from undifferentiated or syncretic states to ever more differentiated ones. This is found in perception, concepts, language, and elsewhere. The categories of higher mental life, such as ego and nonego, perception of individual objects, space, time, number, individual concepts and words, develop (in infants) or evolve (in cultures) from primitive states, an "undifferentiated absolute of self and environment" in Piaget's expression, a "perceptual-conceptual-motivational continuum" after Werner, or "paleologic" perception after Arieti. This is a special case of a general systems-theoretical principle manifest in innumerable biological, behavioral, and social phenomena. It also applies to the reverse.

REGRESSION

Regression is not loss of meristic, higher functions and thus a return to a more primitive and infantile state. Rather, as emphasized by Arieti [1], it is essentially dedifferentiation and disintegration. For this reason, primitive syncretic states such as synesthesia, paleologic thinking, and breakdown of the ego boundary may reappear in regression. Normal differentiation further implies progressive organization within an integrated whole or system. In regression this integration is lost, resulting in splitting of personality, complexes, disturbed ego function, and the like.

CENTRALIZATION

Centralization is another aspect of differentiation. "Leading components" are formed that have a greater triggering influence on the behavior of the total system than have others. This again is a general principle in ontogenesis and evolution, and is found in the development of the nervous system, in encephalization (dominance of the brain), and cerebralization (dominance of the cortex). Its mental counterpart is ego formation over developmental and evolutionary stages. The converse, pathological case is decerebralization, loss of ego identity, splitting of personality, complex formation, mental disintegration, and so forth.

BODY AND MIND

A survey of psychiatric notions in systems theoretical terms soon leads to profound and basic problems such as the time-hallowed question of the *relation of body and mind* [11, 12]. Let me say only this: The classical dualism of matter and consciousness, which was stated first by Descartes and later continued in science and philosophy, has broken down with developments in modern physics, psychology, psychopathology, phenomenology, cultural anthropology, epistemology, and other fields. Body and mind, matter and consciousness are *conceptualizations* which need revision—with the remote goal of a unified science being essentially a theory of systems and embracing both physical and psychical aspects.

Even this brief remark can indicate implications for psychiatry. Matter and mind, body and consciousness are not ultimate realities. Rather, they are conceptualizations to bring order into William James's "buzzing, blooming confusion" of immediate experience. For this reason, there are no rigid metaphysical borderlines between physiological occurrences, unconscious and conscious. Psychosomatic illness is not a mysterious transgression from mind to body and vice versa; physiology—drugs, for example—may influence the mind, and psychotherapy the body. In psychopathology, more ancient forms of experience than the well-established ego boundary and object-subject separation of the modern adult may reappear in the form of fantasmic, animistic, paleologic conceptions [2].

HIERARCHICAL ORDER

To the systems theorist the universe—that is, the totality of experience—appears as a *hierarchical order*, the major levels of which are the inorganic, the living, and the symbolic worlds. The notion *symbolic* appears to define best those activities, products, and interrelationships which separate human from animal behavior and which form the superstructure of human culture and history.

For this reason, human behavior cannot be reduced to animal behavior, to the play of biological drives, conditioning, and the like. We must abandon the *zoomorphic viewpoint* [14] predominant in a large part of psychology and psychiatry which tries to explain the human phenomenon in terms of rat, pigeon, and monkey behavior.

This, again, agrees with recent trends in psychiatry. Mental disease can be grounded in different levels; grossly speaking, it may be organic, psychogenic, or noogenic, if we accept Frankl's term. Consequently, neurosis and psychosis rest only to an extent on biological drives and their suppression. This corresponds, for example, to Kubie's [7] concept of neurosis and to Arieti's [1, 2] concept of schizophrenia. If, say, a successful businessman is suicidally depressed after retirement, or if a schizophrenic believes himself to be the Emperor of China, obviously this has little to do with hunger, sex, or adverse childhood experience in his distant past. Patently, such disturbances are at the symbolic level—loss of value orientation in the businessman's case, confusion of the symbolic world construct in the schizophrenic. Therapy will have to proceed accordingly; for example, rather than "digging up the past" and unveiling old complexes, establishment of new life-goals, "nootherapy" after Frankl, and the like.

SYSTEMS OF VARIOUS ORDERS

No man is an island: Mankind does not consist of isolated individuals but is organized in *systems of various orders*, from small groups such as the family to the largest, called civilizations.

Consequently, psychiatric practice is not confined to the level of the individual. Systems theory gives a theoretical framework and, hopefully, practicable suggestions for therapeutic measures such as family therapy, community programs, and the multifamily therapy presently proposed [8]; similarly, as may be noted, systems analysis has proved useful in education and other facets of normal life.

SYMBOLIC VALUE SUPERSTRUCTURE

Considered at the highest hierarchical level, history is not progress of an amorphous humanity but takes place in sociocultural entities such as groups, classes, nations, and cultures.

Therefore, mental disease, so far as it is disturbance at the symbolic level, is culture-bound. It varies within cultural frameworks as, for example, the well-known changes in neurotic disorders since Freud's time have shown. What appears psychopathological to us may not be so in other societies, as was shown in Ruth Benedict's *Patterns of Culture.* On the other hand, our own supposedly normal behavior may appear pathological when measured against other sociocultural standards, or against an ideal of human behavior. Affluent society may at the same time be a very sick society particularly because, in the midst of plenty, the symbolic value superstructure has broken down.

One good example of this breakdown is the leisure problem. The citizens of old Athens abhorred what we call honest work; indeed, they would have considered the rat race of our competitive society as the worst slavery. But leisure did not give them problems, neurosis, and existential meaninglessness, but made for one of the most creative epochs in history.

Obviously, this overview at best could do no more than sketch a few viewpoints that emerge from general systems theory for consideration of the psychiatrist. The systems approach appears to provide a framework wherein a large number of necessary developments in psychology and psychiatry can be encompassed, and which, hopefully, will lead to further developments. It permits scientific exploration of aspects previously neglected, and is at the same time humanistic, emphasizing what is specific of the human condition. Here I must rest my case and leave matters to the psychiatric practitioner and theorist. Whatever the shortcomings and errors of commission and omission may be, one thing appears to be certain: such reorientation is due in scientific thinking, in medical practice, and in society in general; and we shall hardly be mistaken in assuming that it will be centered in the concept of "systems."

References

1. Arieti, S. *Interpretation of Schizophrenia.* New York: Robert Brunner, 1955.
2. Arieti, S. Contributions to Cognition from Psychoanalytic Theory. In Masserman, J. (Ed.), *Science and Psychoanalysis,* Vol. 8. New York: Grune & Stratton, 1965.
3. Boguslaw, W. *The New Utopians.* Englewood Cliffs, N.J.: Prentice-Hall, 1965.

4. Buckley, W. *Modern Systems Research for the Behavioral Scientist.* Chicago, Ill.: Aldine, 1967.
5. Grinker, R. R. (Ed.), *Toward a Unified Theory of Human Behavior* (2d ed.). New York: Basic Books, 1967.
6. Huizinga, J. *Homo Ludens.* Boston: Beacon, 1955.
7. Kubie, L. The distortion of the symbolic process in neurosis and psychosis. *J. Amer. Psychoanal. Ass.* 1:59, 1953.
8. Laqueur, H. P. General systems theory and multiple family therapy. *American Psychiatric Association, 123rd Annual Meeting.* 1967.
9. Menninger, K., Mayman, M., and Pruyser, P. *The Vital Balance.* New York: Viking, 1963.
10. Mumford, L. *The Myth of the Machine.* New York: Harcourt, Brace & World, 1966.
11. von Bertalanffy, L. The mind-body problem: A new view. *Psychosom. Med.* 24:29, 1964.
12. von Bertalanffy, L. Mind and body re-examined. *J. Hum. Psychol.* 6:113, 1966.
13. von Bertalanffy, L. General System Theory and Psychiatry. In Arieti, S. (Ed.), *American Handbook of Psychiatry,* Vol. 3. New York: Basic Books, 1966.
14. von Bertalanffy, L. *Robots, Men and Minds.* New York: Braziller, 1967.
15. von Bertalanffy, L. *General System Theory. Foundation, Development, Applications.* New York: Braziller, 1968.
16. von Bertalanffy, L. and Rapoport, A. (Eds.) *General Systems.* Washington, D.C.: The Society for General Systems Research. 13 vols., since 1956.

DESIGNATED DISCUSSIONS

KARL A. MENNINGER, M.D.

I consider Dr. von Bertalanffy to be one of my most influential teachers, one who changed many of my opinions in various respects. Dr. von Bertalanffy came to the Menninger Foundation several years ago as Sloan Visiting Professor. Many of von Bertalanffy's ideas have recurred to me over the past years, and I would like to review some of the vivid experiences leading to the present moment.

I recall a meeting of this association nearly fifty years ago at which Southard of Harvard and Meyer of Johns Hopkins were present. Both expressed concern over the rigidity with which Kraepelinian concepts were being applied. With eagerness we seemed to accept Meyer's suggestion that we study reactive processes. The Freudians suggested we study the unconscious mind.

Twenty years later Anna Freud carefully itemized some of the ways in which the effort to cope with life becomes reflected in what we refer to as symptoms. I tried to organize Anna Freud's "defense measures" in my earlier writings and began to ponder how these could fit into the existing system of psychopathology. For some time I saw defenses as "tension-relieving devices," and I employed that term. Arthur Mirsky asked me if I were acquainted with Brody's research into homeostasis and its disturbances. Von Bertalanffy had been influenced by the physiological research of Brody, and I soon sought out Dr. von Bertalanffy. Somewhat later I was also influenced by Harlow Shapley, especially his descriptions of the diverse ways in which galaxies and subgalaxies interact.

It soon became apparent to me that we should revise the way in which we were thinking about states of disturbance we called illness, as well as the various elements or drives within a personality as it strives toward life adjustment. Reminding myself of Samuel Butler's words, "the changing and unchanging conditions within [adjusting] to the changing and unchanging conditions without," I found myself gradually and naturally attracted to what is now referred to as systems theory. Some among us would go back to the concepts of Kraepelin, but psychiatrists should do more than label and pigeon-hole. Psychiatrists are vitally interested in theoretical issues because they are related to our clinical problems. They should assist in whatever way they can with the restoration of more realistic states of balance or function than those states manifested by patients when first seen. Further, I like to think of certain aspects of adjustment as a continuum from health to disease.

Two aspects of general systems theory have always had special appeal for me. The idea of a continuum, no matter what the parameter, is a highly useful concept. The question of one's internal balance is the point which led me to von Bertalanffy's brilliant notions of steady state, in which there is a balance of homeostatic and what might be called heterostatic mechanisms. No one really wants to keep his balance very long. The real problem is to study the process of changes involved. One can assume that there is a circular motion in all human activity, and this is the essential quality of symptom groups we have called illness. In other words I do not want my sick patient to remain as he is. I want him to change, and it is to our good to expedite the changes necessary for each patient.

SILVANO E. ARIETI, M.D.

The comments I am going to make in the limited space at my disposal, though stimulated by the paper reprinted here, are based upon my knowledge of Dr. von Bertalanffy's previous writings, acquaintanceship with his basic ideas, and admiration for the man.

Although the field of general systems is by definition connected with the largest possible number of sciences, when we consider it in relation to psychiatry we must focus on certain aspects of it and run the risk of making it, so to speak, less general. We may see von Bertalanffy's life work in two distinct roles: first, as the work of a man who is dissatisfied with current concepts, with prevailing idolatries—of a man who points out the inadequacies of some of our popular formulations; secondly, we may see von Bertalanffy as the originator of new concepts which are very useful in all sciences which deal with man. I shall briefly take into consideration these two aspects of Dr. von Bertalanffy's contributions.

In his first role Dr. von Bertalanffy has often reminded us that even we psychiatrists or psychologists have, at least in our theoretical formulations, suppressed a large part of human nature. Whether we interpret the psyche according to the stimulus-response formula or as an entity which responds to a genetic code, we see it only in the framework of the doctrine of what he calls the primary reactivity: that is, we see the psyche as an entity which reacts but does not act. Spontaneity, creativity, unpredictability, and the ability to choose and will are all eliminated because they are qualities reputed to be "beyond scientific approach." The quasimachine or quasicomputer model, which is seen as constantly fed from the external world, is not a good representative of the psyche, nor is it a good representative of the anatomophysiological nature of the brain.

Von Bertalanffy has told us that, and I am going to quote him verbatim: "Our usual psychotechnique and behavioral engineering amounts to 'functional decerebration,' that is, to the functional exclusion of the higher cortical centers." Von Bertalanffy tells us that a new image of man is needed, an image which different workers from various orientations have tried to build. This new image envisions man as an active personality system, to some extent as the arbiter of his own destiny, as an entity whose goal is potentially endless—an open system, as something whose evolution follows two principles, the principle of symbolism and the principle of systemic organization. In other words the growth of man depends upon the development of symbolism and the organization of systems.

Von Bertalanffy's second role as the innovator and originator of general systems theory cannot be covered in a few minutes. Today's and tomorrow's sessions will disclose to a large extent its impact, but will not comprehend it in its entirety. As a matter of fact it is impossible to see its potential expansions and ramifications.

I shall mention only one basic point of general systems theory: the principle of equifinality. This principle states that in contrast to equilibriums in closed systems, which are determined by initial conditions, the open system may attain a time-independent state which is independent of initial conditions and determined only by the system parameters.

As I have elaborated elsewhere, this principle is important from many points of view. First of all it reminds us once more of the so-called "genetic fallacy" which has plagued many segments of psychiatry and psychology. The initial conditions, although very important, do not constitute the only factors involved in the development of a normal or abnormal psyche. A developmental approach has undoubtedly added important dimensions to the fields of psychiatry and psychoanalysis, but at times it has been applied only in the sense of giving almost exclusive importance to early conditions such as early childhood. But perhaps even more important is the fact that psychopathological structures are open systems. If they were closed systems, they would follow psychological entropy and would soon disappear. Psychopathological conditions, too, are states of high improbability which are maintained by negative psychological entropy coming from outside the system. This point of view may have important therapeutic implications for our respective disciplines.

3

Living Systems: Basic Concepts*

JAMES G. MILLER, M.D., PH.D.

GENERAL SYSTEMS THEORY is a set of related definitions, assumptions, and propositions which deal with reality as an integrated hierarchy of organizations of matter and energy. General systems behavior theory is concerned with a special subset of all systems, the living ones.

Even more basic to this presentation than the concept of system are the concepts of matter, energy, and information, because the living systems which I shall discuss are made of matter and energy organized by information.

Space and Time

In the most general mathematical sense, a *space* is a set of elements which conform to certain postulates. Euclidean space, for instance, consists of points in three dimensions which are subject to the postulates of Euclid. In a metric space a distance measure is associated with each pair of elements. In a topological space each element has a collection of neighborhoods. The conceptual spaces of mathematics may have any number of dimensions.

More concretely, *physical space* is the extension surrounding a point. It may be thought of as either the compass of the entire universe or some region of such a universe. Classically the three-dimensional geometry of Euclid was considered to describe accurately all regions in physical space. The modern general theory of relativity

* Reprinted, in modified form, with permission from *Behavioral Science* 10: 193–237, 1965.

has shown that physical space-time is more accurately described by a Riemannian geometry of four nonuniformly curved dimensions, three of space and one of time.

My presentation of a general theory of living systems will employ two sorts of spaces in which they may exist, *physical* or *geographical space* and *conceptual* or *abstracted spaces*.

PHYSICAL OR GEOGRAPHICAL SPACE

This will be considered as Euclidean space, which is adequate for the study of all aspects of living systems as we now know them. Among the characteristics and constraints of physical space are the following: (1) The distance from point A to point B is the same as that from point B to point A. (2) Matter or energy moving on a straight or curved path from point A to point B must pass through every intervening point on the path. This is true also of markers (see p. 58) bearing information. (3) In such space there is a maximum speed of movement for matter, energy, and markers bearing information. (4) Objects in such space exert gravitational pull on each other. (5) Solid objects moving in such space cannot pass through one another. (6) Solid objects moving in such space are subject to friction when they contact another object.

The characteristics and constraints of physical space affect the action of all concrete systems, living and nonliving. The following are some examples: (1) The number of different nucleotide bases—configurations in space—which a DNA molecule has determines how many bits of information it can store. (2) On the average, individuals interact more with persons who live near to them in a housing project than with persons who live far away in the project. (3) The diameter of the fuel supply lines laid down behind General Patton's advancing American Third Army in World War II determined the amount of friction the lines exerted upon the fuel pumped through them, and therefore the rate at which fuel could flow through them to supply Patton's tanks. This was one physical constraint which limited the rate at which the army could advance, because they had to halt when they ran out of fuel. (4) The small physical size of Goa in relation to India and its spatial contiguity to India were, in 1961, major determinants in the decision of erstwhile neutralist India to invade and seize it. (5) Today information can flow worldwide almost instantly by telegraph, radio, and television. In the seventeenth century it took weeks for messages to cross an ocean. A government could not send messages to its ambassadors so quickly then as it can now because of the constraints on the rate of movement of the marker bearing the information. Consequently ambassadors of that century had much more freedom of decision than they do now.

Physical space is a common space, for the reason that it is the only space in which all concrete systems, living and nonliving, exist (though some may exist in other spaces simultaneously). Physical space is shared by all scientific observers, and all scientific data must be collected in it. This is equally true for natural science and behavioral science.

Most people learn that physical space exists, which is not true of many spaces I shall mention in the next section. They can give the location of objects in it. A child probably learns of physical space by correlating the spaces presented by at least two sense modalities, such as vision (which may be distorted by such pathologies as astigmatism or aniseikonia), touch, or hearing (which may be distorted by partial or unilateral deafness). Physical space as experienced by an individual is that space which has the greatest commonality with the spaces presented by all his sense modalities.

CONCEPTUAL OR ABSTRACTED SPACES

Scientific observers often view living systems as existing in spaces which they conceptualize or abstract from the phenomena with which they deal. Examples of such spaces are:

1. Pecking order in birds or other animals.
2. Social class space, such as in Warner's scheme (lower-lower, upper-lower, lower-middle, upper-middle, lower-upper, and upper-upper classes).
3. Social distance among ethnic or racial groups.
4. Political distance among political parties of the right and left.
5. The life space of Lewin—the environment as seen by the subject, including the field forces or valences between him and objects in the environment, which can account for his immediately subsequent behavior.
6. Osgood's semantic space as determined by subjects' ratings of words on the semantic differential test.
7. Sociometric space, for example, the rating on a scale of leader-

ship ability of each member of a group by every other member.

8. A space of time costs for various modes of transportation, e.g., travel taking longer on foot than by air, longer upstream than down.
9. A space representing the shortest distances for messages to travel among various points on a telephone network. These may not be the same as the distances among those points in physical space.
10. A space of frequency of trade relations among nations.
11. A space of frequency of intermarriage among ethnic groups.

These conceptual and abstracted spaces do not have the same characteristics and are not subject to the same constraints as physical space. Each has characteristics and constraints of its own. These spaces may be either conceived of by a human being or learned about from others. Interpreting the meaning of such spaces, observing relations, and measuring distances in them ordinarily require human observers. Consequently the biases of individual human beings color these observations. Perhaps pattern-recognition computer programs can someday be written to make such observations with more objective precision.

Social and some biological scientists find conceptual or abstracted spaces useful because they recognize that physical space is not a major determinant of certain processes in the living systems they study. For example, no matter where they enter the body, most of the iodine atoms in the body accumulate in the thyroid gland. The most frequent interpersonal relations occur among persons of like interests or like attitudes rather than among geographical neighbors. Families frequently come together for holidays no matter how far apart their members are. Allies like England and Australia are often more distant from each other in physical space than they are from their enemies.

Scientists who make observations and measurements in any space other than physical should attempt to indicate precisely what the transformations are from their space to physical space. Other spaces are definitely useful to science, but physical space is the only common space in which all concrete systems exist. A scientist who makes observations and measurements in another space, which he or someone else has conceptualized, is developing a special theory. At the same time, however, he is fractionating science unless he or someone else makes an effort to indicate the relationship of the space he is working in to physical space or to other conceptual or abstracted spaces. Any transformation of one space to another is worth carrying out, and science will not be complete and unitary until transformations can be made from any given space to any other [27a]. One can, of course, conceive of spaces that cannot be transformed to other spaces, but it seems unlikely that they will apply to systems in physical space.

Not knowing at the moment how to carry out the transformation from the space in which one is making observations to another space does not prevent one from conducting profitable studies. Many useful observations about heat were made in the space of degrees of temperature before the transformation from that space to the other spaces of the centimeter-gram-second system were known.

Any scientific observations about a designated space which cannot be transformed to other spaces concern a special theory. A general theory such as I shall develop here, however, requires that observations be made in a common space or in different spaces with known transformations. This is essential because one cannot measure comparable processes at different levels of systems, to confirm or disconfirm cross-level hypotheses, unless one can measure different levels of systems or dimensions in the same spaces or in different spaces with known transformations among them. It must be possible, moreover, to make such measurements precisely enough to demonstrate whether or not there is a formal identity across levels.

TIME

This is the fundamental "fourth dimension" of the physical space–time continuum. *Time* is the particular instant at which a structure exists or a process occurs, or the measured or measurable period over which a structure endures or a process continues. For the study of all aspects of living systems as we now know them, for the measurement of durations, speeds, rates, and accelerations, the usual absolute scales of time—seconds, minutes, days, years—are adequate. The modern general theory of relativity, however, makes it clear that, particularly in the very large systems studied in astronomy, time cannot be accurately measured on any absolute scale of succession of events. Its measurement differs with the special reference frame of each particular observer, who has his own particular "clock." A concrete system can move in any direction on the spatial dimensions, but only forward—never backward—on the temporal dimension. The irreversible unidirectionality of time is related to the Second Law of Thermodynamics (pp. 59–60): a system tends to increase in entropy over time. Without new inputs higher in negentropy to the system, this process cannot be reversed in that system, and such inputs always increase the entropy outside the system. This principle has often been referred to as "time's arrow." It points only one way [34].

Matter and Energy

Matter is anything which has mass (m) and occupies space. *Energy* (E) is defined in physics as the ability to do work. The principle of the conservation of energy states that energy can be neither created nor destroyed in the universe, but it may be converted from one form to another, including the energy equivalent of rest-mass. Matter may have (1) *kinetic* energy, when it is moving and exerts a force on other matter; (2) *potential* energy, because of its position in a gravitational field; or (3) *rest-mass* energy, which is the energy that would be released if mass were converted into energy. Mass and energy are equivalent. One can be converted into the other in accordance with the relation that rest-mass energy is equal to the mass times the square of the velocity of light. Because of the known relationship between matter and energy, throughout this article I use the joint term *matter-energy* except where one or the other is specifically intended. Living systems require matter-energy, needing specific types of it, in adequate amounts. Heat, light, water, minerals, vitamins, foods, fuels, and raw materials of various kinds, for instance, may be required. Energy for the processes of living systems is derived from the breakdown of molecules (and, in a few recent cases in social systems, of atoms as well). Any change of state of matter-energy or its movement over space, from one point to another, I shall call *action*. It is one form of process. (The term "action" is here used as in biology and behavioral science rather than as in physics.)

Information

Throughout this presentation *information* (H) will be used in the technical sense first suggested by Hartley [42] and later developed by

Shannon [92] in his mathematical theory of communication. It is not the same thing as *meaning* or quite the same as *information* as we usually understand it. *Meaning* is the significance of information to a system which processes it; it constitutes a change in that system's processes elicited by the information, often resulting from associations made to it on previous experience with it. *Information* is a simpler concept: the degree of freedom that exists in a given situation to choose among signals, symbols, messages, or patterns to be transmitted. The set of all these possible categories (the alphabet) is called the *ensemble* or *repertoire*. The amount of information is measured as the logarithm to the base 2 of the number of alternate patterns, forms, organizations, or messages. (When $m^x = y$, then x is referred to as the logarithm of y to the base m.) The unit is the binary digit, or *bit* of information. It is the amount of information which relieves the uncertainty when the outcome of a situation with two equally likely alternatives is known. Legend says the American Revolution was begun by a signal to Paul Revere from Old North Church steeple. It could have been either one or two lights—"one if by land or two if by sea." If the alternatives were equally probable, the signal conveyed only one bit of information, resolving the uncertainty in a binary choice. (But it carried a vast amount of meaning, meaning which must be measured by other sorts of units than bits.)

Information is the opposite of uncertainty. It is not accidental that the word *form* appears in *information,* since information is the amount of formal patterning or complexity in any system.* Informa-

* In De Beauregard [24] we read that ". . . cybernetics is led to define '*negentropy*' and '*information*' with a sort of subjective doubling, and to admit the possibility of a transition in two senses.

$$\textit{negentropy} \rightleftarrows \textit{information}$$

"Let us note well that the meaning of the word 'information' is not the same in the two senses: in the direct transition *negentropy* → *information*, 'information' signifies acquisition of knowledge; it is the current modern sense, and the corresponding transition appears to be like the elementary process of *observation*. In the reciprocal transition *information* → *negentropy*, 'information' signifies *power of organization;* it is the ancient Aristotelian sense, and the corresponding transition appears to be like the elementary process of *action. To admit, as cybernetics does, reciprocity of the transition negentropy* ⇄ *information, is to admit ipso facto the equivalence of the two meanings, modern and Aristotelian, of the word 'information.'* " [My translation.]

And Zeman [117] writes:

"The Latin word 'informare,' from which is derived the word 'information,' signifies to put in form, to give a form or an aspect, to form, to create, but also

tion theory is a set of concepts, theorems, and measures that were first developed by Shannon for communication engineering and have been extended to other, quite different fields, including theory of organization [93] and theory of music [22].

Von Neumann [110] used the term *marker* to refer to those observable bundles, units, or changes of matter-energy whose patterning bears or conveys the informational symbols from the ensemble or repertoire.* These might be the stones of Hammurabi's day which bore cuneiform writing, parchments, writing paper, Indians' smoke signals, a doorkey with notches, punched cards, paper or magnetic tape, a computer's magnetized ferrite core memory, an arrangement of nucleotides in a DNA molecule, the molecular structure of a hormone, pulses on a telegraph wire, or waves emanating from a radio station. The marker may be static, as in a book or in a computer's memory. Communication of any sort, however, requires that the marker move in space, in relation to the receiver, and this movement follows the same physical laws as the movement of any other sort of matter-energy. The advance of communication technology over the years has been in the direction of decreasing the matter-energy costs of storing and transmitting the markers which bear information. The efficiency of information processing can be increased by lessening the mass of the markers, making them smaller so they can be stored more compactly and transmitted more rapidly and cheaply. Over the centuries engineering progress has altered the mode in markers from

---

to represent, present, create an idea or emotion. It is possible to understand information in general as whatever is put in form or in order. Information signifies the placing of several elements or parties—either material or non-material—into some form, into some classed system—that represents classification of something. Under this general form information is also the classification of symbols and of their relations in a nexus like the organization of the organs and of the functions of a living being or the organization of any social system or any other community in general. Information expresses the organization of a system, which is capable of mathematical description. It does not concern itself with the matter of that system but with the form, which can be the same for very different kinds of matter (black marks of characters on paper, neurons in the brain, ants in an ant nest, etc.).

"If mass is the measure of the effects of gravitation, and of the force of inertia, and energy the measure of movement, information is in the quantitative sense the measure of the organization of the material object. It is evident that with the characteristics of organization is linked not only the matter but also its characteristics relative to space, time, and movement. Matter, space, time, movement, and organization are in mutual connection." [My translation.]

* Christie, Luce, and Macy [21] call the physical form which the communication takes the "symbol design," and the information itself the "symbol contents."

stones bearing cuneiform to magnetic tape bearing electrons, and clearly some limit is being approached. Cuneiform tablets carried approximately of the order of 10^{-2} bits of information per gram; paper with typewritten messages carries approximately of the order of 10^3 bits of information per gram; and electronic magnetic tape storage carries approximately of the order of 10^6 bits of information per gram. If a marker can assume n different states of which only one is present at any given time, it can represent at most $\log_2 n$ bits of information.

The mass of the matter-energy which makes up a system's markers significantly affects its information processing. Bremermann [13] has estimated, on the basis of quantum-mechanical considerations, the minimum amount of energy that can serve as a marker. On the basis of this estimate he concluded that no system, living or nonliving, can process information at a rate greater than 2×10^{47} bits per second per gram of its mass. Suppose that the age of the earth is about 10^9 or 10^{10} years and its mass is less than 6×10^{27} grams. A system the size of the earth, then, could process no more than 10^{93} bits of information in a period equal to the age of the earth. This is true even if the whole system is devoted to processing information, which never happens. It becomes clear that the minimum possible size of a marker is an important constraint on the capacity of living systems when one considers Minsky's [65] demonstration that the number of all possible sequences of moves in a single chess game is about 10^{120}. Thus no earthly system, living or nonliving, could exhaustively review this many alternatives in playing a game. The human retina can certainly see more than a matrix of 100×100 spots, yet a matrix of this size can form 10^{3000} possible patterns. There are, therefore, important practical matter-energy constraints upon the information processing of all living systems exerted by the nature of the matter-energy which composes their markers.

According to Quastler [79] information measures can be used to evaluate any kind of organization, since organization is based upon the interrelations among parts. If two parts are interrelated either quantitatively or qualitatively, knowledge of the state of one must yield some information about the state of the other. Information measures can demonstrate when such relationships exist.

The antecedents of the information concepts include the early work related to thermodynamics of Maxwell [62], Planck [75], Boltzmann and Nabl [10], Helmholtz [108], and Gibbs [36]. Gibbs formulated the Law of the Degradation of Energy, or the Second Law of Thermodynamics. It states that thermodynamic degradation is irrevocable over time, e.g., a burned log cannot be unburned. This law states that "even though there is an equivalence between a certain amount of work and a certain amount of heat, yet in any cyclic process, where a system is restored to its original state, there can never be a net conversion of heat into work, but the reverse is always possible." That is, one cannot convert an amount of heat into its equivalent amount of work, without other changes taking place in the system. These changes, expressed statistically, constitute a passing of the system from ordered arrangement into more chaotic or random distribution. The disorder, disorganization, lack of patterning, or randomness of organization of a system is known as its *entropy* (S). It is the amount of progress of a system from improbable to probable states. The unit in which it is measured empirically is ergs or joules per degree absolute.

It was noted by Wiener [116] and by Shannon [92] that the statistical measure for the negative of entropy is the same as that for information, which Schrödinger [87] has called "negentropy." Discussing this relationship Rapoport [80] says:

> In classical thermodynamics, entropy was expressed in terms of the heat and the temperature of the system. With the advent of the kinetic theory of matter, an entirely new approach to thermodynamics was developed. Temperature and heat are now pictured in terms of the kinetic energy of the molecules comprising the system, and entropy becomes a measure of the *probability* that the velocities of the molecules and other variables of a system are distributed in a certain way. The reason the entropy of a system is greatest when its temperature is constant throughout is because this distribution of temperatures is the *most probable.* Increase of entropy was thus interpreted as the passage of a system from less probable to more probable states.
>
> A similar process occurs when we shuffle a deck of cards. If we start with an orderly arrangement, say the cards of all the suits following each other according to their value, the shuffling will tend to make the arrangement disorderly. But if we start with a disorderly arrangement, it is very unlikely that through shuffling the cards will come into an orderly one.

One evening in Puerto Rico I observed a concrete illustration of how information decreases as entropy progresses. Epiphany was being celebrated according to Spanish custom. On the buffet table of a large hotel stood a marvelous carving of the three kings with their camels, all done in clear ice. As the warm evening went on, they gradually melted, losing their precise patterning or information as entropy increased. By the end of the evening the camels' humps were nearly gone and the wise men were almost beardless.

Since, according to the Second Law, systems tend to increase in entropy over time, they must tend to decrease in negentropy or information. There is, therefore, no principle of the conservation of information as there are principles of the conservation of matter and energy. The total information can be decreased in any system without increasing it elsewhere, but it cannot be increased without decreasing it elsewhere. Making one or more copies of a given informational pattern does not increase information overall, though it may increase the information in the system where the information is copied. Writing an original poem or painting a new picture or composing a new concerto does not create information, but simply selects one of many possible patterns available to the medium. Creating or transmitting such patterns can have great influence on behavior of any receiver of the pattern, but this is an impact of the meaning in the pattern—not the information itself. Of course the information must be transmitted for the meaning to be transmitted.

INFORMATION AND ENTROPY

At least three sorts of evidence suggest that the relationship between information and entropy is more than a formal identity based simply on similar statistical characteristics.

First, Szilard [97] wrote a paper about Maxwell's sorting demon, which had constituted a paradox for physicists since 1871. This is a mythical being* ". . . whose faculties are so sharpened that he can follow every molecule in its course, such a being, whose attributes are still as essentially finite as our own, would be able to do what is at present impossible to us. . . . Now let us suppose that . . . a vessel is divided into two portions, *A* and *B*, by a division in which there is a small hole, and that a being, who can see the individual molecules, opens and closes this hole, so as to allow only the swifter

* Muses [67] points out the amusing fact that all three laws of thermodynamics can be expressed in demoniacal terms: "The first law excludes the existence of a demon who creates energy from nothing [Rothstein (84) calls it Aladdin's demon], the second does the same for Maxwell's demon, while the third disposes of Laplace's demon."

molecules to pass from A to B, and only the slower ones to pass from B to A. He will thus, without expenditure of work, raise the temperature of B and lower that of A, in contradiction to the second law of thermodynamics" [62]. Szilard made important progress in resolving Maxwell's paradox by demonstrating that the demon transforms information into negative entropy. Using thermodynamics and quantum mechanics he calculated the minimum amount of energy required to transmit one bit of information, i.e., the minimum marker. Brillouin [15] carried out comparable calculations of the smallest possible amount of energy used in observing one bit of information. His work was based on the assertion that unless there is light the demon cannot "see" the molecules, and that if light is introduced into the system the entropy in it increases. This supports the Second Law. Like Szilard, Brillouin employed the statistics of thermodynamics and quantum mechanics. It is clear that he believed his work to apply both to microsystems and to macrosystems.* Valentinuzzi and Valentinuzzi [100] have made calculations of the amount of information in various inorganic and organic chemical compounds. They calculated that in order to organize one bit of information in a compound approximately 10^{-12} ergs per bit is required. They sug-

* Brillouin's application of his concepts to large-scale systems is seen in the following passage. He is discussing gas in a container:

"Assume plane parallel walls and an initial situation where a plane wave is moving from one wall to the opposite one; the wave will go on propagating from left to right, be reflected, then come back from right to left, be reflected on the first wall, and so on for a certain time, until viscosity effects finally destroy the wave and change its energy into heat motion. The information contained in the wave persists for a practical length of time before it eventually disappears. The system can be used for storage of information (replace the gas by mercury and you have the mercury delay line memory system) or for communication. The case of electromagnetic waves propagating along a cable is very similar. The wave may be picked up by a receiver at the end of the cable, or it may be reflected and propagate back and forth until it finally dies out by ohmic resistance. At any rate, when information disappears, the whole system goes back to its maximum entropy value.

"Transmission or storage of information is associated with the temporary existence of the system in a state of lower entropy. The decrease in entropy can be taken as a measure of the amount of information." (Reprinted with permission from *The Journal of Applied Physics* [15].)

There has been much discussion pro and con as to whether microscopic physical entropy is the same as the entropy of functionally interdependent macrosystems like living systems. Perhaps at each level of system the principle is the same in that the entropy depends on the number of possible arrangements of the units and the particular arrangement which exists at a given moment, but the relevant units are larger and more complex at each higher level. Among the articles concerning this difficult problem are those by Linschitz [59], Augenstine [7], Baer [8], Branson [11], Linschitz [60], and Morowitz [66].

gested that such methods could be applied to calculations of the amount of information accumulated by living systems throughout growth.

Other relevant material can be found in a discussion by Foster, Rapoport, and Trucco [33] of work by Prigogine [77], De Groot [26], and others on an unresolved problem in the thermodynamics of open systems. They turn their attention to the concept of Prigogine that in an open system (that is one in which both matter and energy can be exchanged with the environment) the rate of entropy production within the system, which is always positive, is minimized when the system is in a steady state. This appears to be a straightforward generalization of the Second Law, but after studying certain electrical circuits they conclude that this theorem does not have complete generality, and that in systems with internal feedbacks (see p. 117), internal entropy production is not always minimized when the system is in a stationary state. In other words, feedback couplings between the system parameters may cause marked changes in the rate of development of entropy. Thus it may be concluded that the "information flow" which is essential for this feedback markedly alters energy utilization and the rate of development of entropy, at least in some such special cases which involve feedback control. While the explanation of this is not clear, it suggests an important relationship between information and entropy.

The other evidence is the work of Pierce and Cutler [74], who calculated the minimum amount of energy used in transmitting one bit of information, the minimum marker, in macrosystems. They arrived at the same value that Szilard [97] and Brillouin [15] independently derived for microsystems. In a communication channel with thermal noise the minimum value was calculated by Pierce and Cutler as 9.56 times 10^{-24} joules per bit per degree absolute. At the body temperature of a human being (37.0° C) for example, this would be 2.96 times 10^{-21} joules per bit. Their approach to this question was to determine how much energy is required to overcome the thermal noise in a channel, which is the unpatterned, random motion of the particles in it. The amount of this noise times the length of the channel determines the amount of energy required to increase the signal above the noise and transmit the information. There are several factors to be considered in this.

Take, for example, a satellite which is sending information. First of all, there is the "housekeeping" energy required to hold the molecules of the system together and keep it operating, maintaining the transmissions along the channel. In a satellite this involves the energy in the atoms and that holding together the molecules, as well as the energy stored in the batteries which operate the transmitter, and so forth. Then the level of thermal noise in the channel must be considered. At lower temperatures this is less, so that less energy is required to transmit information over the noise. It is very much less around absolute zero (which is why cryogenics, the study of phenomena in very cold substances, has developed ways to speed computers by keeping certain of their components very cold). It is therefore necessary to calculate the temperature of any channel above absolute zero and to compute from this a factor by which to multiply the minimal amount of energy required to transmit information at absolute zero. Furthermore, another factor must be allowed for—the lack of efficiency in whatever coding is used, the degree to which the code is less than optimal. Shannon [91] has made calculations of this sort, figuring the upper and lower bounds of the error probability in decoding optimal codes for a continuous channel with an additive Gaussian noise and subject to an average power limitation at the transmitter. Also transmitting systems ordinarily are not optimally efficient, achieving only a certain percentage of the highest possible efficiency. This means that they will need proportionately more energy to accomplish the transmission.

Of course the amount of energy actually required to transmit the information in the channel is a minute part of the total energy in the system, the "housekeeping energy" being by far the largest. For this reason it seems almost irrelevant to calculate the efficiency in terms of the energy required to transmit the information. This can be done only in situations in which the other factors accounting for more of the energy can be held constant, or in which the subsystem directly involved with the transmission is considered while all other subsystems are neglected. That is, such calculations may be important to one studying a single neuron, but when the whole brain or the entire body is considered, so many other "housekeeping" uses of energy appear that the slight changes in energy arising from information transmission may be unrecognizably small.

For such reasons information theorists tend to neglect the calculation of energy costs, so missing an important aspect of systems theory. In recent years systems theorists have been fascinated by the new ways to study and measure information flows, but matter-energy flows are equally important. Systems theory is more than information theory, since it must deal with such matters as the muscular movements of people, the flow of raw materials through societies, or the utilization of energy by brain cells.

Only a minute fraction of the energy used by most living systems is employed for information processing. Nevertheless it may well be possible in specific experimental situations to determine rigorously the minimal amount of energy required to transmit one bit of information, and so to determine for such systems a constant relationship among measures of energy, entropy, and information.

I have noted above that the movement of matter-energy over space, *action*, is one form of process (see p. 56). Another form of process is information processing or *communication*, which is the change of information from one state to another or its movement from one point to another over space.

Communications, while being processed, are often shifted from one matter-energy state to another, from one sort of marker to another. If the form or pattern of the signal remains relatively constant during these changes, the information is not lost. For instance, it is now possible to take a chest x-ray, storing the information on photographic film; then a photoscanner can pass over the film line by line, from top to bottom, converting the signals to pulses on an electrical circuit which represent bits; then those bits can be stored in the core memory of a computer; then those bits can be processed so that contrasts can be systematically heightened; then the resultant altered patterns can be printed out on a cathode ray tube and photographed. The pattern of the chest structures, the information, modified for easier interpretation, has remained largely invariant throughout all this processing from one sort of marker to another. Similar transformations go on in living systems.

One basic reason why communication is of fundamental importance is that informational patterns can be processed over space and the local matter-energy at the receiving point can be organized to conform to, or comply with, this information. As I have already said, if the information is conveyed on a relatively small, light, and compact marker, little energy is required for this process. Thus it is a much more efficient way to accomplish the result than to move the entire amount of matter-energy, organized as desired, from the location of the transmitter to that of the receiver. When Euclid was concerned with the practical problems of laying out real estate on the Nile delta, he demonstrated the truth of this principle. It was much easier for him to solve geometrical problems with symbols than to walk the distances and move the massive blocks of stone which they represented.

Shannon [92] was concerned with the transmission of information in the form of signals or messages from a sender to a receiver over a channel such as a telephone wire or a radio band. These channels always contain a certain amount of unpatterned, random activity, or "noise." This has the statistical character of entropy. In auditory transmission, as in radio, this is heard as noise, and in visual transmission, as in television, it is seen as "snow" on the screen. In order to convey a message, signals in channels must be patterned and must stand out recognizably above the background noise.

Matter-energy and information always flow together. Information is always borne on a marker. Conversely there is no regular movement in a system unless there is a difference in potential between two points, which is negative entropy or information. Which aspect of the transmission is most important depends upon how it is handled by the receiver. If the receiver responds primarily to the material or energic aspect, I shall call it, for brevity, a matter-energy transmission; if the response is primarily to the information, I shall call it an information transmission. For example, the banana eaten by a monkey is a nonrandom arrangement of specific molecules, and thus has its informational aspect, but its use to the monkey is chiefly to increase the energy available to him. So it is an energy transmission. The energic character of the signal light that tells him to depress the lever which will give him a banana is less important than the fact that the light is part of a nonrandom, patterned organization which conveys information to him. So it is an information transmission. Moreover, just as living systems must have specific forms of matter-energy, so they must have specific patterns of information. For example, many species of animals do not develop normally unless they have appropriate information inputs in infancy [9]. Harlow [41] showed, for instance, that monkeys cannot make proper social adjustments unless they interact with other monkeys during a period between the third and six months of their lives.

This treatment of the relationships of information and entropy can be epitomized by Table 3-1. It indicates that there are several pairs

TABLE 3-1
INFORMATION VERSUS ENTROPY

H	$= -$	S
Information		Uncertainty[a]
Negentropy		Entropy
Signal		Noise
Accuracy		Error
Form		Chaos
Regularity		Randomness
Pattern or form		Lack of pattern or formlessness
Order[b]		Disorder
Organization		Disorganization
Regular complexity		Irregular simplicity
Heterogeneity		Homogeneity
Improbability (only one alternative correctly describes the form)		Probability (more than one alternative correctly describes the form)
Predictability (only one alternative correctly describes the form)		Unpredictability (more than one alternative correctly describes the form)

[a] As information about a system increases, uncertainty about it decreases, and so information is the negative of uncertainty. When uncertainty about it is 0, no further information can be received about it [73].

Cf. also Quastler [78], who says that information "is related to such diverse activities as arranging, constraining, designing, determining, differentiating, messaging, ordering, organizing, planning, restricting, selecting, specializing, specifying, and systematizing; it can be used in connection with all operations which aim at decreasing such quantities as disorder, entropy, generality, ignorance, indistinctness, noise, randomness, uncertainty, variability, and at increasing the amount or degree of certainty, design, differentiation, distinctiveness, individualization, information, lawfulness, orderliness, particularity, regularity, specificity, uniqueness."

[b] Burgers [17] points out that any arrangement represents some form of order or pattern of regularity when viewed mathematically. And no form is more important or meaningful than any other. He holds that the distinction between order and disorder is made by the living observer and is not inherent in the physical world as viewed by physicists.

And Schafroth [86] observes:

"It is, in fact, no such a trivial matter to define 'disorder.' Scientists exist who have the habit of piling up papers and books in a seemingly random fashion on their desks, yet know all the time how to find a given thing. If someone brings apparent 'order' to this desk, the poor owner may be unable to find anything. In this case, it is obvious that the apparent 'disorder' is, in fact, order, and vice versa. You will see easily that in this sense the order on the desk can be measured by the information the owner has about its state. This example illustrates that, by trying to define 'disorder' more precisely, we return to the previous definition in terms of 'lack of information.'" (Reprinted with permission from H. Messel, *Selected Lectures in Modern Physics*, Macmillan, 1960.)

of antonyms used in this section, one member of which is associated with the concept of information (H) and the other member of which is associated with its negative, entropy (S). Some of these are precise, technical terms. Others are common-sense words which may be more vague. Noting that such terms as regularity, pattern, and order are listed in the column under *information* one might ask if there is not less rather than more information in a system with highly redundant pattern, order, or regularity. The answer is that knowledge about a small portion of such an arrangement gives much information about the total system, which is not true of randomness, lack of pattern, or disorder.

System

The term *system* has a number of meanings. There are systems of numbers and of equations, systems of value and of thought, systems of law, solar systems, organic systems, management systems, command and control systems, electronic systems, even the New York Central System. The meanings of *system* are often confused. The most general, however, is von Bertalanffy's: A *system* is a set of units with relationships among them [106].* The word *set* implies that the units have common properties. The state of each unit is constrained by, conditioned by, or dependent on the state of other units.† The

* He suggests that a system can be defined much as I define it, as a set of elements standing in interaction, and that this definition is not so vague and general as to be valueless. He believes these systems can be specified by families of differential equations.

† Rothstein [83] deals with the constraints among units of organized systems in terms of entropy and communication as information processing:

"What do we mean by an organization? First of all an organization presupposes the existence of parts, which, considered in their totality, constitute the organization. The parts must interact. Were there no communication between them, there would be no organization, for we would merely have a collection of individual elements isolated from each other. Each element must be associated with its own set of alternatives. Were there no freedom to choose from a set of alternatives, the corresponding element would be a static, passive cog rather than an active unit. We suggest the following general characterization of organization. Consider a set of elements, each associated with its own set of alternatives. We now define a complexion as a particular set of alternatives. There are, of course, as many complexions as there are ways of selecting a representative from each set of alternatives. The set of complexions then has an entropy which is merely the sum of the entropies of the individual sets of alternatives so long as the elements do not interact. Complexion entropy is a maximum for independent elements. Maximal entropy, i.e., zero coupling, will be said to constitute the condition of zero organization.

"On the other hand, it is possible that the coupling between elements is so units are coupled. Moreover, there is at least one measure of the sum of its units which is larger than the sum of that measure of its units.*

---

strong that only one complexion is possible. In this case, the set of complexions has zero entropy and organization is said to be maximal. All elements are then 'cogs.' In general, the interactions in which the organization consists serve to narrow the ensemble of admissible complexions. These interactions are of the nature of correlations, couplings, constraints, orders, or instructions which restrict the choices available to a given element in accordance with choices made from the ensemble of alternatives associated with other elements. The entropy of the set of complexions is thus generally less than the sum of the entropies of the individual sets of alternatives. We define the amount of organization as the excess of this maximum possible value of the complexion entropy over the entropy of the set of complexions calculated with the correlations characterizing the organization taken into account. It is easy to see that organization measures how much information has been introduced into the ensemble of complexions because of the interactions."

Ashby [6] also deals with this. He says, speaking of what organization means as applied to systems, "The hard core of the concept is, in my opinion, that of 'conditionality.' As soon as the relation between two entities *A* and *B* becomes conditional on *C*'s value or state then a necessary component of 'organization' is present. Thus *the theory of organization is partly co-extensive with the theory of functions of more than one variable*."

He goes on to ask when a system is not a system or is not organized: "The converse of 'conditional on,' is 'not conditional on,' so the converse of 'organization' must therefore be, as the mathematical theory shows us clearly, the concept of 'reducibility.' (It is also called 'separability.') This occurs, in mathematical forms, when what looks like a function of several variables (perhaps very many) proves on closer examination to have parts whose actions are *not* conditional on the values of the other parts. It occurs in mechanical forms, in hardware, when what looks like one machine proves to be composed of two (or more) sub-machines, each of which is acting independently of the others. . . .

"The treatment of 'conditionality' (whether by functions of many variables, by correlation analysis, by uncertainty analysis, or by other ways) makes us realize that the essential idea is that there is first a product space—that of the *possibilities*—within which some sub-set of points indicates the actualities. This way of looking at 'conditionality' makes us realize that it is related to that of 'communication'; and it is, of course, quite plausible that we should define parts as being 'organized' when 'communication' (in some generalized sense) occurs between them. (Again the natural converse is that of independence, which represents non-communication.)

"Now 'communication' from *A* to *B* necessarily implies some constraint, some correlation between what happens at *A* and what at *B*. If, for a given event at *A*, all possible events may occur at *B*, then there is no communication from *A* to *B* and no constraint over the possible (*A*, *B*)-couples that can occur. Thus the presence of 'organization' between variables is equivalent to the existence of a constraint in product-space of the possibilities." (Reprinted with permission from H. von Foerster and G. W. Zopf, *Principles of Self-Organization*, 1962, Pergamon Press.)

* Von Foerster [107] points out that, in systems, components or subsystems join in coalitions in which the interacting elements follow a superadditive composition rule: a measure of the sum of its units is larger than the sum of that measure of its units—$\phi(x + y) > \phi x + \phi y$. For instance, if ϕ is the square, then $(x + y)^2 > x^2 + y^2$, for $x^2 + y^2 + 2xy$ is greater than $x^2 + y^2$, by $2xy$. A man with his head is something much more than a man's body plus his separate head.

CONCEPTUAL SYSTEM

UNITS. *Units* of a *conceptual system* are terms, such as words (commonly nouns, pronouns, and their modifiers), numbers, or other symbols, including those in computer simulations and programs.

RELATIONSHIPS. *Relationships* are expressed by words (commonly verbs and their modifiers), or by logical or mathematical symbols, including those in computer simulations and programs, which represent operations like inclusion, exclusion, identity, implication, equivalence, addition, subtraction, multiplication, division, and many others. The language, symbols, or computer programs are all concepts and always exist in one or more concrete systems, living or nonliving (see pp. 70–75). The conceptual systems of science exist in one or more scientific observers, theorists, or experimentalists, or in all of them, whether human or mechanical.

THE OBSERVER. The *observer,* for his own purposes and on the basis of his own characteristics, selects, from an infinite number of units and relationships, a particular set.

VARIABLE. Each member of this set becomes a *variable* of his conceptual system. The observer may select variables from the infinite number of units and relationships which exist in any concrete system or set of concrete systems, or on the other hand he may select variables which have no connection with any concrete system [3, 73]. His conceptual system may be loose or precise, simple or elaborate.

FUNCTION. A correspondence between two variables such that a value of one depends upon a value of the other, as determined by some rule or relation (e.g., plus, multiplied by *n,* greater than) is a *function.* This is a simple conceptual system. Such systems may be very complex, involving many interrelated functions. This sense of

---

This is a recent formulation of the classical view that a system must be viewed as a gestalt, or total configuration. For example, Köhler [51] pointed out many years ago that the pattern of electrical charges on a conductor is a resultant of its particular overall form, and that practically it is impossible to synthesize such a whole from its parts. If some of the charge is withdrawn locally, the entire pattern of distribution of charges over the whole system is altered. Gestalt theory has had an important influence on current systems theory [105], just as physical field theory influenced Gestalt theory. Weiss [112, 113] recognized in an early and classical publication that the behavior of animal organisms is integral system reaction and "not simply the product of a string of component reactions." Recent research on the theory of automata [43] has investigated the necessary and sufficient conditions for decomposing a system into several simpler ones which, when operating in parallel, can produce the same output as the original system.

function is usually a mathematical usage. In a concrete system this word has a different meaning (see p. 83).

THE STATE OF A CONCEPTUAL SYSTEM. This state is the set of values on some scale, numerical or otherwise, which its variables have at a given instant. This state may or may not change over time.

FORMAL IDENTITY. One system may have one or more variables, each of which varies comparably to a variable in another system. If these comparable variations are so similar that they can be expressed by the same function, a *formal identity* or *isomorphism* exists between the two systems. If different functions are required to express the variations, there is a formal disidentity.

RELATIONSHIPS BETWEEN CONCEPTUAL AND OTHER SORTS OF SYSTEMS. A conceptual system may be purely logical or mathematical, or its terms and relationships may be intended to have some sort of formal identity or isomorphism with units and relationships empirically determinable by some operation carried out by an observer, which are selected observable variables in a concrete system or an abstracted system (see pp. 71–79; see also Pask [73]). The observer selects the variables of his conceptual system. As to the many other variables in the concrete or abstracted systems which he has not selected, the observer may either (1) observe that they remain constant, (2) operate on the concrete or abstracted system in order to assure that they remain constant, (3) "randomize them," i.e., assume without proof that they remain constant, or (4) simply neglect them.

Science advances as the formal identity or isomorphism increases between a theoretical conceptual system and objective findings about concrete or abstracted systems.

The chief purpose of this and succeeding articles is to state in prose a conceptual system concerning variables—units and relationships—which have important formal identities or isomorphisms to concrete, living systems.

CONCRETE SYSTEM

A *concrete, real,* or *veridical system* is a nonrandom accumulation of matter-energy, in a region in physical space-time, which is nonrandomly organized into coacting interrelated subsystems or components.

UNITS. The units (subsystems, components, parts, or members) of these systems are also concrete systems [40].

Relationships. Relationships in concrete systems are of various sorts, including spatial, temporal, spatiotemporal, and causal.

Both units and relationships in concrete systems are empirically determinable by some operation carried out by an observer. In theoretical verbal statements about concrete systems nouns, pronouns, and their modifiers typically refer to concrete systems, subsystems, or components; verbs and their modifiers usually refer to the relationships among them.

The observer of a concrete system. The observer, according to Campbell [18], distinguishes a concrete system from nonorganized entities in its environment by the following criteria: (1) physical proximity of its units; (2) similarity of its units; (3) common fate of its units; and (4) distinct or recognizable patterning of its units. He maintains that evolution has provided human observers with remarkable skill in using such criteria for rapidly distinguishing concrete systems. Their boundaries are discovered by empirical operations available to the general scientific community rather than set conceptually by a single observer.

Variable of a concrete system. Any property of a unit or relationship within a system which can be recognized by an observer who chooses to attend to it, which can potentially change over time, and whose change can potentially be measured by specific operations, is a variable of a concrete system. Examples are the number of its subsystems or components, its size, its rate of movement in space, its rate of growth, the number of bits of information it can process per second, or the intensity of a sound to which it responds. A variable is intrasystemic, and is not to be confused with intersystemic variations which may be observed among individual systems, types, or levels.

The state of a concrete system. The state of a concrete system at a given moment is represented by the set of values on some scale which its variables have at that instant. This state always changes over time.

Open system. Most concrete systems have boundaries which are at least partially permeable, permitting sizable magnitudes of at least certain sorts of matter-energy or information transmissions to cross them. Such a system is an *open system*. In open systems entropy may increase, remain in steady state, or decrease.

Closed system. A concrete system with impermeable boundaries through which no matter-energy or information transmissions of any sort can occur is a *closed system*. This is a special case, in which inputs and outputs are zero, of the general case of open systems. No actual concrete system is completely closed, so concrete systems therefore are relatively open or relatively closed. In closed systems entropy generally increases, exceptions being when certain reversible processes are carried on which do not increase it. It can never decrease. Whatever matter-energy happens to be within the system is all there is going to be, and it gradually becomes disordered. A body in a hermetically sealed casket, for instance, slowly crumbles and its component molecules become intermingled. Separate layers of liquid or gas in a container move toward random distribution. Gravity may prevent entirely random arrangement.

NONLIVING SYSTEM. Every concrete system which does not have the characteristics of a living system is a *nonliving system*. This is the general case of such systems, of which living systems are a very special case. Nonliving systems need not have the same critical subsystems (see pp. 105–106) as living systems, though they often have some of them.

LIVING SYSTEM. The *living systems* are a special subset of the set of all possible concrete systems. They are composed of the primitive monerans and protistans as well as the higher plants and animals. They all have the following characteristics:

1. They are open systems.
2. They maintain a steady state of negentropy even though entropic changes occur in them as they do everywhere else. This they do by taking in inputs of matter-energy higher in complexity of organization or in negative entropy, i.e., lower in entropy, than their outputs. Thus they restore their own energy and repair breakdowns in their own organization. Commenting on a remark of Schrödinger [87] that "What an organism feeds upon is negative entropy," Rapoport writes:* "Now certain processes associated with life seem to go counter to this current [of increasing entropy]. Within living organisms proteins are synthesized as well as broken down or denatured; gradients are established as well as neutralized; specific structures are created; behavior tends to go from random to specific in many learning processes, and on the grand scale biological evolution seems to

* Rapoport, A. Personal communication, 1961.

give rise to ever-increasing ordered complexity in the development of higher species." Walling off living systems to prevent exchanges across their boundaries results in what Brillouin [14] calls "death by confinement." The Second Law is an arrow pointing along the one-way road of the inevitable forward movement which we call time [34]. For walled-off living systems this means that entropy will ultimately increase, the disorganization resulting in inevitable death, but the law does not state the rate at which dissolution approaches. It might even be zero; the Second Law has no time limit.

3. They have more than a certain minimum degree of complexity.

4. They contain genetic material composed of deoxyribonucleic acid (DNA), presumably descended from some primordial DNA common to all life, or have a charter, or both. One or both of these is the template—the original "blueprint" or "program"—of their structure and process from the moment of their origin (see p. 109).

5. They are largely composed of protoplasm (containing water and proteins, constructed from about a score of amino acids and other characteristic organic compounds) and its derivatives.

6. They contain a decider, the essential critical subsystem which controls the entire system, causing its subsystems and components to coact, without which there is no system (see p. 105).

7. They also contain certain other specific critical subsystems, or they have symbiotic or parasitic relationships (see pp. 104–105) with other living or nonliving systems which carry out the processes of any such subsystem they lack.

8. These subsystems are integrated together to form actively self-regulating, developing, reproducing unitary systems, with purposes and goals (see pp. 121–125).

9. They can exist only in a certain environment. Any change in their environment of such variables as temperature, air pressure, hydration, oxygen content of the atmosphere, or intensity of radiation, outside a relatively narrow range which occurs on the surface of the earth, produces stresses to which they cannot adjust (see p. 110). Consequently they die.

Totipotential system. A living system which is capable of carrying out all critical subsystem processes necessary for life is *totipotential.*

Partipotential system. A living system which does not itself carry out all critical subsystem processes is *partipotential.* It is a special case of which the totipotential system is the general case. A partipotential system must interact with other systems that can carry out the processes which it does not, or it will not survive. Partipotential systems must be parasitic on or symbiotic with other systems, either living or nonliving, to supply the missing processes.

FULLY FUNCTIONING SYSTEM. A system is *fully functioning* when it is carrying out all the processes of which it is capable.

PARTIALLY FUNCTIONING SYSTEM. A system is *partially functioning* when it is carrying out only some of the processes of which it is capable. If it is not carrying out all the critical subsystem processes it cannot survive, unless it is parasitic on or symbiotic with some other system which supplies the missing other processes. Furthermore, it must do its own deciding or it is not a system.

ABSTRACTED SYSTEM

UNITS. The units of *abstracted systems* are relationships abstracted or selected by an observer in the light of his interests, theoretical viewpoint, or philosophical bias. Some relationships may be empirically determinable by some operation carried out by the observer, but others are not, being only his concepts.

RELATIONSHIPS. The relationships mentioned above are observed to inhere and interact or coact in concrete, usually living, systems. In a sense, then, these concrete systems are the relationships of abstracted systems. The verbal usages of theoretical statements concerning abstracted systems are often the reverse of those concerning concrete systems: the nouns and their modifiers typically refer to relationships and the verbs and their modifiers (including predicates) to the concrete systems in which these relationships inhere, interact, or coact. These concrete systems are empirically determinable by some operation carried out by the observer. A theoretical statement oriented to concrete systems typically would say "Lincoln was President," but one oriented to abstracted systems, concentrating on relationships or roles, would very likely be phrased "The Presidency was occupied by Lincoln."*

An abstracted system differs from an *abstraction*, which is a concept (like those that make up conceptual systems) representing a

* Cervinka [20] very precisely distinguishes, at the group level, between a concrete system which the author calls a "socius," that is, a single person in a group together with all his relationships, and a "groupoid," an abstracted system, which is a pattern of attachments of a single kind of relation selected by an observer, which interrelates a set of people.

class of phenomena all of which are considered to have some similar "class characteristic." The members of such a class are not thought to interact or be interrelated, as are the relationships in an abstracted system.

Abstracted systems are much more common in social science theory than in natural science.

Since abstracted systems usually are oriented toward relationships rather than toward the concrete systems which have those relationships, spatial arrangements are not usually emphasized. Consequently their physical limits often do not coincide spatially with the boundaries of any concrete system, although they may. Speaking of system hierarchies, Simon [94] says:

> There is one important difference between the physical and biological hierarchies, on the one hand, and social hierarchies, on the other. Most physical and biological hierarchies are described in spatial terms. We detect the organelles in a cell in the way we detect the raisins in a cake—they are "visibly" differentiated substructures localized spatially in the larger structure. On the other hand, we propose to identify social hierarchies not by observing who lives close to whom but by observing who interacts with whom. These two points of view can be reconciled by defining hierarchy in terms of intensity of interaction, but observing that in most biological and physical systems relatively intense interaction implies relative spatial propinquity. One of the interesting characteristics of nerve cells and telephone wires is that they permit very specific strong interactions at great distances. To the extent that interactions are channeled through specialized communications and transportation systems, spatial propinquity becomes less determinative of structure.*

There are other reasons why abstracted systems are sometimes preferred to concrete. Functionalists may resist the use of space-time coordinates because they seem static. But one must have such coordinates in order to observe and measure process. Subjectivists may resist such coordinates because their private experience does not seem to be presented to them in external space-time. But where else do their inputs arise?

Parsons [71] has attempted to develop general behavior theory using abstracted systems. An interesting colloquy at a conference on unified theory conducted by Grinker [39] spells out ways in which a

* Reprinted with permission from Herbert A. Simon, *Proceedings of the American Philosophical Society* 106:467, 1962.

theory developed around abstracted systems differs from one using concrete systems. Ruesch, Parsons, and Rapoport are speaking:

RUESCH: Previously I defined culture as the cumulative body of knowledge of the past, contained in memories and assumptions of people who express this knowledge in definite ways. The social system is the actual habitual network of communication between people. If you use the analogy of the telephone line, it corresponds to actual calls made.* The society is the network—the whole telephone network. Do you agree with these definitions?

PARSONS: No, not quite. In the limiting conception a society is composed of human individuals, organisms; but a social system is not, and for a very important reason, namely, that the unit of a partial social system is a role and not the individual.

RAPOPORT: The monarch is not an individual, but is a site into which different individuals step. Is that your unit of the social system?

PARSONS: Yes. A social system is a behavioral system. It is an organized set of behaviors of persons interacting with each other: a pattern of roles. The roles are the units of a social system. We say, "John Jones is Mary Jones' husband." He is the same person who is the mail carrier, but when we are talking about the mail carrier we are abstracting from his marriage relationship. So the mail carrier is not a person, just a role. On the other hand, the society is an aggregate of social subsystems, and as a limiting case it is that social system which comprises all the roles of all the individuals who participate.†

What Ruesch calls the social system is something concrete in space-time, observable and presumably measurable by techniques such as those of natural science. To Parsons the system is abstracted from this, being the set of relationships which are the form of organization. To him the important units are classes of input-output relationships of subsystems rather than the subsystems themselves.

Grinker [39] accurately described this fundamental, but not irresolvable, divergence when he said:

Parsons stated that . . . [action] is not concerned with the internal structure of processes of the organism, but is concerned with the organism as a unit in a set of relationships and the other terms of that relationship, which he calls situation. From this point of view the system is a system of relationship in action, it is neither a physical

* Ruesch appears to confuse structure and process in this sentence.

† Reprinted with permission from *Toward a Unified Theory of Human Behavior* edited by Roy R. Grinker, Sr., ("Boundary Relations Between Sociocultural and Personality Systems" Presentation by Talcott Parsons), Basic Books, Inc., Publishers, New York, 1956.

organism nor an object of physical perception. On the other hand, some of us consider that the foci or systems which are identified in a living field must be considered as being derived through evolution, differentiation and growth from earlier and simpler forms and functions and that within these systems there are capacities for specializations and gradients. Sets of relationships among dimensions constitute a high level of generalization that can be more easily understood if the physical properties of its component parts and their origins and ontogenetic properties are known.*

ABSTRACTED VERSUS CONCRETE SYSTEMS

To some it may appear that the distinction between concrete and abstracted systems† is something like the difference between saying "A has the property *r*" and saying "*r* is a property of A." This translation is logically trivial. In empirical work, however, there can be an important difference between discovering that A has the property *r* and finding an A which has the property *r*.

It is possible to assert connections in abstracted systems among all sorts of entities, like or unlike, near together or far apart, with or without access to each other in space—even Grandpa's moustache, Japanese haiku poetry, and the Brooklyn Bridge—depending upon the particular needs of a given project. How and why this is done will determine whether the results are trivial, like a sort of intellectual "Rube Goldberg apparatus," or whether they are functional.

A science of abstracted systems certainly is possible and under some conditions may be useful. When Euclid was developing geometry, with its practical applications to the arrangement of Egyptian real estate, the solid lines in his figures originally represented the borders of land areas or objects. Sometimes, as in Figure 3-1, he would use dotted "construction lines" to help in the conceptualization of a geometric proof. The dotted line did not correspond to any actual border in space. Triangle ABD would be shown to be congruent to triangle CBD, and therefore the angle BAD was equal to the angle BCD. After the proof was completed, the dotted line might well be erased, since it did not correspond to anything real and was useful only for the proof. Such construction lines, representing relationships

* Reprinted with permission from *Toward a Unified Theory of Human Behavior* edited by Roy R. Grinker, Sr. ("Boundary Relations Between Sociocultural and Personality Systems" Presentation by Talcott Parsons), Basic Books, Inc., Publishers, New York, 1956.

† Levy [58] makes a very similar distinction between what he calls "concrete structure" and "analytic structure."

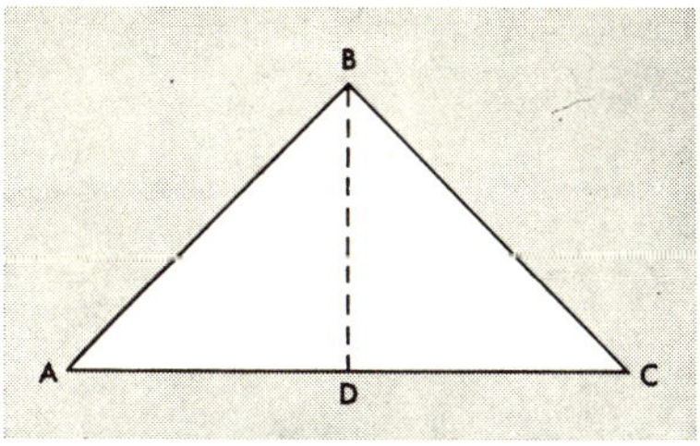

FIGURE 3-1. A EUCLIDEAN FIGURE.

among real lines, were used in the creation of early forms of abstracted systems.

If the diverse fields of science are to be unified, it would help if all disciplines were oriented either to concrete or to abstracted systems. Moreover, it is vital for science to distinguish clearly between them. To use both kinds of systems in theory leads to unnecessary problems. It would be best if one type of system or the other were generally used in all disciplines. Past tradition is not enough excuse for continuing to use both. Since one can conceive of a relationship between any concrete system and any other, one can conceive of many abstracted systems which do not correspond to any reality. The existence of such systems is often asserted in science, and frequently empirical studies show there really are no such systems.

Confusion of abstracted and concrete systems has resulted in the contention that the concept of system is logically empty because one cannot think of anything or any collection of things which could not be regarded as a system [16]. What is *not* a concrete system? Any set of subsystems or components in space-time which do not interact or coact, which do not have relationships in terms of the variables under consideration, is not a concrete system. Physicists call it a *heap*. My heart and your stomach, together, are not a concrete system; the arrangements of cells in your fingernails and in your brown felt hat are not a concrete system; the light streaming through my study window and the music floating out from my phonograph are not a concrete system. All the coal miners in Wales were not a concrete system until they were organized into an intercommunicating, coacting trade union. Sherlock Holmes assumed red-haired men in general were not a concrete system, but when he got evidence that some of them were interacting he deduced the existence of an organized Red-Headed League.

When abstracted systems are used it is essential that they be distinguished from abstractions. Is *culture* an abstraction, the class of all stored and current items of information which are shared in common by certain individuals who are members of a group, organization, or society, as revealed by similarities of those persons' customary behavior or of their artifacts—art objects, language, or writing [53]? Or does the term *culture* imply interactions among those items of information, so representing an abstracted system? Or, to take another example, is an individual's *personality* merely a class of traits as represented by repeated similar acts, gestures, and language of a person, or does it imply interactions among these traits, which would be an abstracted system? Terms such as *culture* or *personality* can be useful in behavioral science to refer to commonalities among people or among characteristics of a single person, but they must be used unambiguously as *either* an abstraction *or* an abstracted system.

No scientist, in social science or any other field, will change his traditional procedures without reason. There are, however, a number of simple, down-to-earth, practical reasons why theorists should focus upon concrete systems rather than abstracted systems:

1. In the first place it is easier. Our sense organs reify, distinguishing objects from their environment. Since childhood concrete objects (mamma, cup) have been the nouns of most of our sentences, and words representing relationships or changes in relationships (loves, runs) the verbs. We are used to seeing the world as a collection of concrete objects in space-time, and these objects naturally draw our attention. Relationships are unnecessarily complicated for our ordinary thinking processes. We are used to putting things into the framework of space and time. It helps us orient them accurately to other things. Movies which jump around in time puzzle us. We are confused when the action of a novel or play skips about in space from one place to another. For general theory embracing biological and social aspects of life and behavior, conceptualizations referring to concrete systems in space-time enable us to profit from a lifetime of experience in thinking that way. Abstracted systems are usually at best inconvenient and clumsy conceptual tools. Spatial propinquity or accessibility to information transmitted over physical channels are essential for all social interactions, except for interactions based on mutual agreements remembered from past interactions. Even then,

spatial contact in the past is essential. Spatial orientation, therefore, is important for both biological and social science: (a) It is a significant fact about cellular function that deoxyribonucleic acid (DNA) is found only within the space bounded by the nuclear membrane, while ribonucleic acid (RNA) is found on both sides of the membrane and can cross it. (b) The location of pain sensory tracts near the central canal of the human spinal cord explains why pain sensation is halted, in the bodily regions to which those tracts lead, when the disease process of syringomyelia widens the central canal until it transects the tracts. (c) The wings of the ostrich are of inadequate size to carry its large weight, so it must run rather than fly. (d) Strodtbeck and Hook [96] showed that the spatial positions of jury members around the table significantly affected their behaviors. (e) The groups which make up organizations interact most frequently and most effectively when they are close in space. Powell [76] found that differences in proximity of houses in two Costa Rican villages were associated with differences between the two villages in the frequency of visiting among families. In the village in which the houses were close to each other, 53 percent of the visiting occurred daily; in the more open village only 34 percent was daily. (f) It is well recognized by sociologists, economists, and political scientists that many sorts of behaviors are different in rural regions from those in urban areas. (g) International relations are often affected by the spatial locations and geographical characteristics of nations and the relationships of their land masses, their bodies of water, and the seas around them. The histories of the Panama Canal or the Suez Canal, of Switzerland or Cyprus attest to such geopolitical factors.

2. Variations in the units of systems appear to contribute as much or more to the total variance in the systems than variations in their relationships, although of course the total system variance arises from both, plus interactions between the two. Any cell in a given location at a given time, any ruler of a given nation in a given period, receives comparable matter-energy and information inputs. But they may act quite differently. If their inputs or relationships vary, of course their actions vary. Process of systems is explained only when we take account of both units and relationships—of cells and the internal environment around them, of the leader and the *Zeitgeist*.*

* I cannot accept Kaplan's overemphasis on relationships when he writes [49], "The inclusion of the set of essential rules in the state description of political or social action systems reflects the belief that the most important descriptive aspects

3. Theory which deals with concrete systems avoids the sort of confusion that arises when theory in social science or other areas appears to assume that actions, roles, or relationships carry on a life of their own, independent of other aspects of the people or other concrete systems whose processes they are. Or that information is transmitted from person to person without markers to bear it. When the head of a brokerage firm who is also a Sunday school teacher, in his role as chairman of the board, connives with the bookkeeper in an embezzlement, the chairman takes the Sunday school teacher right along with him into jail. They are aspects of the process of a concrete system in a suprasystem.

4. If a surgeon does not cut along planes of cleavage, he may become confused about spatial relations as he gets farther and farther into a region such as, for instance, the pelvis. Not only is it harder for him to reconstruct firm muscles when he sews up again, but it is more difficult for him to conceptualize the relationship between different structures. Behavioral scientists, if they deal with abstracted systems, easily forget the intrasystem relationships in concrete systems which influence processes within and between those systems. Consequently their understanding of the phenomena they study is often incomplete and inaccurate.

5. If the social sciences were to formulate their problems, whenever possible, in the way which has proved most convenient for the natural sciences over centuries, unification of all the sciences would be accelerated.

---

of these systems are represented in those general relationships which are independent of the specific role occupants. No matter how important labeling was to the Tarquins, sociological and political analysis of the Roman Kingdom must be directed to the social and political relationships between rulers and led rather than to the fact that a particular family, the Tarquins, was incumbent in that role. Political theory aspires to discover why such a system arose, how it operated, and why it declined. Political theory assumes that had any other family the same attributes and opportunities, the same kind of system would have arisen. Essential rules permit the investigation of types rather than of particulars."

In a later paragraph, Kaplan [49] appears to modify his view, admitting that: "An actor may fail to do something he has the capability to do if he is unaware of his capabilities. He may attempt something he is unable to do if he overestimates his capabilities." He may also not be motivated to do it, not think of it, have religious taboos against it, or act in an unexpected fashion for many other personal reasons related to his particular situation as an individual system. In 1960 many Protestants did not want a Catholic President of the United States because they thought in some decisions his personal biases might prevail. And in not supporting Federal aid for parochial schools, indeed they may have, though in a way the electorate did not expect. Kennedy may have "leaned over backward" away from the view of most of the Catholic hierarchy.

ABSTRACTED VERSUS CONCEPTUAL SYSTEMS

Because some of the relationships in abstracted systems are selected by scientific observers, theorists, or experimentalists (or all three), it is possible that they might be confused with conceptual systems, since both units and relationships of conceptual systems are so selected. The two kinds of systems differ in that some units or relationships, or both, of every abstracted system are empirically determined, and this is not true of any conceptual system.

All three meanings of *system* are useful in science, but confusion results when they are not differentiated. In my following statements the single word *system*, for brevity, will always mean *concrete system*. The other sorts of systems will always be referred to as *conceptual system* or *abstracted system*.

Structure

The *structure* of a system is the arrangement of its subsystems and components in three-dimensional space at a given moment of time [112]. This may remain relatively fixed over a long period or it may change from moment to moment, depending upon the characteristics of the process in the system. This process halted at any given moment—as when motion is frozen by a high-speed photograph—would reveal the three-dimensional spatial arrangement of the system's components as of that instant. When anatomists study structure they use dead, often fixed, material in which no further activity can be expected to occur. Similarly historians study the relationships among units of a society at a given period. These are studies of structure.

Process

All change over time of matter-energy or information in a system is *process*. If the equation describing a process is the same no matter whether the temporal variable is positive or negative, it is a *reversible* process; otherwise it is *irreversible*. Process includes the ongoing *function* of a system, reversible actions succeeding each other from moment to moment. This usage should not be confused with the mathematical usage of *function* defined earlier (pp. 70–71). Process also includes *history*, less readily reversed changes such as mutations, birth, growth, development, aging, and death; changes that commonly follow trauma or disease; and the changes resulting from learning which is not later forgotten. Historical processes alter both the structure and the function of the system. I have said "less readily reversed" instead of "irreversible" (although many such changes are in fact irreversible) because structural changes sometimes can be reversed: a component which has developed and functioned may atrophy and finally disappear with disuse; a functioning part may be chopped off a hydra and regrow. History, then, is more than the passage of time. It involves also accumulation in the system of residues or effects of past events (structural changes, memories, and learned habits). A living system carries its history with it in the form of altered structure, and consequently of altered function also. So there is a circular relation among the three primary aspects of systems—structure changes momentarily with functioning, but when such change is so great that it is essentially irreversible, a historical process has occurred, giving rise to a new structure.*

I have differentiated carefully between structure and process because often this is not done. Leighton [56] has shown that the meanings of *structure* and *function* (or *process*) are not always clearly distinguished. He contends that what is meant by structure in the study of societies is what is ordinarily called *function* in the study of bodily organs. He lists components of a sociocultural unit such as a town as: "family, including extended families; neighborhoods; associations; friendship groups; occupational associations; institutions such as those concerned with industry, religion, government, recreation, and health; cultural systems; socioeconomic classes; and finally societal roles." In my terminology not all of these are structural components of any living system, social or otherwise (see pp. 100–103).

Leighton says [56]:

> Components such as these and their arrangement in relation to each other are often called "structure" by sociologists and anthro-

* Gerard [35] uses for structure, function, and history, the terms *being, behaving*, and *becoming*. Life is fundamentally process, exquisitely controlled change over time. Merton [63] notes the recent increase in emphasis on function at all levels of systems. Science is increasingly concerned with dynamics. But all systems have both structure and process, and all general theories must deal with both.

pologists. This usage of the term parallels that of psychiatrists and psychologists when they speak of the 'structure' of personality in referring to the relationships of such components as the id, ego, and superego. In both instances the word means process. It stands for patterned events which tend to occur and recur with a certain amount of regularity. Hence, when one says that the structure of a community or a personality has such and such characteristics, he is, in effect, talking about an aspect of function.

It seems to me that "structure" as a term can be troublesome when one is trying to grasp and analyze the nature of sociocultural and psychological phenomena. This is probably not the case with those authors whose names are associated with the term, but in my experience it does confuse people new to the field, especially those from other disciplines trying to master the concepts and develop an understanding of both personality and sociocultural processes. Hence some impressions on the reasons for these difficulties may be worth recording.

The meaning attributed to "structure" by sociologists, anthropologists, psychologists, and psychiatrists is one that is limited, denotative, and reasonably clear. Trouble arises from the fact that connotative meanings are carried over from other contexts in which the word has markedly different significance. For example, the usage with reference to personality and society is dynamic, while in anatomy, in architecture, and in many everyday contexts the word refers to the static aspect of things. A structure is not something which keeps coming back in a regular flow of movement like a figure in a dance; it is something which just sits there like a chair.

Another and more important connotation is that of substance. The overwhelming force of the word in everyday usage is of an entity which can be seen and felt. It is—relative to other experiences in living—something directly available to the senses. This common meaning is also found in many sciences, particularly biology. When one speaks of the structure of the heart he is talking about visible-palpable substance, not the rhythmical contractions. The latter are an aspect of its functioning. Yet it is precisely the analogue in behavior of these contractions, this regular functional process, that is meant when one speaks of "structure" in a society. The brain offers another example. Its "structure" consists in the arrangements that can be seen with and without the aid of instruments such as the microscope—cerebellum, medulla oblongata, layers of the cortex, and so on. The recurrent electrical events called brain waves are not considered structure, but rather a manifestation of functioning. Again, however, they are the kind of phenomena which in discussions of society are called "structure." The closest analogue in the community of the anatomical use of "structure" is the arrangement of streets, houses, and other buildings.

A further point is this: in common terms, and also in biology, "structure" is for the most part a *description* of observed nature, whereas in discussions of personality and society it is usually an *inference from* observed nature. No one, for instance, has ever seen a class system in the same sense in which the layers of the body can be seen—skin, fascia, muscles, etc.*

Leighton suggests that the term *structure* is so misleading it should perhaps not be used.† He continues by saying that Hughes [47] in a personal communication suggested to him that " 'Structure' refers to configurations which pre-exist other processes that are the focus of our attention—namely the 'functions.' " Then he quotes von Bertalanffy [105] who says:

The antithesis between *structure* and *function*, *morphology* and *physiology* is based upon a static conception of the organism. In a machine there is a fixed arrangement that can be set in motion but can also be at rest. In a similar way the preestablished structure of, say, the heart is distinguished from its function, namely rhythmical contraction. Actually this separation between a pre-established structure and processes occurring in the structure does not apply to the living organism. For the organism is the expression of an everlasting orderly process, though, on the other hand, this process is sustained

* Reprinted with permission from *My Name Is Legion* by Alexander H. Leighton, M.D., Basic Books, Inc., Publishers, New York, 1959.

† But Turner, the historian, has distinguished structure and process quite distinctly. (Turner, R. Personal communication, 1960.)

"As the scientific understanding of phenomena has developed, they have come more and more to be dealt with in terms of two general concepts: (1) structure and (2) process.

"The concept *structure* predicates that a phenomenon consists of identifiable parts organized in functional relations; i.e., the parts work together as a whole. . . . In each instance the part is a whole, and the whole is a part. Where this relation appears not to exist may be taken as indicating a limit of man's knowledge rather than as an end of the relation. Therefore, it may be held that within the boundaries of observation, all phenomena enter into a structure of one kind or another.

"The concept *process* predicates that a structure under the play of external forces and through its own energy undergoes action or acts so that change affects it. By processes, therefore, structures are broken down and built up, and all structures may be conceived as having existence in terms of some process.

"Together the concepts *structure* and *process* may be seen as exhibiting the static and dynamic aspects of phenomena. In some phenomena the static aspect may appear more significant, while in others the dynamic aspect may seem decisive; actually they should be viewed together, each as a manifestation of the other. Structure, however enduring, exists in terms of process, and process, no matter how slowly or rapidly it operates, always moves through structure. Structure and process are correlative, not opposing, aspects of phenomena.

"To study human affairs in terms of the concepts *structure* and *process* would seem, therefore, to be the scientific way to an understanding of them."

by underlying structures and organized forms. What is described in morphology as organic forms and structures, is in reality a momentary cross-section through a spatio-temporal pattern.

What are called structures are slow processes of long duration, functions are quick processes of short duration. If we say that a function such as a contraction of a muscle is performed by a structure, it means that a quick and short process wave is superimposed on a long-lasting and slowly-running wave.

My terminology avoids this semantic morass. I agree with Leighton that the family, various groups, associations, and institutions are part of the structure of a town or other concrete system. The cultural systems, societal roles, and social and economic classes which Leighton refers to are, however, abstractions, relationships, or abstracted systems, unless it is demonstrated, as it sometimes is, that their members coact as organized, concrete systems.* Structure is the static arrangement of a system's parts at a moment in three-dimensional space. Process is dynamic change in the matter-energy or information of that system over time. The two are entirely different and need not be confused.

Type

If a number of individual living systems are observed to have similar characteristics they often are classed together as a *type*. Types are abstractions. Nature presents an apparently endless variety of living things which man, from his earliest days, has observed and classified —first, probably, on the basis of their threat to him, their susceptibility to capture, or their edibility, but eventually according to categories which are scientifically more useful. Classification by species is applied to organisms, plants or animals, or to free-living cells, because of their obvious relationships by reproduction. These systems are classified together by taxonomists on the basis of likeness of structure and process, genetic similarity and ability to interbreed, and local coaction, often including, in animals, ability to respond appropriately to each other's signs. The individual members of a given species are commonly units of widely separated concrete systems. The reason the species is not a concrete system is that, though all its mem-

* Hearn [44] makes an interesting effort to apply general systems behavior theory to the field of social work, but in places (cf. pp. 59–62) he confuses structure and function.

bers *can* interbreed and coact, they do so only locally and there is no overall species organization. Of course at some time in the past their ancestors did, but that may have been long ago. Complete isolation of one local set of members of a species from other local sets, after a time, may lead to the development of a new species because mutations occur in one local interbreeding set which are not spread to others of the species.

There are various types of systems at other levels of the hierarchy of living systems besides the cell and organism levels, each classed according to different structural and process taxonomic differentia. There are, for instance, primitive societies, agricultural societies, and industrial societies. There are epithelial cells, fibroblasts, red blood cells, and white blood cells, as well as free-living cells. Biological interbreeding as a way of transmitting a new system's template, which is a specialized form of information processing (see p. 109), does not occur at certain levels. At these levels—like the organization or society—it may well be, however, that the template, the "charter" information which originally "programmed" the structure and process of all individual cases of a particular type of system, had a common origin with all other templates of that type.

Types of systems often overlap one another along a given variable. Within one animal species, for instance, there may be individuals which are larger than many members of another species which on the average is much larger. Primitive societies in general have been less populous than agricultural societies, but there have been exceptions. Rank-ordering of types is also different depending upon the variable. The rabbit, though larger, seems less intelligent than the rat. He has much larger ears—more like those of a horse in size—but a very much shorter and better upholstered tail.

Level

The universe contains a hierarchy of systems, each higher *level* of system being composed of systems of lower levels.* *Atoms* are com-

* This concept is not a product of our times. It developed long ago. For instance, in the middle of the nineteenth century, Virchow [103] wrote that the scope of the life sciences must include the cellular, tissue, organism, and social levels of living organization. In modern times the concept of hierarchical levels of systems is, of course, basic to the thought of von Bertalanffy and other general systems theorists [cf. von Bertalanffy, 106]. Even some scientists not explicitly of such persuasion, who

posed of *particles; molecules*, of atoms; *crystals* and *organelles*, of molecules. About at the level of crystallizing *viruses*, like the tobacco mosaic virus, the subset of living systems begins. Viruses are neces-

---

have perhaps been skeptical in the past, recognize value in such an approach. For example, Simon [94] writes: "A number of proposals have been advanced in recent years for the development of 'general systems theory' which, abstracting from properties peculiar to physical, biological, or social systems, would be applicable to all of them. We might well feel that, while the goal is laudable, systems of such diverse kinds could hardly be expected to have any nontrivial properties in common. Metaphor and analogy can be helpful, or they can be misleading. All depends on whether the similarities the metaphor captures are significant or superficial.

"It may not be entirely vain, however, to search for common properties among diverse kinds of complex systems. The ideas that go by the name of cybernetics constitute, if not a theory, at least a point of view that has been proving fruitful over a wide range of applications. It has been useful to look at the behavior of adaptive systems in terms of the concepts of feedback and homeostasis, and to analyze adaptiveness in terms of the theory of selective information. The ideas of feedback and information provide a frame of reference for viewing a wide range of situations, just as do the ideas of evolution, of relativism, of axiomatic method, and of operationalism."

He goes on to assert that "hierarchic systems have some common properties that are independent of their specific content. . . .

"By a hierarchic system, or hierarchy, I mean a system that is composed of interrelated subsystems, each of the latter being, in turn, hierarchic in structure until we reach some lowest level of elementary subsystem. In most systems in nature, it is somewhat arbitrary as to where we leave off the partitioning, and what subsystems we take as elementary. Physics makes much use of the concept of 'elementary particle' although particles have a disconcerting tendency not to remain elementary very long. Only a couple of generations ago, the atoms themselves were elementary particles; today, to the nuclear physicist they are complex systems. For certain purposes of astronomy, whole stars, or even galaxies, can be regarded as elementary subsystems. In one kind of biological research, a cell may be treated as an elementary subsystem; in another, a protein molecule; in still another, an amino acid residue.

"Just why a scientist has a right to treat as elementary a subsystem that is in fact exceedingly complex is one of the questions we shall take up. For the moment, we shall accept the fact that scientists do this all the time, and that if they are careful scientists they usually get away with it." (Reprinted with permission from Herbert A. Simon, *Proceedings of the American Philosophical Society* 106:467, 1962.)

Leake [54] sees value in the concept of levels for contemporary theory about biological organization. He writes:

"Life begins with complex macromolecules such as genes and viruses, and here the principles of physics and chemistry directly apply. Macromolecules may be organized and integrated with many other chemical materials to form cells, which at Virchow's time were thought to be the basic units of life. Cells, however, may be organized into tissues or organs, with specific integrations serving their specific functions. These tissues and organs may further be integrated into organisms, constituting individuals such as human beings. Human beings, and indeed many other organisms, are capable of further integration and organization into societies. These societies in turn may be integrated with a more or less limited ecological environment." (Reprinted with permission from C. D. Leake, *Science* 134:2069–2079, December 29, 1961. Copyright 1961 by the American Association for the Advancement of Science.)

The view is also well stated by de Chardin [25]:

"The existence of 'system' in the world is at once obvious to every observer of nature, no matter whom.

sarily parasitic on cells, so cells are the lowest level of living systems. *Cells* are composed of atoms, molecules, and multimolecular organelles; *organs* are composed of cells aggregated into *tissues; organisms,* of organs; *groups* (*e.g.*, herds, flocks, families, teams, tribes), of organisms; *organizations,* of groups (and sometimes single individual organisms); *societies,* of organizations, groups, and individuals; and *supranational systems,* of societies and organizations. Higher levels of systems may be of mixed composition, living and nonliving. They include *planets, solar systems, galaxies,* and so forth. It is beyond my competence and the scope of this article to deal with the characteristics—whatever they may be—of systems below and above those levels

---

"The arrangement of the parts of the universe has always been a source of amazement to men. But this disposition proves itself more and more astonishing as, every day, our science is able to make a more precise and penetrating study of the facts. The farther and more deeply we penetrate into matter, by means of increasingly powerful methods, the more we are confounded by the interdependence of its parts. Each element of the cosmos is positively woven from all the others: from beneath itself by the mysterious phenomenon of 'composition,' which makes it subsistent throughout the apex of an organized whole; and from above through the influence of unities of a higher order which incorporate and dominate it for their own ends.

"It is impossible to cut into this network, to isolate a portion without it becoming frayed and unravelled at all its edges.

"All around us, as far as the eye can see, the universe holds together, and only one way of considering it is really possible, that is, to take it as a whole, in one piece." (From *Phenomenon of Man* by P. T. de Chardin. Copyright 1955 by Editions du Seuil. Copyright © 1959 by Wm. Collins Sons & Co. Ltd., London, and Harper & Row, Publishers, Incorporated, New York. Reprinted by permission of the publishers.)

Kaplan [49] has applied the concept of a hierarchy of systems to international relations: "The same variables will be used at different system levels. The international system is the most inclusive system treated by this book. National and supranational systems are subsystems of the international system. They may, however, be treated separately as systems, in which case inputs from the international system would function as parameters. This holds also for subsystems of nation states and even for personality systems."

The Panel on Basic Research and Graduate Education of the President's Science Advisory Committee of the United States in 1960 appeared also to recognize value in a general systems approach [89]. They wrote: ". . . we suggest that there is great promise in such an emerging subject as a general study of complex systems in action, within which such very large questions as the communication sciences, cognition, and large parts of biology itself might conceivably be treated as special cases."

A textbook of psychology has been written [23] which embodies a conceptualization of a hierarchy of living systems like that I advance in the present article.

And there is widespread scientific and popular implicit recognition of hierarchical levels of living systems. As one instance out of many, six banners in one of the halls of the United Nations Palais des Nations in Geneva depict six levels of social organization. They say: Family, Village, Clan, Medieval State, Nation, and Federation.

which include the various forms of life, although others have done so [68, 69]. This article, in presenting general systems behavior theory, is limited to the subset of living systems—cells, organs, organisms, groups, organizations, societies, and supranational systems.

It would be convenient for theorists if the hierarchical levels of living systems fitted neatly into each other like Chinese boxes. The facts are more complicated, as my discussion of subsystems and components indicates (see pp. 100–108). I have distinguished seven levels of living systems for analysis here, but I do not argue that there are exactly these seven, no more and no less. For example, one might conceivably separate tissue and organ into two separate levels. Or one might maintain that the organ is not a level, since there are no totipotential organs. Likewise, until at least one truly totipotential supranational system comes into being, with its own independent decider, there is, strictly, no such level.

What are the criteria for distinguishing any one level from the others? They are derived from a long scientific tradition of empirical observation of the entire gamut of living systems. This extensive experience of the community of scientific observers has led to a consensus that there are certain fundamental forms of organization of living matter-energy. Indeed the classical division of subject-matter among the various disciplines of the life or behavioral sciences is implicitly or explicitly based upon this consensus. Observers recognize that there are in the world many similar complexly organized accumulations of matter-energy, each identified by the characteristics I have already mentioned (see pp. 70–73): (1) physical proximity of its units; (2) similarity of its units; (3) common fate of its units; and (4) distinct or recognizable patterning of its units.

The distinctive patterning for any given level includes the following aspects:

1. Multiple constituent units which are systems of the sort characteristic of the next lower level; i.e., just as molecules are made up of two or more atoms and atoms are composed of two or more particles, so groups are made up of two or more organisms and organs are composed of two or more cells.

2. A boundary subsystem around its exterior, over which there is less transmission of matter-energy and information than there is within the system or within its environment.

3. A decider subsystem which receives information from all parts of the system and from the environment, makes decisions, and transmits command information which controls a significant part of the process of the units of the system.

4. Several or all the other critical subsystems carrying out their processes (see pp. 105–106), or parasitic or symbiotic interactions (see pp. 104–105) with other systems to carry out these processes.

5. A characteristic size and a characteristic duration of survival. These are not entirely dependable aspects like the others, however, because the levels overlap somewhat in both size and duration. In general a system at each higher level is larger and exists longer than the average system at the next lower level. But some mammalian nerve cells are several feet long while some lower multicellular organisms are only a few micra in diameter. Or the Bank of England, an organization, has survived much longer than many societies.

INTERSYSTEM GENERALIZATION

A fundamental procedure in science is to make generalizations from one system to another on the basis of some similarity between the systems which the observer sees and which permits him to class them together. For example, since the nineteenth century, the field of "individual differences" has been expanded, following the tradition of scientists like Galton in anthropometry and Binet in psychometrics. In Figure 3-2, states of separate specific individual systems on a specific structural or process variable are represented by I_1 to I_n. For differences among such individuals to be observed and measured, of course, a variable common to the type, along which there are individual variations, must be recognized (T_1). Physiology depends heavily, for instance, upon the fact that individuals of the type (or species) of living organism called cats are fundamentally alike, even though minor variations from one individual to the next are well recognized.

Scientists may also generalize from one type to another (T_1 to T_n).

$$\underbrace{I_1 \; . \; . \; . \; . \; I_n}$$
$$\underbrace{T_1 \; . \; . \; . \; . \; T_n}$$
$$L_1 \; . \; . \; . \; . \; L_n$$

FIGURE 3-2. INDIVIDUAL, TYPE, LEVEL.

An example is cross-species generalization, which has been commonly accepted only since Darwin. It is the justification for the patient labors of the white rat in the cause of man's understanding of himself. Rats and cats, cats and chimpanzees, chimpanzees and human beings are similar in structure, as comparative anatomists know, and in function, as comparative physiologists and psychologists demonstrate.

The amount of variance among species is greater than among individuals within a species. If the learning behavior of cat Felix is compared with that of mouse Mickey, we would expect not only the sort of individual differences which are found between Mickey and Minnie Mouse, but also greater species differences. Cross-species generalizations are common, and many have good scientific acceptability, but in making them, interindividual and interspecies differences must be kept in mind. The learning rate of men is not identical to that of white rats, and no man learns at exactly the same rate as any other.

The third type of scientific generalization indicated in Figure 3-2 is from one level to another. The basis for such generalization is the assumption that each of the levels of life, from cell to society, is composed of systems of the previous lower level. These cross-level generalizations will, ordinarily, have greater variance than the other sorts of generalizations, since they include variance among types and among individuals. But they can be made, and they can have great conceptual significance.

That there are important uniformities, which can be generalized about, across all levels of living systems is not surprising since, from cell to society, they all presumably have arisen from the same primordial template (see p. 109) or genes, diversified by evolutionary change. All are composed of the same carbon-hydrogen-nitrogen constituents, most importantly a score of amino acids organized into similar proteins and protoplasms. All are equipped to live in a water-oxygen world rather than, for example, on the methane and ammonia planets so dear to science fiction. Also they are all adapted to relatively narrow ranges of temperature, pressure, radiation, and other physical variables [45].

A formal identity among concrete systems is demonstrated by a procedure composed of three logically independent steps: (1) recognizing an aspect of two or more systems which has comparable status in those systems; (2) hypothesizing a quantitative identity between them; and (3) empirically demonstrating that identity within a certain range of error by collecting data on a similar aspect of each of the two or more systems being compared. Thus a set of observations at one level of behaving systems can be associated with findings at another, to support generalizations that are far from trivial. It may be possible to use the same conceptual system to represent two very different sorts of concrete systems, or to make models of them with the same mathematical constructs. It may even be possible to make useful generalizations which apply to all living systems at all levels. Such formal identities should include terms which also state specifically the intertype and interindividual disidentities. The identification and confirmation of these formal identities is a matter for empirical study.

In order to make it easier to recognize similarities that exist in systems of different types and levels, it is helpful to use general systems terms. These words are carefully selected according to the following criteria:

1. They should be as acceptable as possible when applied at all levels and to all types of living systems. For example, *sense organ* is one word for the subsystem that brings information into the system at the level of organisms, but *input transducer* is also satisfactory, and it is a more acceptable term for that subsystem at the society level (e.g., a diplomat, foreign correspondent, or spy) or in an electronic system. Consequently I use it. I select terms which refer to a commonality of structure or process across systems. Such a usage may irritate some specialists used to the traditional terminology of their fields. After all, one of the techniques we all use to discover whether a person is competently informed in a certain field is to determine through questioning whether he can use its specialized terminology correctly. A language which intentionally uses words that are acceptable in other fields is, of necessity, not the jargon of the specialty. Therefore whoever uses it may be suspected of not being informed about the specialty. The specialist languages, however, limit the horizons of thought to the borders of the discipline. They mask important intertype and interlevel generalities which exist and make general theory as difficult as it is to think about snow in a language that has no word for it. Since no single term can be entirely appropriate to represent a structure or process at every level, readers of general systems literature must be flexible, willing to accept a word to which they are not accustomed, so long as it is precise and accurate, if the

term is useful in revealing cross-type and cross-level generalities. I do not wish to create a new vocabulary but to select, from one level, words which are broadly applicable, and to use them in a general sense at all levels. This is done recognizing that these terms have synonyms or near synonyms which are more commonly employed at certain levels. Actually, it is impossible with the current usages of scientific language always to use general systems terms rather than type-specific and level-specific words. If that were done, the discussion would appear meaningless to experts in the field.

2. The terms should be as neutral as possible. It is preferable that they should not be associated exclusively with any type or level of system, with biological or social science, with any discipline, or with any particular school or theoretical point of view.

What are some examples of the sort of general systems terms I shall use? For a structure, *ingestor*. This is the equivalent of a number of different words used at the various levels; for example: cell—aperture in the cell membrane; organ—hilum; organism—mouth; group—the family shopper; organization—the receiving department; society—the dock workers of the country. For a process, *moving*. This is a close equivalent of: cell—contraction; organ—peristalsis; organism—walking; group—hiking; organization—moving a factory; society—nomadic wandering; supranational system—migration (but it is questionable whether any supranational system has ever done this).

All systems at each level have certain common characteristics which differentiate them from systems at other levels. Such differentia include material composition, degree of spatial cohesiveness of subsystems over time, type of boundaries, amount of mobility, average duration as a system, details of actions, and so forth. For example, the small group level is characterized by a number of components which are individual organisms. These often move actively about within the group boundary, and frequently and easily disperse to reunite at a later time. Systems at the organism level have subsystems with much less mobility in relation to each other, more fixed spatial relationships, and more readily observable boundaries. In fact, the striking differences in mobility are one of the chief reasons why many scientists have found it difficult to recognize their fundamental similarity as systems.

Within each level, systems display type differences and individual differences. No two specific organisms, even two peas, are exactly alike. No two groups have exactly identical compositions or interactions.

The importance of interindividual, intertype, or interlevel formal identities among systems, and what makes them of absorbing interest, is that very different structures, if they can be shown to carry out similar processes, may well turn out to perform in ways which are so alike that they can be very precisely described by the same formal model. Conversely, it may perhaps be shown as a general principle that subsystems with comparable structures but extremely different processes may have quantitative similarities as well.

In one of the succeeding articles I shall present numerous hypotheses about cross-level formal identities concerned with either structure or process. They are the warp of general systems behavior theory. The woof are the disidentities, differences among the levels. One such difference is represented by the various units of size employed at the various levels—cells are small, supranational systems are large. At different levels, also, typical systems characteristically endure for different lengths of time, and there are many other systematic cross-level differences in density, diffusion rates, growth rates, and so on. The ultimate task in making predictions about living systems is to learn the quantitative characteristics of the general, cross-level formal identities on the one hand and the type and individual differences on the other, combining both in a specific prediction.

It is important to follow one procedural rule in systems theory, in order to avoid confusion. Every discussion should begin with an identification of the level of reference, and the discourse should not change to another level without a specific statement that this is occurring.* Systems at the indicated level are called systems. Those at the level above are *suprasystems*, and at the next higher level, *suprasuprasystems*. Below the level of reference are *subsystems*, and below them *subsubsystems*. For example, if one is studying a cell, its organelles are the subsystems, and the tissue or organ is its suprasystem,

* Herbert [46] makes it clear that one should make the level of reference explicit. He says that often, in writing on group research, for instance, an author will change his level of reference from the leader (organism) to the group and back to a group member (organism) again without explicitly referring to the change. This produces confusing conceptual ambiguity.

unless it is a free-living cell whose suprasystem includes other living systems with which it interacts.*

I have stated that a measure of the sum of a system's units is larger than the sum of that measure of its units (see pp. 68–69 and the footnote on p. 70). Because of this, the more complex systems at higher levels manifest characteristics, more than the sum of the characteristics of the units, not observed at lower levels. These characteristics have been called *emergents*, and it is contended that significant aspects of these systems will be neglected if they are described only in terms and dimensions used for the subsystems.

Braynes, Napalkov, and Svechinskiy [12] have pointed out that the remarkable capabilities of both the computer and the human brain derive from the complex way in which the elements are combined. Individual nerve cells, and parts of the computer, have less functional scope. I agree that certain original aspects—new patterns of structure and process—are found at higher levels which are not seen at lower ones. For these new qualities new terms and dimensions are needed. But that is no reason for a complete new conceptual system. It makes for scientific unity and parsimony simply to add to the concepts needed at lower levels. Moreover, it is vital to be clear about the nature of emergents. One electronic system—a wire connecting the poles of a battery—may be able only to conduct electricity and heat the wire. Add several tubes, condensers, resistors, and controls, and it can become a radio, capable of receiving sound messages. Add dozens of other components, including a picture tube and several more controls, and it can become a television set which can receive sound and a picture. And this is not just more of the same. These are emergent capabilities the first system did not have, emergent from its special design

* Illustrative of the similarities between the approach outlined here and recent thinking about electronic system design is the following statement by Goode [38] concerning the need to identify the level of reference:

"Confusion . . . arises from consideration of the level of design. System design may be done:

"1. At the *set* level: that is, a radar, an ignition system, a navigation set. Any of these may be designed on a system engineering basis, given a need and the necessary analysis of requirements.

"2. At the *set of sets* level: thus an airplane, a telephone exchange, a missile system, each is itself a set of sets and is subject to system design.

"3. At the *set of sets of sets* level: thus an overall weapon system, a telephone system, an air traffic system, represent such sets of sets of sets."

In a similar analysis Malcolm [61] distinguishes eight hierarchical levels in a large weapon system: system, subsystem, component, assembly, subassembly, unit, unit component, and part.

of much greater complexity. But there is nothing mystical about the colored merry-go-round and racing children on the TV tube—it is the output of a system which can be completely explained by a complicated set of differential equations such as electrical engineers write, including terms representing the characteristics of all the set's components. There should be opposition only to a concept of emergence (like that held early, and later rejected, by some Gestalt psychologists [Köhler, 52]) that maintains there is some intangible character of the whole, greater than the sum of the parts, which is not susceptible to the ordinary methods of scientific analysis.

Echelon

This concept may seem superficially similar to the concept of level, but it is distinctly different. Many complex living systems, at various levels, are organized into two or more echelons. It may be that all levels are not—for example, there is no evidence that cells are. In systems with multiple echelons, the decider, an information-processing subsystem, is so organized that certain decisions (usually certain types of decisions) are made by one component of that subsystem and others by another. These components are hierarchically arranged. Each is an *echelon.* All echelons are within the same system boundary and interact similarly with the other critical subsystems of the system. Ordinarily each echelon is made up of components of the same level as those which make up every other echelon in that system. Characteristically the decider component at one echelon gets information from multiple sources at its echelon. The number of sources does not matter except that, if there is only a single one, the echelon structure is unnecessary. Such redundant structure, when it exists, often appears incongruous—ludicrous (e.g., both an admiral and a captain in a small nation which owns only one ship) or tragic (e.g., the commanding general of a division which has lost all its men in battle, except one platoon, sharing the command of that platoon with its lieutenant).

After the decider makes a decision on the basis of the information received, it is transmitted, through a single component which may or may not be the same as the decider, upward to the next higher echelon, which goes through a similar process, and so on to the top echelon. Here a final decision is made and then command information is transmitted downward to lower echelons. Characteristically information is abstracted or made more general as it proceeds upward from echelon to echelon and it is made more specific or detailed as it proceeds downward. If a given component does not decide but only passes on information, it is not functioning as an echelon. In some cases of decentralized decision-making, certain types of decisions are made at lower echelons and not transmitted to higher echelons in any form, while information relevant to other types of decisions is transmitted upward. If there are multiple parallel deciders, without a hierarchy that has subordinate and superordinate deciders, there is not one system but multiple ones.

Suprasystem

The *suprasystem* of any living system is the next higher system in which it is a component or subsystem. For example, the suprasystem of a cell or tissue is the organ it is in; the suprasystem of an organism is the group it is in at the time. Presumably every system has a suprasystem except the "universe." The suprasystem is differentiated from the *environment*. The immediate environment is the suprasystem minus the system itself. The entire environment includes this plus the suprasuprasystem and the systems at all higher levels which contain it. In order to survive the system must interact with and adjust to its environment, the other parts of the suprasystem. These processes alter both the system and its environment. It is not surprising that living systems characteristically adapt to their environment, and in return mold it. The result is that, after some period of interaction, each in some sense becomes a mirror of the other (see p. 110). For example, Emerson [32] has shown how a termite nest, an artifact of the termites as well as part of their environment, reveals to inspection by the naturalist, long after the termites have died, much detail about the social structure and function of those insects. Likewise a pueblo yields to the anthropologist facts about the life of the Indians who inhabited it centuries ago. Conversely, living systems are shaped by their environments. Sailors' skins are weathered and cowboys' legs are bowed. As Tolman [98] pointed out, each of us carries with him a cognitive map of the organization of his environment, of greater or lesser accuracy—stored information and memories which are essential for effective life in that environment.

Subsystem and Component

In every system it is possible to identify one sort of unit, each of which carries out a distinct and separate process, and another sort of unit, each of which is a discrete, separate structure. The totality of all the structures in a system which carry out a particular process is a *subsystem.* A subsystem, thus, is identified by the process it carries out. It exists in one or more identifiable structural units of the system. These specific, local, distinguishable structural units are called *components* or *members* or *parts.* I have referred to these subsystems and components in my definition of a concrete system as "a nonrandom accumulation of matter-energy, in a region in physical space-time, which is nonrandomly organized into coacting, interrelated subsystems or components" (see p. 71). There is no one-to-one relationship between process and structure. One or more processes may be carried out by two or more components. Every system is a component, but not necessarily a subsystem, of its suprasystem.

The concept of subsystem process is related to the concept of *role* used in social science [57]. Organization theory usually emphasizes the functional requirements of the system which the subsystem fulfills, rather than the specific characteristics of the component or components that make up the subsystem. The typical view is that an organization specifies clearly defined roles (or subsystem functions) and human beings "fill them" [111]. But it is a mistake not to recognize that characteristics of the component—in this case the person carrying out the role—also influence what occurs. A role is more than simple "social position," a position in some social space which is "occupied." It involves interaction, adjustments between the component and the system. It is a multiple concept, referring to the demands upon the component by the system, to the internal adjustment processes of the component, and to how the component functions in meeting the system's requirements. The adjustments it makes are frequently compromises between the requirements of the component and the requirements of the system.

It is conceivable that some systems might have no subsystems or components, although this would be true only of an ultimate particle.* The components of living systems need not be alive. Cells, for example, are composed of nonliving molecules or complexes of molecules. Systems of less than a certain degree of complexity, as I have said, cannot have the characteristics of life (see p. 74).

Often the distinction between process units and structural units, between subsystems and components, is not clearly recognized by scientists. This results in confusion. For example, the "organ" of most physiologists is a process unit, while the "organ" of many anatomists is a structural unit. Yet the same word is used for both.

Sometimes confusion is avoided by giving a unit of a system both a structural name and a title referring to the process or role it carries out. Elizabeth Windsor is a structural name and her process title is Queen.

It is notoriously hard to deduce process from structure, and the reverse is by no means easy. Thomas Wharton, the seventeenth century anatomist, demonstrated how delightfully wrong one can be in determining a subsystem's process from its structure. After carefully examining the thyroid gland, he concluded† that it has four purposes:

* The criticism has been made that it is impossible to prove or disprove the statement that every system has subsystems [16]. This is not really a problem in relation to concrete systems. If one found a system, however small, in which the distribution of matter-energy was entirely homogeneous and without parts, one could say that it did not have subsystems.

† Wharton [114]: "Usus harum glandularum primus ac praecipuo videtur, superfluas quasdam nervi recurrentis humidiates excipere, et in venosum genus denuo deducere per lymphaeductus susos;

"2. Cartilagines, alioquin frigidiores, quibus affigitur, calore suo fovere. Est enim copiosis arteriis perfusa et sanguine abundat, unde commode vicinis partibus calorem impertit;

"3. Laryngis lubricationi suis halitibus conducere adeoque vocem laeviorem, canoram, suavioremque reddere;

"4. Ad colli rotunditatem et ornatum multum contribuere, implent enim vacua spacia circa laryingem, partesque eius protuberantes fere in laevorem ac planitiem deducunt: praesertim in foeminis, quibus ab hanc causam majores obtigerunt, eorumque colla aequaliora ac venustiora reddunt."

Translation:

"It seems that the first and most important use of these glands is to extract the superfluous moisture from the recurrent nerve, and, through their lymphatic ducts, to lead it again into the venous system;

"2. To warm the colder cartilages to which [the thyroid] is attached. In fact it has numerous arteries and is richly supplied with blood, and therefore it can properly distribute warmth to the adjacent regions;

"3. To lubricate the larynx and thus to make the voice lighter, melodious, and sweeter;

"4. To contribute to the roundness and embellishment of the neck, filling the empty space around the larynx, and smoothing its protuberant parts; especially in women, who for this reason were endowed with larger [thyroid glands], the neck is made smoother and more beautiful."

(1) to serve as a transfer point for the superfluous moisture from the nerves through the lymphatic ducts to the veins which run through the gland; (2) to keep the neck warm; (3) to lubricate the larynx, so making the voice lighter, more melodious, and sweeter; and (4) to round out and ornament the curve of the neck, especially in women.

Such confusion about the process carried out by a structure can exist at any level: a lively argument still persists as to whether during President Woodrow Wilson's illness he was the nation's chief executive and decision maker, or whether it was his wife, or his physician, Dr. Cary T. Grayson. Everyone who has ever served on a committee knows that Cohen may be the chairman, but Kelly can be the leader, or vice versa.

In defining *system,* I indicated that the state of its units is constrained by, conditioned by, or dependent upon the state of other units. That is, the units are coupled (pp. 68–69). Some systems and components are also constrained by their suprasystems and subsystems. The form of allocation of process to structure determines the nature of the constraint or dependency in any given system. It is the nature of organization that each subsystem and component has some autonomy and some subordination or constraint, from lower level systems, other systems at the same level, and higher level systems. Conflicts among them are resolved by adjustment processes (see pp. 120–121).

The way living systems develop does not always result in a neat distribution of exactly one subsystem to each component. The natural arrangement would appear to be for a system to depend on one structure for one process. But there is not always such a one-to-one relationship. Sometimes the boundaries of a subsystem and a component exactly overlap, are congruent. Sometimes they are not congruent. There can be (1) a single subsystem in a single component; (2) multiple subsystems in a single component; (3) a single subsystem in multiple components; or (4) multiple subsystems in multiple components.

Systems differ markedly from level to level, type to type, and perhaps somewhat even from individual to individual, in their *patterns of allocation* of various subsystem processes to different structures. Such a process may be (1) localized in a single component; (2) combined with others in a single component; (3) dispersed laterally to other components in the system; (4) dispersed upward to the suprasystem or above; (5) dispersed downward to subsubsystems or below;

or (6) dispersed to other systems external to the hierarchy it is in. Which allocation pattern is employed is a fundamental aspect of any given system. For a specific subsystem function in a specific system one strategy results in a more efficient process than another. In all probability there are general systems principles as to which are the optimal sorts of structures to carry out specific processes. Possible examples are: Structures which minimize the distance over which matter-energy must be transported or information transmitted are the most efficient. If multiple components carry out a process, the process is more difficult to control and less efficient than if a single component does it. If one or more components which carry out a process are outside the system, the process is more difficult to integrate than if they are all in the system. Or if there are duplicate components capable of performing the same process, the system is less vulnerable to stress and therefore is more likely to survive longer, because if one component is inactivated, the other can carry out the process alone.

The following sorts of subsystems and other contents exist in living systems or are associated with them:

LOCAL SUBSYSTEM

If the boundary of a subsystem is congruent with the boundary of a component, and all its parts are contiguous in space, it is a *local subsystem,* limited to one component. The system in this case is dependent on only one component for the process.

COMBINED SUBSYSTEM

If the boundary of a subsystem is not congruent with the boundary of a component, and the subsystem is located in a smaller region than the component, sharing it with one or more other subsystems, it is a *combined subsystem.* The system in this case is dependent on part of one component for the process.

LATERALLY DISPERSED SUBSYSTEM

If the boundary of a subsystem is not congruent with the boundary of a component, and the subsystem is located in a larger region, including more than one component of the system, it is a *laterally dispersed subsystem.* In this case the system is dependent on multiple components for the process. To coordinate these components there must be a sufficient degree of communication among the parts so that they are able to coact.

JOINT SUBSYSTEM

At times a subsystem may be simultaneously a part of more than one local concrete system—for example, when one person plays the fourth position at two bridge tables or when a yeast cell is budding into two. A *joint subsystem* usually coacts with only one system at a given level at any one moment, though its relationships fluctuate rapidly. In this case the system is dependent for the process on a component it shares with another system.

UPWARDLY DISPERSED SUBSYSTEM

If the subsystem boundary is not congruent with a component boundary, but the process is carried out by a system at a higher level, it is an *upwardly dispersed subsystem.* In this case the system is dependent on a suprasystem for the process.

DOWNWARDLY DISPERSED SUBSYSTEM

If the subsystem is not congruent with any component, but the process is carried out by a subsubsystem at a lower level, it is a *downwardly dispersed subsystem.* In this case the system is dependent on a subsubsystem for the process.

OUTWARDLY DISPERSED SUBSYSTEM

If the boundary of a subsystem is not congruent with the boundary of a component, but the process is carried out by another system, living or not, it is an *outwardly dispersed subsystem.* If the other system performs the process in exchange for nothing, *parasitism* exists (see pp. 92 and 105). If it carries out the process in exchange or economic trade-off for some reward or service which constitutes a cost to the first system, *symbiosis* exists (see pp. 92 and 105). In either case the system is dependent for the process upon another system, at the same or at another level. By definition we shall not call it parasitism or symbiosis if the dependence is on the system's suprasystem or systems at higher levels which include it, or on a subsubsystem or systems at lower levels included in it. A person may be parasitically or symbiotically dependent on cells of another person (e.g., blood transfusion recipients) or on organs of another (e.g., kidney transplant recipients) or on another organism (e.g., a blind man with a leader dog) or on another group than his own family (e.g., the "Man Who Came to Dinner") or on another organization than his own (e.g., a visiting professor) or on another nation (e.g., a foreign tourist). Such assistance is required for all partipotential systems and all totipotential ones which are not functioning fully. If they did not have this aid they would not survive.

When a member of a family goes away to college he ceases to be a subsystem of the local concrete family group and becomes parasitic or symbiotic on the college organization. He may keep in sufficient touch through the use of the telephone or by mail to coordinate his plans with the family's and play a part in family interactions. The family may spend a large part of its existence in dispersed form, coming together only for reunions. The group can be coordinated by information flows so that all members convene at the same time. If the information flows break down, the group may cease to exist. Foreign secret agents who are dispersed into other social systems are sometimes detected because their secret radio messages or other information transmissions are monitored and their participation in another system discovered. The coordination and mutual influence require information flow, and the agent must communicate if he is to follow directives of his government and also send back intelligence to it.

CRITICAL SUBSYSTEMS

Certain processes are necessary for life and must be carried out by all living systems that survive or be performed for them by some other system. They are carried out by the following *critical subsystems:* reproducer, boundary, ingestor, distributor, converter, producer, matter-energy storage, extruder, motor, supporter, input transducer, internal transducer, channel and net, decoder, associator, memory, decider, encoder, and output transducer. Of these only the decider is essential, in the sense that a system cannot be parasitic or symbiotic with another system for its deciding. A living system does not exist if the decider is dispersed upwardly, downwardly, or outwardly.

Since all living systems are genetically related, have similar constituents, live in closely comparable environments, and process matter-energy and information, it is not surprising that they should have comparable subsystems and relationships among them. All systems do not have all possible kinds of subsystems. They differ individually, among types, and across levels, as to which subsystems they have and the structures of those subsystems. But all living systems either have a complete complement of the critical subsystems carrying out the functions essential to life or are intimately associated with and effectively interacting with systems which carry out the missing life functions for them.

Often there are structural cues as to which are the critical subsystems. Natural selection has wiped out those species whose critical subsystems were vulnerable to stresses in the environment. Those have survived whose critical subsystems are either duplicated (like the kidney) or especially well protected (like the brain suspended in fluid in a hard skull or the embryo suspended in amniotic fluid in the uterus). So structural characteristics may reveal secrets of process.

INCLUSION

Sometimes a part of the environment is surrounded by a system and totally included within its boundary. That is an *inclusion.* Any living system at any level may include plant, animal, or nonliving components. The amoeba, for example, ingests both inorganic and organic matter and may retain particles of iron or dye in its cytoplasm for many hours. A surgeon may replace an arteriosclerotic aorta with a plastic one and the patient may live comfortably with it for years. To the two-member group of one dog and one cat an important plant component is often added—one tree. An airline firm may have as an integral component a computerized mechanical system for making reservations which extends into all its offices. A nation includes many sorts of vegetables, minerals, buildings, and machines, as well as its land.

The inclusion is a component or subsystem of the system if it carries out or helps in carrying out a critical process of the system; otherwise it is part of the environment. Either way the system, to survive, must adjust to its characteristics. If it is harmless or inert it can often be left undisturbed. But if it is potentially harmful—like a pathogenic bacterium in a dog or a Greek in the giant gift horse within the gates of Troy—it must be rendered harmless or walled off or extruded from the system or killed. Because it moves with the system in a way the rest of the environment does not, it constitutes a special problem. Being inside the system it may be a more serious or more immediate stress than it would be outside the system's protective boundary. But also, the system that surrounds it can control its physical actions and all routes of access to it. For this reason international law has developed the concept of extraterritoriality to provide freedom of action to ambassadors and embassies, nations' inclusions within foreign countries.

An employee, an officer, or a stockholder of a company is certainly a component in that system. But is a client who enters the organization's boundary to buy or a customer who goes into a theater to see a movie? If a shopper simply wanders into a store, looks at a television set on display, and then wanders out, probably he was just an inclusion. But if a significant interaction occurs or a contract, implicit or explicit, is agreed to (as when a customer buys a ticket to enter the theater or hires a lawyer to represent him) the customer or client is an inclusion (not a component) and he is at the same time another system in the environment of the organization or firm, interacting with it in the suprasystem.

ARTIFACT

An *artifact* is an inclusion in some system, made by animals or man. Spider webs, bird nests, beaver dams, houses, books, machines, music, paintings, and language are artifacts. They may or may not be *prostheses*, inventions which carry out some critical process essential to a living system. An artificial pacemaker for a human heart is an example of an artifact which can replace a pathological process with a healthy one. Insulin and thyroxin are replacement drugs which are human artifacts. Chemical, mechanical, or electronic artifacts have been constructed which carry out some functions of all levels of living systems.

Living systems create and live among their artifacts. Beginning presumably with the hut and fire, the pot and the vase, the plow and the wheel, mankind has constructed tools and devised machines. The Industrial Revolution of the nineteenth century, capped by the recent harnessing of atomic energy, represents the extension of man's matter-energy processing ability, his muscles. A new Industrial Revolution, of even greater potential, is just beginning in the twentieth century, with the development of information and logic-processing machines, adjuncts to man's brain. These artifacts are increasingly becoming prostheses, relied on to carry out critical subsystem processes. A chimpanzee may extend his reach with a stick; a man may extend his cognitive skills with a computer. Today's prostheses include input transducers which sense the type of blood cells that pass before them and identify missiles that approach a nation's shores; photographic, mechanical, and electronic memories which can store masses of information over time; computers which can solve problems, carry out logical and mathematical calculations, make decisions, and control other machines; electric typewriters, high speed printers, cathode ray tubes, and photographic equipment which can output information. An analysis of many modern systems must take into account the novel problems which arise at man-machine interfaces.

Music is a special sort of human artifact, an information-processing artifact [22, 64]. So are the other arts. So is language. Whether it be a natural language or the machine language of some computer system, it is essential to information processing. Often stored only in human brains and expressed only by human lips, it can also be recorded on nonliving artifacts like stones, books, and magnetic tapes. It is not of itself a dynamic system that can change. It changes only when man changes it. As long as it is used it is in flux, because it must remain compatible with the ever-changing living systems that use it. But the change emanates from the users, and without their impact the language is inert. The artifactual language used in any information transmission in a system determines many essential aspects of that system's structure and process [115]. Scientists sometimes neglect to distinguish between living systems and their artifacts. Because artifacts are the products of living systems, they often mirror aspects of their producers and thus have systems characteristics of their own. Termites' nests are highly complex artifacts. So are the pots and jewelry of primitive tribes. So are music and language. These systems have important systems characteristics which can be studied for themselves alone as well as to understand the living systems that produced them. But they are not living systems. Systems theory, for instance, can be applied to linguistics because language is a system produced by living systems. But in itself it is not a living system, although the living systems that produce and use it may change it over time.

Transmissions in Concrete Systems

All process involves some sort of transmission among subsystems within a system, or among systems. There are *inputs* across the boundary into a system, *internal processes* within it, and *outputs* from it.

Each of these sorts of transmissions may consist of either (1) some particular form of matter; (2) energy, in the form of light, radiant energy, heat, or chemical energy; or (3) some particular pattern of information. The terms *input* and *output* seem preferable to *stimulus* and *response*, which are used in some of the behavioral sciences, because the former terms make it easy to distinguish whether the transmission is of matter, energy, or information, which the latter terms often conceal.

The *template*, genetic input or charter, of a system is the original information input that is the program for its later structure and process, which can be modified by later matter-energy or information inputs from its environment. Von Neumann [109] called it an *instruction.*

Steady State

When opposing variables in a system are in balance, that system is in equilibrium with regard to them. The equilibrium may be static and unchanging or it may be maintained in the midst of dynamic change. Since living systems are open systems, with continually altering fluxes of matter-energy and information, many of their equilibria are dynamic and are often referred to as *flux equilibria* or *steady states*. These may be *unstable*, in which a slight disturbance elicits progressive change from the equilibrium state—like a ball standing on an inverted bowl; or *stable*, in which a slight disturbance is counteracted so as to restore the previous state—like a ball in a cup; or *neutral*, in which a slight disturbance makes a change, but without cumulative effects of any sort—like a ball on a flat surface.

All living systems tend to maintain steady states (or homeostasis) of many variables, keeping an orderly balance among subsystems which process matter-energy or information. Not only are subsystems usually kept in equilibrium, but systems also ordinarily maintain steady states with their environments and suprasystems, which have outputs to the systems and inputs from them. This prevents variations in the environment from destroying systems. The variables of living systems are constantly fluctuating, however. A moderate change in one variable may produce greater or lesser alterations in other related ones. These alterations may or may not be reversible (see pp. 83–84).

STRESS, STRAIN, AND THREAT

There is a *range of stability* for each of numerous variables in all living systems. It is that range within which the rate of correction of deviations is minimal or zero, and beyond which correction occurs. An input or output of either matter-energy or information which, by lack or excess of some characteristic, forces the variables beyond the range of stability, constitutes a *stress** and produces a *strain* (or strains) within the system. Input lack and output excess both produce the same strain—diminished amounts in the system. Input excess and output lack both produce the opposite strain—increased amounts. Strains may or may not be capable of being reduced, depending upon their intensity and the resources of the system. The totality of the strains within a system resulting from its template program and from variations in the inputs from its environment can be referred to as its *values* (see pp. 121–122). The relative urgency of reducing each of these specific strains represents its *hierarchy of values.*

Stress may be anticipated. Information that a stress is imminent constitutes a *threat* to the system. A threat can create a strain. Recognition of the meaning of the information of such a threat must be based on previously stored (usually learned) information about such situations. A pattern of input information is a threat when—like the smell of the hunter on the wind, a change in the acidity of fluids around a cell, a whirling cloud approaching the city—it is capable of eliciting processes which can counteract the stress it presages. Processes—actions or communications—occur in systems only when a stress or a threat has created a strain which pushes a variable beyond its range of stability. A system is a constantly changing cameo and its environment is a similarly changing bas-relief, and the two at all times fit each other. That is, outside stresses of threats are mirrored by inside strains. Matter-energy storage and memory also mirror the past environment, but with certain alterations (see pp. 99–100).

LACK STRESS. Ordinarily there is a standard range of rates at which each sort of input enters a system. If the input rate falls below this range, it constitutes a *lack stress.*

* The most extensive work applying the concept of stress to living systems, at the organism level exclusively, has been done by Selye [90].

Excess stress. If the input rate goes above this range, it is an *excess stress.*

Matter-energy stress. There are various ways for systems to be stressed. One class of stresses is the *matter-energy stresses,* including: (1) matter-energy input lack or underload—starvation or inadequate fuel input; (2) input of an excess or overload of matter-energy; and (3) restraint of the system, binding it physically (this may be the equivalent of [1] or [2]).

Information stress. Also there are *information stresses,* including: (1) information input lack or underload, resulting from a dearth of information in the environment or from improper function of the external sense organs or input transducers; (2) injection of noise into the system, which has an effect of information cut-off, much like the previous stress; (3) information input excess or overload. Informational stresses may involve changes in the rate of information input or in its meaning.

The Le Chatelier principle in closed and open systems. Le Chatelier [55] stated his principle, which applies to nonliving systems and possibly also to living systems, as follows:

"Every system in chemical equilibrium undergoes, upon the variation of one of the factors of the equilibrium, a transformation in such a direction that, if it had produced itself, would have led to a variation of opposite sign to the factor under consideration." A common statement of it is: "A stable system under stress will move in that direction which tends to minimize the stress." That is, a compensatory force will develop which will tend to minimize the effect of stress; it will be exerted opposite to the stress, and it is usually accompanied by changes in other related, subsidiary variables. By this we mean system variables not primarily and directly affected by the applied stress.*

* Students of living systems at all levels have maintained that some principle like that of Le Chatelier operates to maintain variables in them in steady states. For instance, Pareto [70] described what happens to a social system if its state is artificially modified. He said: "At once a reaction will take place, tending to restore the changing form to its original state as modified by normal change."

In his early article on organisms as systems, Weiss [112, 113] wrote that "A *system* may therefore be defined as any complex which tends to retain its constancy as a whole relative to the outer world during and not withstanding the

This principle or theorem was originally stated after a consideration of the thermodynamics of closed systems, but it has been adapted for open systems by Prigogine [77]. Furthermore, a related theorem has been developed by Prigogine concerning steady states in open systems. He has stated that, for a fairly general class of cases, such steady states approach minimum entropy production. It is possible for entropy not to increase in such systems, and they are able to maintain steady states.

Figure 3-3 represents one possible model for such a system in steady state. If a ping-pong ball is held in a kitchen strainer, it is possible to blow horizontally through a straw at the ball. The faster the stream of air moves, the higher the ball will rise in the strainer, until finally it passes a critical point and goes over the edge. Then a change of state results.

Vertical downward forces (G) tend to return the ball as close as possible to the equilibrium point. Something is minimized in such systems, and it appears to be the rate of change of entropy produc-

---

changes taking place in its parts. . . . This tendency to retain constancy denotes that the systemic state represents the stable, i.e., unequivocal condition of the complex in the total field of external conditions. . . . The maintenance of constancy by the system as a whole despite alternatives of some of its parts can only be achieved by reactive changes of opposite sign in the remaining parts."

Writing about the concept he named *homeostasis,* Cannon [19] stated the proposition "that a certain degree of constancy in a complex system is itself evidence that agencies are acting or are ready to act to maintain that constancy . . . when a system remains steady it does so because any tendency toward change is met by increased effectiveness of the factor or factors which resist the change." And he went on to propose a cross-level generality: "Are there not general principles of stabilization? May not the devices developed in the animal organism for preserving steady states illustrate methods which are used, or which could be used, elsewhere? Would not a comparative study of stabilizing processes be suggestive? Might it not be useful to examine other forms of organization—industrial, domestic, or social—in the light of organization of the body?" Buck [16] has wondered whether a "nonhomeostatic" system might exist. My answer is that it could not, by the definition of concrete system used here. At least one variable must be maintained in steady state. Locating a concrete living system which is not homeostatic would create serious doubts about my whole conceptual approach.

Schumpeter [88] has used equilibrium theory extensively in economics. Ashby [1] has applied similar steady state theory to living systems generally, including brains. Easton [29] has considered its applicability to social systems generally. Kaplan [49] has applied it to political systems.

Kempf [50] notes that every level of organization (atomic, molecular, crystalline, enzymatic, protoplasmic, cellular, organismic, and social, among others) has components which interact in space and time, which function to maintain internal and external equilibria in the presence of stresses capable of disturbing them. If they did not, they would not continue to exist.

Gifford [37] has applied steady state theory to cities as systems.

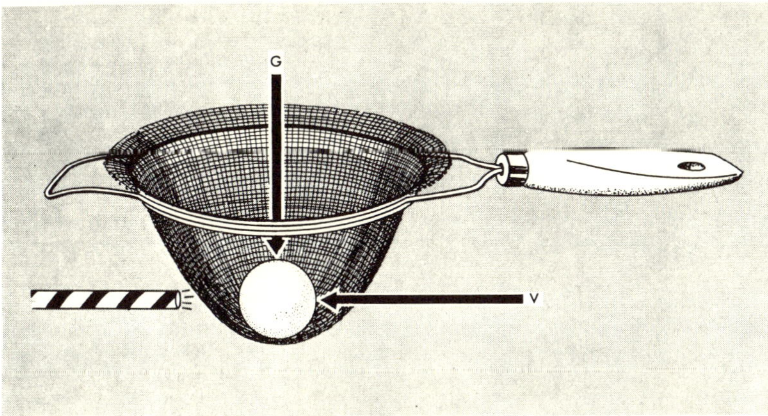

FIGURE 3-3. A STEADY STATE IN AN OPEN SYSTEM.

tion. The single variable (V) which, according to Le Chatelier's principle, tends to return the ball as close as possible to the equilibrium point, is equal and opposite in effect to the stream of air coming in. Within the system this variable or equilibratory force tends to operate at the expense of certain other associated variables related to adjustment processes of the system. There are, of course, fluctuations in these variables over time. Systems which maintain stability over long periods of time apparently tend to reduce the costs involved in the activation of these associated variables.*

ADJUSTMENT PROCESSES

Those processes of subsystems which maintain steady states in systems, keeping variables within their ranges of stability despite stresses, are *adjustment processes*. In some systems a single variable may be influenced by multiple adjustment processes. As Ashby [1] has pointed out, a living system's adjustment processes are so coupled that the system is ultrastable. This characteristic can be illustrated by the example of an army cot. It is made of wires, each of which would break under a 300-pound weight, yet it can easily support a sleeper of that weight. The weight is applied to certain wires, and as it becomes greater, first nearby links and then those farther and farther away take up part of the load. Thus a heavy weight which would break any of the component wires alone can be sustained. In

* This may be a general statement of the much misunderstood and often mystically used Principle of Least Effort of Zipf [118].

a living system, if one component cannot handle a stress, more and more others are recruited to help. Eventually the entire capacity of the system may be involved in coping with the situation.

FEEDBACK. The term *feedback* means that there exist two channels carrying information such that Channel B loops back from the output to the input of Channel A and transmits some portion of the signals emitted by Channel A [86] (see Fig. 3-4). These are tell-tales or monitors of the outputs of Channel A. The transmitter on Channel A is a device with two inputs, formally represented by a function with two independent variables, one the signal to be transmitted on Channel A and the other a previously transmitted signal fed back on Channel B. The new signal transmitted on Channel A is selected to decrease the strain resulting from any error or deviation in the feedback signal from a criterion or comparison or reference signal indicating the state of the output of Channel A which the system seeks to maintain steady. This provides control of the output of Channel A on the basis of actual rather than expected performance.

The feedback signals have a certain probability of *error*. They differ in the *lag* in time which they require to affect the system. Their lag may be minimal, so that each one is fed back to the input of the main channel before the next signal is transmitted. Or their lag may be longer and several signals may be transmitted before they arrive to affect the decision about what signal to transmit next. Feedback signals also differ in their *gain* or extent of corrective effect. When the signals are fed back over the feedback channel in such a manner that they increase the deviation of the output from a steady state, *positive feedback* exists. When the signals are reversed, so that they

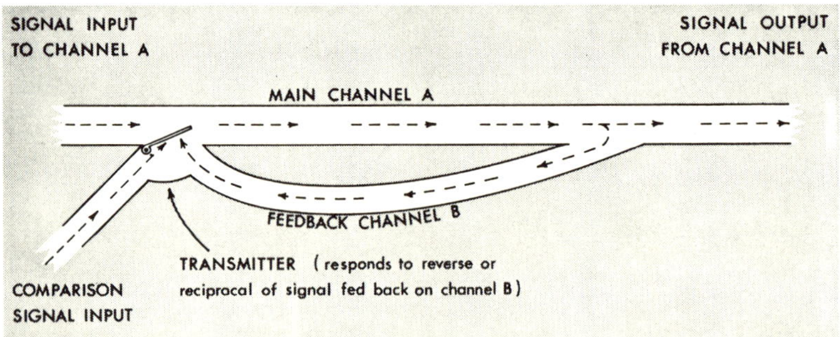

FIGURE 3-4. NEGATIVE FEEDBACK.

decrease the deviation of the output from a steady state, it is *negative feedback*. Positive feedback alters variables and destroys their steady states. Thus it can initiate system changes. Unless limited, it can alter variables enough to destroy systems. Negative feedback maintains steady states in systems. It cancels an initial deviation or error in performance. As Ashby [2] says: ". . . the importance of feedback as a necessary method for the correction of error is now accepted everywhere."

Cybernetics, the study of methods of feedback control, is an important part of systems theory. It has led to the recognition of certain formal identities among various sorts of nonliving and living systems.* In a complex system, control is achieved by many finely adjusted, interlocking processes involving tranmissions of matter-energy and information [31].

There are many such systems, living and nonliving. An automatic tracking device is one nonliving example. By means of such a device aircraft-to-aircraft fire-control systems may be set up that keep guns or missiles pointed accurately at a maneuvering target in spite of the motion of the plane in which they are mounted [99].

Steady states in all living systems are controlled by negative feedbacks. A living system is self-regulating because in it input not only affects output but output often adjusts input. The result is that the system adapts homeostatically to its environment. Elkinton and Danowski [31] point out how complex these physiological self-regulating servomechanisms of mammalian organisms are. They illustrate this by the example of bodily water balance:

> The output of water in excess of electrolyte controlled by the antidiuretic hormone in the kidney, produces a rise in extra-cellular electrolyte concentration. The rise in this concentration feeds back to the osmoreceptors in the hypothalamus to stimulate the production of antidiuretic hormone (ADH) in the supraoptico-hypophyseal system, and so the error in output of water tends to be corrected. At the same time this system is linked to regulation of intake through thirst. Hypertonicity of extra-cellular fluid with resultant cellular dehydration stimulates thirst and increased intake of water as well as the production of ADH. Thus both intake and output are regulated to minimize error in water content of the body.

* These formal identities have been repeatedly pointed out by the pioneers of cybernetics, Wiener [116] and Ashby [4]. That they are also recognized in the Soviet Union is indicated by a statement of Veltistov [101].

They go on to describe the further relationships between intake of sodium, appetite, and water balance:

> It is tempting to consider the possibility of describing all these linked servomechanisms in the organism in terms of control of energy exchange with the environment. Thus the total body content of solids and fluids is maintained in the healthy adult at a constant level with oscillation about a mean. . . . The dynamics of the body fluids are one aspect of the integrated function of the organism by which a steady state is maintained with the aid of exogenous energy ultimately derived from the sun.*

Vickers [102] describes adjustment processes of living systems in terms of feedbacks which correct deviations of systems from desirable states as follows:

> The problem for *R* [a regulating process] is to choose a way of behaving which will neutralize the disturbance threatening the maintenance of *E* [a desirable state]. Success means initiating behavior which will reduce the deviation between the actual course of affairs and the course which would be consonant with *E*; or at least preventing its nearer approach to the limit of the unacceptable or the disastrous.
>
> This decision is a choice between a limited number of alternatives. Men and societies have only a finite number of ways of behaving, perhaps a much smaller number than we realize; and the number actually available and relevant to a given situation is far smaller still. It is thus essential to regard these decisions as the exercise of restricted choice.
>
> These decisions are of four possible kinds. When the usual responses fail, the system may alter itself, for instance by learning new skills or reorganizing itself so as to make new behaviors possible; it may alter the environment; it may withdraw from the environment and seek a more favorable one; or it may alter *E*. These are possible, if at all, only within limits; and all together may prove insufficient.
>
> It remains to ask how men and societies choose from among these alternatives, when choose they must. In brief, the answer is "by experience."

At every level of living systems numerous variables are kept in a steady state, within a range of stability, by negative feedback controls. When these fail, the structure and process of the system alter

* Reprinted with permission from J. R. Elkinton and T. S. Danowski, *The Body Fluids: Basic Physiology and Practical Therapeutics*, © 1955. The Williams & Wilkins Co., Baltimore, Md. 21202, U.S.A.

markedly—perhaps to the extent that the system does not survive. Feedback control always exhibits some oscillation and always has some lag. When the organism maintains its balance in space, this lag is caused by the slowness of transmissions in the nervous system, but is only of the order of hundredths of seconds. A social institution, like a firm, may take hours to correct a breakdown in an assembly line, days or weeks to correct a bad management decision. In a society the lag can sometimes be so great that, in effect, it comes too late. General staffs often plan for the last war rather than the next. Governments receive rather slow official feedbacks from the society at periodic elections. They can, however, get faster feedbacks from the press, other mass media, picketers or demonstrators. Public opinion surveys can accelerate the social feedback process. The speed and accuracy of feedback have much to do with the effectiveness of the adjustment processes it mobilizes.

There are various different types of feedback.

Internal Feedback. Such a feedback loop never passes outside the boundary of the system. An example is the temperature-control mechanism of mammals.

External Feedback. Such a loop passes outside the system boundary; for instance, when a patient asks a nurse to fetch him an extra blanket for his bed.

Loose Feedback. Such a loop permits marked deviations from steady state, or error, before initiating corrections. In a democratic country, for instance, an elected official usually remains in office for his entire term even though his constituency disapproves of his actions.

Tight Feedback. Such a loop rapidly corrects any errors or deviations. An illustration is a tightrope-walker's balance control.

From a study of electronic systems which carry out some sort of adaptive control, Kazda* has listed five functional types of feedback. Each of these types, and combinations of them, can be found among the complexly adaptive living systems. They are:

Passive Adaptation. Achieves adaptation without changing system variables, but by altering environmental variables. Examples: a heater controlled by a thermostat; a snake's temperature control.

Input Signal Adaptation. Adapts to changes in characteristics of

* Kazda, Louis F. Personal communication, 1961.

the input signal by altering system variables. Examples: automatic radio volume control; iris of the eye.

Extremum Adaptation. Self-adjusts for a maximum or minimum of some variable. Examples: a computer which minimizes passenger waiting time for a battery of elevators; a department store buyer who purchases as cheaply as possible articles which he thinks his store can sell for the best profit.

System Variable Adaptation. Bases self-adjustment on measurement of system variables. Examples: an automatic train dispatcher; a political system which counts votes to determine policies.

System Characteristic Adaptation. Self-adjustment based on measurements made on the output of the system. Examples: an autogyro; a student who practices speaking a foreign language by listening to recordings of his own speech.

POWER. In relation to energy processing, *power* is the rate at which work is performed, work being calculated as the product of a force and of the distance through which it acts. The term also has another very different meaning. In relation to information processing, *power* is control, the ability of one "master" system to influence in a specific direction the decision of a "slave" system at the same or another level, to elicit compliance from it. The system influenced may be the system itself—a man may be his own master; it may be some subsystem or component of it; it may be its suprasystem; or it may be some external system at any level. Characteristically, in hierarchies of living systems, each level has a certain autonomy and to a degree is controlled by levels above and below it. A mutual "working agreement" thus is essential.

How is power or control exerted? A system transmits an information output, a command signal or message. Such a message has certain specific characteristics: (1) It has an *address*—it includes information indicating to what specific receiver system or systems it is transmitted, those which are to be influenced. If the channel on which it is transmitted does not branch, simply sending it on that channel gives the address information. If the channel branches, the address indicates the appropriate routing at branching points. (2) It has a *signature*—it includes information indicating what system transmitted it. If it travels on a channel that has only one transmitter, its presence on that channel gives the signature information. Simply having a form that can be uniquely produced by only one system can give the information. Or it may have specific signature symbols added to the content. (3) It contains evidence that the transmitter is a *legitimate* or appropriate source of command information to influence decisions of the receiver. In some systems commands of a certain sort are complied with regardless of the source. For example, thyroid cells respond to thyrotropic hormone regardless of whether it comes from the pituitary gland of that system or is an intravenous injection. Telephone information operators respond to requests for telephone numbers regardless of who makes them. In such systems the form of the command carries its own evidence of legitimacy. In other systems the message must include the title of the transmitter or other evidence of its legitimacy, along with the context of the command. (4) It is often literally in the *imperative mood,* styled as a command, but even when it is not couched in this form, it implies expectation of compliance. (5) The primary content of the message *specifies an action* the receiver is expected to carry out. It reinforces one alternative rather than others in a decision the receiver is constrained to make.

Why can such a message elicit compliance? At lower levels because the electrical or chemical form of what is transmitted sets off a specific reaction. At higher levels because the receiving system is part of a suprasystem that can transmit rewarding and punishing inputs to it. The receiver has learned that, because the signature indicates that the message is from a legitimate source capable of influencing some part of the suprasystem to make such inputs, there is a certain significant probability of receiving such rewards or punishments, depending on how it responds. This is why legitimacy of the source is important—it indicates that the message is from a transmitter which has an established relationship with the suprasystem and can therefore influence the receiver through it. This fact helps to determine values and purposes or goals of the system, motivating it to act in compliance with the command (pp. 121–122). Mrs. Martin, for example, can command Mrs. Wrenn's support in the women's club election because Mrs. Martin is on the committee which selects the girls to be invited to serve at the annual Christmas party and Mrs. Wrenn has a daughter who wants to serve. Consequently Mrs. Martin has "fate control" over Mrs. Wrenn, being able to influence her actions. Power among nations frequently depends on ability to trade off with other countries, a nation which can offer favorable trade inducements or foreign aid often gaining a measure of control over others.

Measures of power are joint functions of: (1) the percentage of acts of a system which are controlled, i.e., changed from one alternative to another; (2) some measure of how critical the acts controlled are to the system; (3) the number of systems controlled; and (4) the level of systems controlled, since control of one system at a high level may influence many systems at lower levels.

Certain differences among systems influence how power is wielded. As I have noted above (see pp. 103–105), systems can be either *local* or *dispersed.* Transmitting commands throughout dispersed systems requires more energy than in local systems because the components are farther apart and the markers must be dispatched over longer channels.

Systems, also, may be either *cohesive* or *noncohesive.* They are cohesive if the parts remain close enough together in space, despite any movement of the system, to make possible transmission of coordinating information along their channels. Otherwise they are noncohesive.

Systems, also, may be either *integrated* or *segregated.* If they are integrated, they are centralized, the single decider of the system exercising primary control. If they are segregated, there are multiple deciders, each controlling a subsystem or component. The more integrated a system is, the more feedbacks, commands, and information relevant to making the central decisions and implementing them flows among its parts. Therefore the more one part is likely to influence or control another. A system is more likely to be integrated if it is local rather than dispersed. Integration, of course, requires less energy in local than in dispersed systems. The degree of integration of a system is measured by a joint function of: (1) the percentage of decisions made by the system's central decider; (2) the rate at which the system accurately processes information relevant to the central decisions, without significant lag or restriction of the range of messages; and (3) the extent to which conflict among systems and components is minimized.

Conflict. In branching channels or networks commands may come to a receiver simultaneously from two or more transmitters. If these messages direct the receiver to do two or more acts which it can carry out successfully, simultaneously or successively, there is no problem. If they direct the receiver to carry out two or more actions which are incompatible—because they cannot be done simultaneously or because doing one makes it impossible later to do the other—a special sort of strain, *conflict*, arises. The incompatible commands may arise from two or more systems at the same level or at different levels. For example, two subsystems may demand more energy input and the system may be unable to meet the demands. (Jean Valjean could not provide the bread to feed his whole family.) Or two systems are in competition for a desired input, but there is not enough for both. (An embryo develops with stunted legs because the blood supply to the lower part of the body is partially blocked.) Or a system makes demands which threaten the existence of its suprasystem. (The great powers demand a veto on all significant actions of the United Nations.) An effective system ordinarily resolves such conflicts by giving greater compliance to the command with higher priority in terms of its values (see p. 110). But it may resolve the conflict by many sorts of adjustment processes.

Purpose and goal. By the information input of its charter or genetic input, or by changes in behavior brought about by rewards and punishments from its suprasystem, a system develops a preferential hierarchy of values that gives rise to decision rules which determine its preference for one internal steady state value rather than another. This is its *purpose*. It is the comparison value which it matches to information received by negative feedback in order to determine whether the variable is being maintained at the appropriate steady state value. In this sense it is normative. The system then takes one alternative action rather than another because it appears most likely to maintain the steady state. When disturbed, this state is restored by the system by successive approximations, in order to relieve the strain of the disparity recognized internally between the feedback signal and the comparison signal. Any system may have multiple purposes simultaneously.

A system may also have an external *goal*, such as reaching a target in space, or developing a relationship with any other system in the environment. Or it may have several goals at the same time. Just as there is no question that a guided missile is zeroing in on a target, so there is no question that a rat in a maze is searching for the goal of food at its end, or that the Greek people under Alexander the Great were seeking the goal of world conquest. As Ashby [5] notes, natural selection permits only those systems to continue which have goals that enable them to survive in their particular environments. The external goal may change constantly, as when a hunter chases a moving fox or a man searches for a wife by dating one girl after another, while the internal purpose remains the same [27].

It is not difficult to distinguish purposes from goals, as I use the terms: an amoeba has the purpose of maintaining adequate energy levels and therefore it has the goal of ingesting a bacterium; a boy has the purpose of keeping his body temperature in the proper range and so he has the goal of finding and putting on his sweater; Switzerland had the purpose in 1938 of remaining uninvaded and autonomous and so she sought the goal of a military organization which could keep all combatants outside her borders or disarm them if they crossed them.

A system's hierarchy of values determines its purposes as well as its goals. The question is often asked of the words *goal* and *purpose*, as it is of the word *value*, whether they are appropriately defined as whatever is *actually* preferred or sought by the system, or as what *should* be preferred or sought. I shall use it in the former sense, unless I indicate that the latter sense is being employed. When the latter meaning is used, I shall not imply that the norm as to what the goal should be is established in any absolute way, but rather that it is set by the system's suprasystem when it originates its template, or by rewards and punishments. Ashby [6] has said that "there is no property of an organization that is good in any absolute sense; all are relative to some given environment, or to some given set of threats and disturbances, or to some given set of problems." A system is adjusted to its suprasystem only if it has an internal purpose or external goal which is consistent with the norm established by the suprasystem. Since this is not always true, it is important to distinguish the two notions of the actual and the normative.

The reason it is important to a receiver whether a command signal is transmitted from a legitimate source (see p. 119) is that, if it is legitimate, it can influence the suprasystem to make reward and punishment inputs to the receiver and so potentially can alter both its process and its goals.

It is necessary to distinguish two meanings of the term *purpose*. One is function or role (see p. 100) of the system in the suprasystem, and the other, independent concept is the internally determined control process of the system which maintains one of its variables at a given steady state value. Rosenblueth, Wiener, and Bigelow [82], in their early paper on cybernetics, saw rudimentary purposive behavior in some nonliving systems, a torpedo which can home to a moving target. The concept of purpose has been made suspect to most scientists by teleological formulations which suggest that living systems strive for mystical ends which are not clearly formulated. These formulations are from the viewpoint of the scientific observer. On this topic, Rothstein [85] has written that:

> . . . one would not introduce the notion of purpose unless the system were only partially specified. With complete specification the "stimulus" is specified, likewise the action of the regulator and ditto the response of the system. It is only when an ensemble of possible stimuli is considered and *no information is available* to predict *a priori* which of the ensemble will materialize that one is motivated to introduce the concept of purpose.
>
> . . . One can say the initial state causes the final state, or that the final state is the purpose of the initial state. In this form one can object that the concept of purpose has been reduced to an empty play on words. However, consider an experimenter interested in producing some particular situation. In many cases he sets up an initial configuration from which the desired situation will ensue because of the laws the system obeys. The final situation is the goal or purpose of the experimenter, which has determined his choice of initial conditions. In this sense, we can call his purpose the cause of the initial condition. For completely defined physical systems, there is thus no logical distinction between cause and purpose as either determines the other. Meaningful distinctions are only possible in terms of considerations extrinsic to the system. It now follows that physics is as incapable of finding a purpose or goal of the whole universe as it is of finding its origin or cause.

Rothstein believes that the next-to-last sentence is true of systems in general.

But if *purpose* is defined not in terms of the observer but in terms of specific values of internal variables which systems maintain in steady states through corrective actions, then the concept is scientifically useful. Sommerhoff [95] reinterprets *purpose* in concepts of modern physics, maintaining that the notion concerns a certain future event, a "focal condition" (in my terms, a *goal*). This focal condition, he says, is a determinant of a "directive correlation." Such a correlation is characteristically found between processes in living systems and in their environments. Variables in them are so "geared" or interrelated that, within certain ranges, they will at a later time only bring about the focal condition. Such a situation requires that there be some prior state of affairs which gives rise jointly both to the processes in the system and to those in its environment. Feedback is one way such joint causation can be accomplished. Sommerhoff believes that this sort of process explains such phenomena of living systems as adaptation of individuals and species to their environments, coordination and regulation of internal system processes, repair of systems after trauma, and various sorts of behavior including learning, memory, and decision-making. For example, one cannot distinguish between products which are put out by a system and wastes which are excreted without knowing the purpose of the system internally and its related goals in the suprasystem. This is graphically demonstrated by the following "Ballad of the Interstellar Merchants" [30]:

> Among the wild Reguleans
> we trade in beer and hides
> for sacks of mMomimotl leaves
> and carcasses of brides.
>
> They love 'em and they leave 'em,
> once affection's been displayed,
> to the everloving merchants
> of the Interstellar Trade.
>
> Chorus: Don't throw that bride away, friends
> don't turn that carcass loose.
> What's only junk on Regulus
> is gold on Betelgeuse. . . .

Engineers must know the purposes which a machine is to fulfill, what steady state values its variables are to have, before they begin to design it. This may or may not be related to some purpose or function in the suprasystem. Occasionally comics have built apparatuses with wheels, cogs, gears, pistons, and cams that merely operate, without any useful function in the suprasystem, or gadgets that function only to turn themselves off. If one is to understand a system, know what it is to optimize, or measure its efficiency (i.e., the ratio between the effectiveness of its performance and the costs involved), one must learn its expected function or purpose in the suprasystem. The charter of a group, organization, society, or supranational system describes this. Biologists, however, have a difficult time defining the functions of a cell, organ, or organism, except in terms of the survival of the system itself, or of the organism of which it is a part, or of its particular type.

Such facts as that a normal sea urchin can develop either from a complete egg or from a half egg led Driesch [27] to embrace vitalism, the doctrine that phenomena of life cannot be explained in natural science terms. This sort of *equifinality*, he contended, could be explained only by some mystical vitalism. Equifinality means that a final state of any living system may be reached from different initial conditions and in different ways.* But this is exactly what all cybernetic systems do, living and nonliving.

Von Bertalanffy [104, 106] has opposed Driesch's views on the basis of an analysis of living systems as open systems. The steady states of open systems depend upon system constants more than environmental conditions, so long as the environment has a surplus of essential inputs. Within a wide range of inputs the composition of living tissue, for example, remains relatively constant. Of course —and von Bertalanffy does not always make this clear—inputs outside the "normal" range may destroy the system or affect its structure and functioning. Each separate system, moreover, has its own history, different from others of its kind, and therefore any final state is affected by the various preceding genetic and environmental influences which have impinged upon the system. All organisms do not develop into perfect adulthood, and presumably each single cell may have slightly different characteristics as a result of its history. These limitations upon von Bertalanffy's principle do not destroy its importance. The obvious purposive activities of most living systems, which have seemed to many to require a vitalistic or teleological interpretation, can be explained as open system characteristics by means of this principle. Some open physical systems also have this characteristic.

* A closely related concept, *finalism*, was described by Rignano [81].

COSTS. All adjustment processes have their *costs,* in energy of nonliving or living systems, in material resources, in information (including, in social systems, a special form of information often conveyed on a marker of metal or paper—money), or in time required for an action. Any of these may be scarce. (Time is a scarcity for mortal systems.) Any of these is valued if it is essential for reducing strains. The costs of adjustment processes differ from one to another and from time to time. They may be immediate or delayed, short-term or long-term. Systems constantly make economic decisions which increase efficiency by improving performance and decreasing costs. How efficiently a system adjusts to its environment is determined by what strategies it employs in selecting adjustment processes and whether they satisfactorily reduce strains without being too costly. This decision process can be analyzed by a mathematical approach to economic decisions, game theory. This is general theory concerning the best strategies for weighing "plays" against "pay-offs," selecting actions which will increase profits while decreasing losses, increase rewards while decreasing punishments, improve adjustments of variables to appropriate steady state values, or attain goals while diminishing costs. Relevant information available to the decider can improve such decisions. Consequently it is valuable. But there are costs to obtaining such information. Hurley [48] has developed a mathematical theory on how to calculate the value of relevant information in such decisions. This depends on such considerations as whether it is tactical (about a specific act) or strategic (about a policy for action); whether it is reliable or unreliable, overtly or secretly obtained, accurate, distorted, or erroneous.

Conclusions

My analysis of living systems uses concepts of thermodynamics, information theory, cybernetics, and systems engineering, as well as the classical concepts appropriate to each level. The purpose is to produce a description of living structure and process in terms of input and output, flows through systems, steady states, and feedbacks, which will clarify and unify the facts of life.

In such fundamental considerations it would be surprising if many new concepts appear, for countless good minds have worked long on these matters over many years. Indeed, new, original ideas should at first be suspect, though if they withstand examination they should be welcomed. My intent is not to create a new school or art form, but to discern the pattern of a mosaic which lies hidden in the cluttered, colored marble chips of today's empirical facts. I may assert, along with Pascal [72], "Let no man say that I have said nothing new—the arrangement of the material is new. In playing tennis, we both use the same ball, but one of us places it better. I would just as soon be told that I have used old terms. Just as the same thoughts differently arranged form a different discourse, so the same words differently arranged form different thoughts."

References

1. Ashby, W. R. *Design for a Brain.* New York: Wiley, 1954. Pp. 153–158, 210–211.
2. Ashby, W. R. Cybernetics. In G. W. T. H. Fleming and A. Walk (Eds.), *Recent Progress in Psychiatry.* New York: Grove, 1959. Vol. III, p. 94.
3. Ashby, W. R. *Design for a Brain* (2d. ed. rev.). New York: Wiley, 1960. P. 16.
4. Ashby, W. R. *An Introduction to Cybernetics.* London: Chapman, 1961.
5. Ashby, W. R. *Cybernetics Today and Its Future Contribution to the Engineering-Sciences.* New York: Foundation for Instrumentation, Engineering and Research, 1961. Pp. 6–7.
6. Ashby, W. R. Principles of the Self-Organizing System. In H. von Foerster and G. W. Zopf (Eds.), *Principles of Self-Organization.* New York: Pergamon, 1962. Pp. 255–257, 266.
7. Augenstine, L. Information and Thermodynamic Entropy. In H. Quastler (Ed.), *Information Theory in Biology.* Urbana, Ill.: University of Illinois Press, 1953. Pp. 16–20.
8. Baer, R. M. Some General Remarks on Information Theory and Entropy. In H. Quastler (Ed.), *Information Theory in Biology.* Urbana, Ill.: University of Illinois Press, 1953. Pp. 21–24.
9. Beach, F. A., and Jaynes, J. Effects of early experience upon the behavior of animals. *Psychol. Bull.* 51:239, 1954.
10. Boltzmann, L., and Nabl, J. Kinetische Theorie der Materie. *Encyklopädie der Mathematischen Wissenschaften,* Vol. V, Part I. Leipzig: Teubner, 1903.
11. Branson, H. R. A Definition of Information from the Thermodynamics of Irreversible Processes. In H. Quastler (Ed.), *Infor-*

mation Theory in Biology. Urbana, Ill.: University of Illinois Press, 1953. Pp. 25–40.

12. Braynes, S. N., Napalkov, A. V., and Svechinskiy, V. B. *Uchenyye Zapiski* (*Problemy Neyrokebernetiki*). [Scientific Notes (On the Problem of Neurocybernetics).] Moscow: Publishing House of the Academy of Medical Sciences U.S.S.R., 1959.
13. Bremermann, H. J. Optimization Through Evolution and Recombination. In M. C. Yovits, G. T. Jacobi, and G. D. Goldstein (Eds.), *Self-Organizing Systems*. Washington: Spartan, 1962. Pp. 93–106.
14. Brillouin, L. Life, thermodynamics, and cybernetics. *Amer. Sci.* 37:554, 1949.
15. Brillouin, L. Maxwell's demon cannot operate: Information and entropy, I and II. *J. Appl. Physics* 22:334, 1951.
16. Buck, R. C. On the Logic of General Behavior Systems Theory. In H. Feigl and M. Scriven (Eds.), *Minnesota Studies in the Philosophy of Science*. Vol. I, *The Foundations of Science and the Concepts of Psychology and Psychoanalysis*. Minneapolis: University of Minnesota Press, 1956. Pp. 224–226, 235.
17. Burgers, J. M. On the emergence of patterns of order. *Bull. Amer. Math. Soc.* 69:1, 1963.
18. Campbell, D. T. Common fate, similarity, and other indices of the status of aggregates of persons as social entities. *Behav. Sci.* 3:14, 1958.
19. Cannon, W. B. *Wisdom of the Body*. New York: Norton, 1939. Pp. 287, 293.
20. Cervinka, V. A dimensional theory of groups. *Sociometry* 11:100, 1948.
21. Christie, L. S., Luce, R. D., and Macy, J., Jr. *Communication and Learning in Task-Oriented Groups*. Cambridge, Mass.: Massachusetts Institute of Technology Technical Report No. 231, May 13, 1952.
22. Cohen, J. E. Information theory and music. *Behav. Sci.* 7:137, 1962.
23. Coleman, J. C. *Personality Dynamics and Effective Behavior*. Chicago: Scott, Foresman, 1960.
24. De Beauregard, O. C. Sur l'équivalence entre information et entropie. *Sciences* 11:51, 1961.
25. De Chardin, P. T. *The Phenomenon of Man*. New York: Harper, 1959. Pp. 43–44.
26. De Groot, S. R. *Thermodynamics of Irreversible Processes*. New York: Interscience, 1952.
27. Deutsch, K. W. *The Nerves of Government*. New York: Free Press, 1963. Pp. 186–187.

27a. Deutsch, K. W., and Isard, W. A. Note on a generalized concept of effective distance. *Behav. Sci.* 6:308, 1961.
28. Driesch, H. A. *The Science and Philosophy of the Organism.* London: Adam & Charles Black, 1908.
29. Easton, D. Limits of the equilibrium model in social research. *Behav. Sci.* 1:86, 1956.
30. Eberhart, S. S. The ballad of the interstellar merchants. *Galaxy* 23:92, 1964.
31. Elkinton, J. R., and Danowski, T. S. *The Body Fluids: Basic Physiology and Practical Therapeutics.* Baltimore: Williams & Wilkins, 1955. Pp. 24, 26.
32. Emerson, A. E. The Organization of Insect Societies. In W. C. Allee, A. E. Emerson, O. Park, T. Park, and K. P. Schmidt (Eds.), *Principles of Animal Ecology.* Philadelphia: Saunders, 1949. Pp. 419–435.
33. Foster, C., Rapoport, A., and Trucco, E. Some unsolved problems in the theory of non-isolated systems. *Gen. Syst.* 2:9, 1957.
34. Franklin, W. S. On entropy. *Physical Rev.* 30:766, 1910.
35. Gerard, R. W. Becoming: The Residue of Change. In S. Tax (Ed.), *Evolution After Darwin.* Vol. II, *The Evolution of Man.* Chicago: University of Chicago Press, 1960. P. 255.
36. Gibbs, J. W. *Elementary Principles of Statistical Mechanics.* New Haven: Yale University Press, 1902.
37. Gifford, J. V. *The Nature of the City as a System.* Mimeographed report. Dept. of City Planning, City and County of San Francisco, Calif., 1962.
38. Goode, H. H. Intracompany systems management. *I.E.E.E. Trans. Engin. Manag.* EM-7:14, 1960.
39. Grinker, R. (Ed.) *Toward a Unified Theory of Human Behavior.* New York: Basic Books, 1956. Pp. 328, 371.
40. Hall, A. D., and Fagan, R. E. Definition of system. *Gen. Syst.* 1:18, 1956.
41. Harlow, H. F., and Harlow, M. K. Social deprivation in monkeys. *Sci. Amer.* 207:137, 1962.
42. Hartley, R. V. L. Transmission of information. *Bell Syst. Tech. J.* 7:535, 1928.
43. Hartmanis, J. Symbolic analysis of a decomposition of information processing machines. *Inform. Control* 3:154, 1960.
44. Hearn, G. *Theory Building in Social Work.* Toronto: University of Toronto Press, 1958.
45. Henderson, L. J. *The Fitness of the Environment: An Inquiry into the Biological Significance of the Properties of Matter.* Boston: Beacon, 1958.
46. Herbert, P. G. Situation dynamics and the theory of behavior systems. *Behav. Sci.* 2:13, 1957.

47. Hughes, C. C. Personal communication to A. H. Leighton. In A. H. Leighton, *My Name Is Legion*. New York: Basic Books, 1959. P. 223.
48. Hurley, W. V. *A Mathematical Theory of the Value of Information*. Report 63-3. New York: Port of New York Authority, Engineering Dept., Research and Development Division, May, 1963.
49. Kaplan, M. A. *System and Process in International Politics*. New York: Wiley, 1957. Pp. 6–9, 11, 12.
50. Kempf, E. J. Basic biodynamics. *Ann. N.Y. Acad. Sci.* 73:869, 1958.
51. Köhler, W. *Die Physischen Gestalten in Ruhe und im stationaren Zustand*. Braunschweig: Vieweg, 1921.
52. Köhler, W. *Gestalt Psychology*. New York: Liveright, 1929. Pp. 187–223.
53. Kroeber, A. L., and Kluckhohn, C. Culture: A critical review of concepts and definitions. *Pap. Peabody Museum* 47: No. 1, 1952.
54. Leake, C. D. The scientific status of pharmacology. *Science* 134:2069, 1961.
55. Le Chatelier, H. Recherches expérimentales et théoriques sur les équilibres chimiques. *Annales des Mines*, Huitième Série, Mémoires, XIII, Paris: Dunod, 1888. P. 200.
56. Leighton, A. H. *My Name Is Legion*. New York: Basic Books, 1959. Pp. 204, 221–222.
57. Levinson, D. J. Role, personality, and social structure in the organizational setting. *J. Abnorm. Soc. Psychol.* 58:170, 1959.
58. Levy, M. J. *The Structure of Society*. Princeton: Princeton University Press, 1952. Pp. 88–90.
59. Linschitz, H. Information and Physical Entropy. In H. Quastler (Ed.), *Information Theory in Biology*. Urbana, Ill.: University of Illinois Press, 1953. Pp. 14–15.
60. Linschitz, H. The Information Content of a Bacterial Cell. In H. Quastler (Ed.), *Information Theory in Biology*. Urbana, Ill.: University of Illinois Press, 1953. Pp. 251–262.
61. Malcolm, D. G. Reliability maturity index (RMI)—an extension of PERT into reliability management. *J. Industr. Engin.* 14:3, 1963.
62. Maxwell, J. C. *Theory of Heat*. London: Longmans, Green, 1871. Pp. 308–309.
63. Merton, R. K. *Social Theory and Social Structure* (rev. ed.). Glencoe, Ill.: Free Press, 1957. Pp. 46–47.
64. Meyer, L. B. Meaning in music and information theory. *J. Aesthet. Art. Crit.* 15:412, 1957.

65. Minsky, M. Steps toward artificial intelligence. *I.R.E. Proc.* 49:3, 1961.
66. Morowitz, H. J. Some order-disorder considerations in living systems. *Bull. Math. Biophys.* 17:81, 1955.
67. Muses, C. A. Foreword. In J. Rothstein, *Communication, Organization, and Science*. Indian Hills, Colo.: Falcon's Wing Press, 1958. Pp. xcii–xciii.
68. Neyman, J., and Scott, E. L. On a mathematical theory of populations conceived as conglomerations of clusters. *Gen. Syst.* 3:180, 1958.
69. Neyman, J., Scott, E. L., and Shane, C. D. Statistics of images of galaxies with particular reference to clustering. *Gen. Syst.* 3:193, 1958.
70. Pareto, V. *The Mind and Society*. New York: Harcourt, Brace, 1935. Vol. IV, p. 1435.
71. Parsons, T., and Shils, E. A. (Eds.) *Toward a General Theory of Action*. Cambridge, Mass.: Harvard University Press, 1951.
72. Pascal, B. *Pascal's Pensées*. (Orig. ed., 1670). Translated by H. F. Stewart. New York: Pantheon, 1950. Pp. 358–359.
73. Pask, G. *An Approach to Cybernetics*. London: Hutchinson, 1961. Pp. 26–27, 77.
74. Pierce, J. R., and Cutler, C. C. Interplanetary Communications. In F. I. Ordway, III (Ed.), *Advances in Space Science*. New York: Academic, 1959. Vol. I, pp. 68–69.
75. Planck, M. *Treatise on Thermodynamics*. New York: Dover, 1945.
76. Powell, R. M. Sociometric analysis of informal groups—their structure and function in two contrasting communities. *Sociometry* 15:367, 1952.
77. Prigogine, I. *Introduction to Thermodynamics of Irreversible Processes*. Springfield, Ill.: Thomas, 1955. Pp. 82, 84.
78. Quastler, H. (Ed.) *Information Theory in Biology*. Urbana, Ill.: University of Illinois Press, 1953. P. 41.
79. Quastler, H. Information Theory Terms and their Psychological Correlates. In H. Quastler (Ed.), *Information Theory in Psychology*. Glencoe, Ill.: Free Press, 1955. Pp. 159–160.
80. Rapoport, A. What is information? *Synthese* 9:157, 1953.
81. Rignano, E. The concept of purpose in biology. *Mind* 40:335, 1931.
82. Rosenblueth, A., Wiener, N., and Bigelow, J. Behavior, purpose and teleology. *Phil. Sci.* 10:18, 1943.
83. Rothstein, J. *Communication, Organization, and Science*. Indian Hills, Colo.: Falcon's Wing Press, 1958. Pp. 34–36.
84. Rothstein, J. Physical demonology. *Methodos* 11:99, 1959.

85. Rothstein, J. Thermodynamics and some undecidable physical questions. *Phil. Sci.* 31:40, 1964.
86. Schafroth, M. R. The Concept of Temperature. In H. Messel (Ed.), *Selected Lectures in Modern Physics*. London: Macmillan, 1960.
87. Schrödinger, E. *What Is Life?* New York: Macmillan, 1945. P. 72.
88. Schumpeter, J. A. *Business Cycles*. New York: McGraw-Hill, 1939.
89. Seaborg, G. T. (Chairman). Panel on Basic Research and Graduate Education of the President's Science Advisory Committee. Scientific progress and the federal government. *Science* 132:1802, 1960.
90. Selye, H. *The Stress of Life*. New York: McGraw-Hill, 1956.
91. Shannon, C. E. Probability of error for optimal codes in a Gaussian channel. *Bell Syst. Tech. J.* 38:611, 1959.
92. Shannon, C. E., and Weaver, W. *The Mathematical Theory of Communication*. Urbana, Ill.: University of Illinois Press, 1949.
93. Simon, H. A. *The New Science of Management*. New York: Harper, 1960.
94. Simon, H. A. The architecture of complexity. *Proc. Amer. Phil. Soc.* 106:467, 1962.
95. Sommerhoff, G. *Analytical biology*. London: Oxford University Press, 1950.
96. Strodtbeck, F. L., and Hook, L. H. The social dimensions of a twelve-man jury table. *Sociometry* 24:397, 1961.
97. Szilard, L. Über die Entropieverminderung in einem thermodynamischen System bei Eingriffen intelligenter Wesen. *Phys.* 53:840, 1929. (Translated by A. Rapoport and M. Knoller as: On the increase of entropy in a thermodynamic system by the intervention of intelligent beings. *Behav. Sci.* 9:301, 1964.)
98. Tolman, E. C. Cognitive maps in rats and men. *Psychol. Rev.* 55:189, 1948.
99. Truxal, J. G. *Automatic Feedback Control System Synthesis*. New York: McGraw-Hill, 1955. P. 2.
100. Valentinuzzi, M., and Valentinuzzi, M. E. Information content of chemical structures and some possible biological applications. *Bull. Math. Biophys.* 24:11, 1962.
101. Veltistov, Y. E. USSR improves computers. *Ogonec* 36:14, 1962.
102. Vickers, G. Is adaptability enough? *Behav. Sci.* 4:219, 1959.
103. Virchow, R. Atome und Individuen, *Vier Reden über Leben und Kranksein*. Berlin, 1862. (Translated by L. J. Rather as: Atoms and Individuals. In *Disease, Life, and Man. Selected*

Essays by Rudolph Virchow. Stanford: Stanford University Press, 1958. Pp. 120–141.)

104. von Bertalanffy, L. Der Organismus als physikalisches System betrachtet. *Naturwissenschaften* 28:521, 1940.
105. von Bertalanffy, L. *Problems of Life: An Evaluation of Modern Biological Thought*. New York: Wiley, 1952. Pp. 134, 147–151, 189–194.
106. von Bertalanffy, L. General systems theory. *Gen. Syst.* 1:1, 1956.
107. von Foerster, H. Communication amongst automata. *Amer. J. Psychiat.* 118:865, 1962.
108. von Helmholtz, H. L. F. *Abhändlungen zur Thermodynamik*. Leipzig: Engelmann, 1902.
109. von Neumann, J. The general and logical theory of automata. In L. A. Jeffress (Ed.), *Cerebral Mechanisms in Behavior*. New York: Wiley, 1951. Pp. 25–31.
110. von Neumann, J. *The Computer and the Brain*. New Haven: Yale University Press, 1958. Pp. 6–7.
111. Weber, M. *The Theory of Social and Economic Organization*. (Translated by A. M. Henderson and T. Parsons.) New York: Oxford University Press, 1947.
112. Weiss, P. A. In R. W. Gerard (Ed.), Concepts of biology. *Behav. Sci.* 3:92, 1958.
113. Weiss, P. Animal behavior as system reaction. *Gen. Syst.* 4:1, 1959. (Translated and revised from Tierisches Verhalten als "Systemreaktion," *Biol. Gen.* 1:167, 1925.)
114. Wharton, T. *Adenographia: Sive Glandularum Totius Corporis Descriptio*. Amsterdam: Ravestein, 1659. Pp. 110–111.
115. Whorf, B. L. *Language, Thought, and Reality*. New York: Wiley, 1956.
116. Wiener, N. *Cybernetics*. New York: Wiley, 1948. P. 76.
117. Zeman, J. Le sens philosophique du terme "l'information." *Docum. France* 3:19, 1962.
118. Zipf, G. K. *Human Behavior and the Principle of Least Effort*. Cambridge, Mass.: Addison-Wesley, 1949.

4

Symbolism and General Systems Theory *

ROY R. GRINKER, SR., M.D.

THE SUBJECT of symbolism is a vast area which embraces practically all of human psychology. The specific focus here will be on the comparison of Freudian metapsychology and general systems theory as they conceptualize symbolism. Both, however, are global theoretical abstractions covering a variety of subtheories at varying distances from empirical data, so that some choices must be made. For this purpose I have chosen, not arbitrarily, topological theory of psychoanalysis and transactional theory as they apply to symbolism. Certainly this is not the only possible comparison, but the easiest.

Topological theory identifies mental processes with reference to their degree of awareness or to their relationship with consciousness. Refinement or at least later modification of psychoanalytic theory includes the allocation of psychological functions to defined "structures" (id, ego, and superego) which vary in their topological positions. The so-called unconscious, the preconscious, and the conscious represent levels of awareness and phases of ontogenesis from least and earliest to most and latest. Between the unconscious and preconscious levels of awareness a conceptual barrier activates and maintains repression, while what is preconscious can by effort and recall become conscious.

General systems theory [8, 9] includes concepts of integration and processes by which integration is maintained in all open living systems inseparable from their environments. Each system is composed

* Presented at the 123rd annual meeting of the American Association for the Advancement of Science, Washington, D.C., December 29, 1966, as part of a symposium on Symbolism, a joint program of the American Academy of Psychoanalysis and the Society for General Systems Research.

of parts or subsystems under control and regulation within specified gradients. Information exchange occurs among subsystems, and between systems, at their interfaces by means of reverberating circular transactions. I propose to consider how symbolic systems develop from an integration of component subsystems and how they function in transaction with reference to their topological positions.

Symbolic Transformation

From the evolutionary point of view Langer [6] states that symbolic transformation derived from simpler animal activities begins with the development of language. I, on the other hand, do not subscribe to the words of the ancient Rabbi who said of the man-made Golem that it was not human because it had no language. All humans can symbolize in pictures, images, and actions without words, especially while dreaming.

Recent work with the deaf indicates that language and operational thinking are not necessarily related. Persons born deaf have no verbal language at beginning school age and never develop normal vocabulary or an English language syntactical structure. Although there are many crippling consequences of deafness, depending on early experiences and social factors at home, school, and community, the structure of thinking according to Furth [2] is not necessarily affected. In confirmation of Furth's experimental studies, Vernon's [7] investigation of thirty-three independent research reports involving over 8000 deaf persons showed that (1) there is no functional relationship between verbal language and cognition, (2) verbal language is not the mediating symbol system of thought, and (3) there is no relationship between concept formation and level of verbal language development.

Langer [5] states that symbolic thinking rests on the capacity for abstract seeing or visual imagery representative of visible things. Since similar images are stimulated from many sources, there is a tendency to generalize, and the visual sign becomes a symbol whereby one may maintain an idea or thought of an object in its absence. In Gibson's [3] terms, symbols are coded knowledge about the environment in contrast with signs of the environment. However, symbols may be developed from signs derived from any sense perception. They represent an economical process which saves man from repetitive information overloading and maintains his psychological functions in the face of restricted sensory inputs. We are thus able to scan a wide variety and range of perceptions, with relative independence of the current sensory input in the process of recognizing universals.

The components of the symbolic system begin early, but at every stage learning from experience is important in shaping its quality. The vital organs functioning spontaneously and the perceptual systems search actively for releasing stimuli. As the infant develops an image of its body as a whole, some of his autonomic activities are projected or experienced as if on the surface. In fact the body surface and the subsequent body image are not only related to biological needs and satisfactions but, as Kubie [4] stated, become the roots of the abstractions or symbols of I and not-I. Deficiencies in these subsystems have been considered highly significant as precursors of the schizophrenic process and childhood autism.

Levels of Awareness

The transposition of the biological strata of the symbol system into dim unconscious awareness occurs, but how we do not know. The strata reappear during functional regressions or during electrical stimulation of the temporal lobe. Eating, elimination, and phallic sensations are incorporated into symbols which are detached from their signs and applied to social process in feeling, thinking, and dreaming. In fact, psychoanalytic theory concerning phases of libidinal development utilizes body signs to qualify psychological symbolic processes. In psychological regression symbolism becomes less generalized and closer to signs. These may become so concrete that body language may become the core of psychosomatic symptoms.

In addition to somatic perceptions as signs, other perceptive subsystems either consciously or subliminally perceived become part of the unconscious psychological organization. Most important are visual images which at first are representatives of things seen. These become complex combinations combining motivational wishes and satisfactions as well as frustration of needs and denied experiences. Time, logic, consistency, and clear differentiation are missing, so that only primitive symbols develop.

The preconscious functions with a high degree of freedom, processing internal associations of symbols and automatizing consciously learned activities. Although highly developed preconscious symbols are largely visual condensations, it is within this process that much imagination, creative fantasy, and dreaming occur. For such, words are not necessary. In fact, when we translate preconscious concepts into language, our thinking is slowed since we have to choose the correct and applicable verbal symbol for a vast array of fast-moving images.

Finally, conscious processes, although primarily verbal, may take on other forms which have not been well studied. Nevertheless the word, picture, geometrical figure, mathematical symbol, or artistic movements and sounds may achieve universality, in part at least, and be utilized in the transmission of information, facts, experiences, attitudes, and emotions. Symbol systems transmitting information to the young may be called the genes of culture.

The ontogenic phases of symbol systems progress through learning of various types such as spontaneous self-action or reflexes which are reinforced, imprinting by the influence of releasing mechanisms, primary identification or imitation, and by secondary identification or memory images of significant interpersonal transactions. Combinations or symbol systems specific for the individual comprise his psychological patterning, his cognitive organizations, or his life-style.

The vagueness and fluidity of the internal transactions of the symbolic system constitute its constant dynamic qualities. There seem to be constant reverberating communication and temporary interruptions of transactions among all phases in every direction. There may be deficiencies in any subsystem, often leading to compensations by another. Threshold values within these transactions may cut off some or all of the flow of information among the parts, resulting in splitting or detours in the flow of information. For example, bodily experiences may not reach consciousness, and the preconscious may be blocked so that conscious feelings, imagination, and creativity may be impoverished. Conscious logical and intellectual functioning may be inhibited, and emotional acting-out may dominate behaviors. Finally the symbolic system may develop a life of its own apart from internal or external references, as in the psychoses. In fact, Kubie states that neurosis and creativity have in common "a disturbance in the relationship of the symbolic process to whatever it represents."

Phases of the Symbolic System

In summary we may view topological psychoanalytic theory in terms of general systems theory as follows:

1. The symbolic system has developed from a system of signs by an evolutionary jump-step, resulting in preconscious and conscious process as distinctly (?) human phenomena.
2. There are ontological phases of learning from body signs to visual imagery to primitive symbols to creative thinking, but the flow of information among these phases persists in all directions throughout life.
3. There are flexible transactional operations among these parts so that all are involved in all forms of thinking.
4. All the phases or parts of the symbolic system are in transactional relationship with reality and inner experiences.
5. A disintegration of optimum or effective relations among parts of the symbolic systems may lead to breaking off of transactions (repression), and thereby to distorted thinking and behavior or to temporary acceleration of creativity.

I have tried to speculate how one aspect of general systems theory and a part of psychoanalytic theory interdigitate not by analogy or tautology but by an attempt to superimpose concepts of process in a sketchy manner. General systems theory includes in its global concepts isomorphism from cell to society. As yet there has been an almost exclusive emphasis on semiotics and statistics characteristic of science. On the other hand, psychoanalysis has been more concerned with the individual and sometimes universal meaning of symbols as characteristic of the humanities. Perhaps someday these two approaches may coalesce.

This is not to state that a principle unifying the biological and psychological aspects of development may, as we once hoped, be on the horizon. As Elsasser [1] states, neither reductionism nor vitalism is applicable to living systems. Rather, he concludes that organisms are semiautonomous, dynamic systems characterized by hierarchies of order and a high degree of individuality and with incredible variation. It is in symbol systems that we are confronted with the maximum of these characteristics.

References

1. Elsasser, W. M. *Atom and Organism—A New Approach to Theoretical Biology*. Princeton, N.J.: Princeton University Press, 1966.
2. Furth, H. G. *Thinking Without Language*. New York: Free Press, 1966.
3. Gibson, J. J. *On Perception*. Boston: Houghton Mifflin, 1966.
4. Kubie, L. S. *Neurotic Distortion of the Creative Process*. Laurence, Kansas: University of Kansas Press, 1958.
5. Langer, S. K. *Philosophy in a New Key*. New York: Mentor, 1948.
6. Langer, S. K. *Philosophical Sketches*. New York: Mentor, 1964.
7. Vernon, McC. Relationship of language to the thinking process. *Arch. Gen. Psychiat.* (Chicago) 16:325, 1968.
8. von Bertalanffy, L. A Biologist Looks at Human Nature. In S. J. Beck and H. Molish (Eds.), *Reflexes to Intelligence: A Reader in Clinical Psychology*. Glencoe, Ill.: Free Press, 1959.
9. von Bertalanffy, L. General Systems Theory in Psychiatry. In S. Arieti (Ed.), *American Handbook of Psychiatry*, Vol. III. New York: Basic Books, 1966.

5

A General Systems Theory Based on Human Communication

JURGEN RUESCH, M.D.

THE CREATION of a new science occurs in four distinct phases [16]:

Phase 1. The recognition of the existence of a new set of problems.
Phase 2. The collection of observations that lead to new generalizations.
Phase 3. The creation of separate organizations and institutions that facilitate the development of new methods and theories.
Phase 4. The integration of mature disciplines into fields concerned with similar problems, methods, and language.

The humanities entered Phase 4 at the time of the Renaissance, the physical sciences arrived at the same point at the beginning of the twentieth century, and the behavioral and social sciences will probably reach it some time in the future. Special theories are characteristic of Phases 1 and 2; unified theories emerge when several fields converge, as in Phase 4; and general systems theories come into their own when groups of sciences—for example, the physical sciences [12]—share their theoretical models and rules of evidence. To talk about a general systems theory of behavior, therefore, may be somewhat premature [32, 33]. Nonetheless, as a result of the convergence of various behavioral disciplines, the foundations for a general systems theory are being laid at the present time. These are rooted in four fundamental approaches to human behavior.

The *behavioral and social sciences* are concerned with individuals and groups. Their concepts revolve around psychological topics such as personality, identity, learning, action, perception, and speech [3], or around social topics such as organization, rules, roles, status, values, and power [14]. These disciplines are searching for facts that have universal validity, the truth of which should hold for longer periods of time. For these reasons, the behavioral and social sciences focus upon structure and the more time-enduring qualities of human behavior.

The *information sciences* are message-oriented, not people-oriented, and are dedicated to the search for facts that can be quantified and have universal validity [21]. Not bound to any particular structure, the engineer is able to pursue a message from its source to its destination, regardless of how many times the message may be transformed and recoded, passing on its way through numerous people or machines [2]. More formally speaking, the science of cybernetics, or steersmanship, is made up of information theory; filtering, detection, and prediction theory; feedback and servomechanism theory; and the theories governing design and programming of computers [24]. The application of these theories has been successfully tested in the areas of automata [10], genetic control [7], nervous system functioning [11], humoral functioning [15], interpersonal communication [26], and mass communication [19].

The *managerial disciplines* specialize in organization, communication, and social operations [18]. If the behavioral and social scientists are structure-oriented and communications engineers message-oriented, the managerial specialists are task-oriented. Professionals such as the lawyer, clergyman, teacher, military man, business manager, personnel officer, and government agent all are concerned with people in groups; but being charged with specific operations and goals, these professionals always deal with a concrete problem [23]. The knowledge they are looking for is place-, time-, person-, group-, and situation-bound. Only ultraspecific information will enable them to institute relevant action with some promise of success. However, knowledge in action is not as universally valid as knowledge that is not subjected to this test.

The *health sciences* are concerned with the malfunctioning of people, singly and in groups [8]. In addition to using scientific information provided by the behavioral scientist, the communications engineer, and the medical scientist, the physician uses knowledge derived from the field of organization and management to maintain a state of health. The psychiatrist's concern with psychological typologies of character and personality has recently been extended to include typologies of situations and groups [26]; and his interest in drugs and somatic therapies is complemented by a concern for institutional resources, the organizational framework, and prevention [5].

These four approaches to human behavior contain certain ingredients common to cells, organ systems, individuals, groups, and societies. They should prove useful for the construction of a future general systems theory of human behavior.

Four Basic Postulates

SUBJECTIVITY AND OBJECTIVITY

Eddington [9] once said that man stands halfway between the mass of the atoms and the mass of the stars. Whenever man remains in this optimal position—the ratio of the masses being about 10^{23}—then the mutual influence between observer and observed can be disregarded, and the relationship is called objective. When the ratio of the masses is 1 (or 10^0), as in a relationship of two people, then the two entities influence each other, and the relationship is subjective. When the ratio of the masses is in the range of 10^3 to 10^7, we deal with a semiobjective relationship characteristic of the social sciences [29].

POSTULATE 1. In a general systems theory the ratio of the masses of the observer to the observed must remain of the same magnitude. No generalizations made in the framework of $\frac{10^0}{10^{-23}}$ or $\frac{10^0}{10^{23}}$ can be compared, for example, to the universe of generalizations made in the framework of $\frac{10^0}{10^4}$.

PATTERNS AND PROBABILITIES

The physical sciences—and this includes some of the biological sciences and experimental psychology—tend to measure in succession one variable aspect of a complex situation after another, while keeping all other features constant. The behavioral and social sciences, in contrast, cannot, during the period of measurement, control the parameters of their field. Therefore the methodology is different [6]. First, typical patterns are described. Second, the data are sorted according to these patterns. Third, the frequency of occurrence of each pattern is established. Fourth, predictions are made about the probability of occurrence of the pattern in similar situations over a given period of time. Once the nature of the pattern is known and the probability of occurrence established, a scientific tool is available to deal with a variety of transactions.

POSTULATE 2. A general systems theory must make provisions for a multivariate approach to human behavior, allowing for the assessment of patterns and the establishment of probabilities.

ACTION AND SYMBOLIC BEHAVIOR

Whenever a human observer perceives an event, two distinctly different aspects are recorded [27]. The contractions of human muscles or the movements of physical particles in nature have a material impact that is judged by the criteria of the physical or biological sciences. Simultaneously these events, if perceived by an observer, have symbolic or representational qualities. These aspects are judged by the criteria derived from the information sciences and the humanities.

POSTULATE 3. A general systems theory of human behavior must make provisions for the dual nature of human observations. On the one hand, action occurring in a physical field has material consequences; on the other hand, signaling behavior occurring in a social field has consequences only if it is perceived and changes the informational body of the receiver.

RATES OF CHANGE

In the behavioral sciences the events that have a slow rate of change serve as frames of reference for the measurement or the observation of events with a faster rate of change. The terms *slow* and *fast* thus refer to changes relative to the time scales of the observer [28]. Thus the next larger entity in terms of mass, space, and time scales is considered more stable than the next smaller one. The cell generally is thought to have a faster rate of change than the organ, the individual a faster rate than society, and the planet a faster rate than the solar system [29].

POSTULATE 4. In a general systems theory of human behavior, stability is attributed to the next larger entity, which thus becomes the field. Variability is attributed to the next smaller entity, which becomes the variable in terms of the mass and time scales of the human observer.

These four postulates reflect the special conditions that surround the observation of behavior by a human observer, whose ways of perceiving may be analyzed by means of the notions of set, isolate, and heterogeneous pattern (see Fig. 5-1).

Closest to human experience is the set. A set is made up of two or more constituent components perceived as being set apart from other events [13]. This percept is usually given a denotative label from which we infer its separateness in the experience of the human being—e.g., hat and coat (clothing), brother and sister (siblings), handshake and smile (gesture).

An isolate is an abstraction of the constituents that make up a set. Although when studied these isolates may turn out to be percepts in their own right, they usually constitute mere ways of analyzing structure. Into this category fall all scientific measurements such

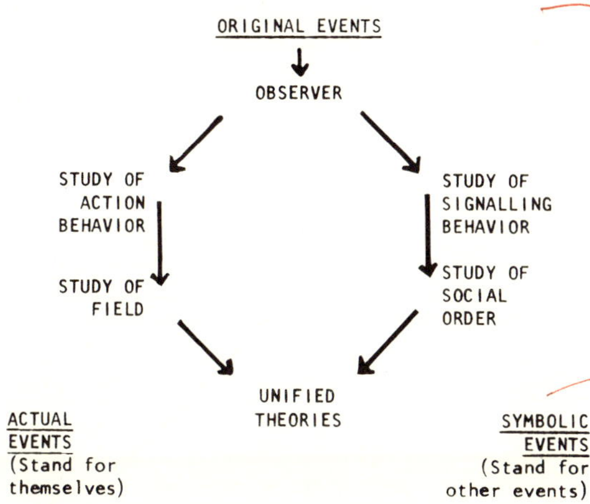

FIGURE 5-1. UNIFIED THEORIES OF BEHAVIOR.

as height and weight, theoretical constructs such as roles and rules, and similar abstractions.

A social pattern emerges when set is related to set in a given context. The combination of sets into patterns occurs according to rules and principles that derive from nature or the man-made order. The organization of space and the apportioning of time, work, and play are pertinent examples.

Pathology and abnormal behavior usually are perceived as deviations from accepted patterns. In 1957 Ruesch [25] suggested that pathology is not characterized by the emergence of new components (or isolates) but by deviations in the organization of patterns. Deviations in intensity and timing relative to the accepted norms for a given situation are features that characterize psychological and social pathology. This fact has been incorporated in the following ditty:

Too much, too little;
Too early, too late;
At the wrong place
Is pathology's fate.

Conventional Approaches to Behavior

THE STRUCTURE SPECIALTIES

Traditionally those interested in biology or medicine have isolated the entity they wish to study, focusing mostly upon the time-enduring features. Table 5-1 shows the approximate scheme of the structure specialists [30].

TABLE 5-1
SCHEME OF THE STRUCTURE SPECIALISTS

The Observer	The Observed Entity	Observations
Medical scientist	Cell	Boundaries
Biological scientist	Organ	Internal structure and function
Behavioral scientist	Organism	Maintenance processes
Social scientist	Group	Growth and decline of entity
	Society	Reversible and irreversible changes
	Nation	

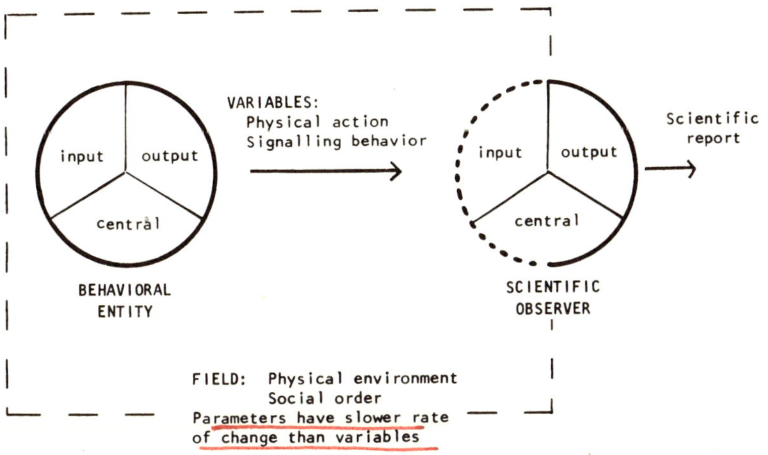

FIGURE 5-2. ENTITY AND FIELD.

DISTORTIONS. Because of isolation for purposes of study, the interconnection of an entity to other entities of a smaller, equal, or larger order frequently has been neglected. For the same reason, greater stability and autonomy are attributed to an isolated entity than it actually possesses. In focusing upon the time-enduring features, many of the interactional and communicative aspects with the surroundings are lost (see Fig. 5-2).

TABLE 5-2
SCHEME OF THE FIELD SPECIALISTS

The Observer	The Observed Fields
Naturalists	The natural environment
Earth scientists	The man-made environment
Social scientists	The experimental environment

Observations

Physical Environment	Resources	Social Environment
Geographic location	Time	People
Climate	Space	Institutions
Flora	Money	Government
Fauna	Raw materials	Language
Architecture	Energy sources	Organization
	Technical equipment	

THE FIELD SPECIALTIES

While the structure specialists study entities, the field specialists select the environment as a subject of study [17]. Obviously the environment is likewise made up of entities, but the scientist's attention is focused upon the larger interconnections. Because of diurnal, biological, or other cyclic variations, the field specialists tend to deal with statistical envelopes or ranges rather than with averages, as shown in Table 5-2.

DISTORTIONS. The field specialists attribute statistical mass effects to the environment. But when a person interacts with another person, or a dog, this interaction becomes ultraspecific. The exchange processes between an entity, and the surroundings therefore are based upon mass effects only when considered from the view of the superhuman observer.

THE SYMBOLIC SPECIALTIES

The engineers join the humanities in the study of communication [22]. In focusing upon symbolic behavior, every discipline in essence studies input and output of a given entity (see Table 5-3).

TABLE 5-3
SCHEME OF THE SYMBOLIC SPECIALISTS

The Observer	The Observed Entity	Observations
Linguist	Signals	Language
Cultural anthropologist	Symbols	Speech
Art historian	Noise	Bodily movements
Novelist	Codes	Written or printed words or figures
Psychologist	Messages	Art objects
Psychiatrist		Theatre
Engineer		Music
		Architecture
		Photographs

DISTORTIONS. From the study of signaling behavior, inferences are made about conjectural internal processes which at the human level are called thinking, feeling, motivation, and the like.

THE SOCIAL ORDER SPECIALTIES

Messages travel in circuits. While the engineers build the communication machines and connect them by networks in which signals can travel, the social scientists are concerned with the human beings who use the machines and communicate in face-to-face encounters. Spoken, written, or printed words; symbolic movements; dress; and many other symbolic expressions cannot be contained in discrete channels as in the case of impulses traveling in electronic circuits. Nonetheless, symbolic behavior is neither random nor universal but is governed by explicit and implicit social rules [4]. The social field in which communication and interaction take place is a system that the participants have to master in order to express themselves and the observers have to understand in order to interpret the messages. The scheme is shown in Table 5-4.

TABLE 5-4
SCHEME OF THE SOCIAL ORDER SPECIALISTS

The Observer	The Observed Field	Observations
Social scientists Stylists Designers Propagandists Advertisers	The social order in which symbolic behavior occurs, the rules governing communication, and the attribution of meaning	Organizational structure Power structure Rules Roles Metacommunication Networks

DISTORTIONS. Culture and society usually define the ideal norm and the tolerable range of symbolic deviations. The assumption is made that every participant knows these norms and rules, but in practice one finds that one and the same symbol may be used to refer to different events. Therefore social controls are needed to restrict the meaning of symbols.

TYPOLOGIES OF ENTITIES

When scientists superimpose multiple individual observations and project these onto a screen, they arrive at an artificial typology exemplified in character types, diagnostic categories, or popular stereotypes [1]. Statistically speaking, measures of central tendency are frequently used to describe typologies, as shown in Table 5-5.

TABLE 5-5
SCHEME OF THE ENTITY TYPOLOGISTS

The Observer	The Observed Entity	Observations
Physician	The "disease"	Value orientations (future, past)
Psychologist	The "occupation"	Role types (husband, wife)
Psychoanalyst	The "personality"	Class types (lower, middle)
Psychiatrist	The "constitution"	Personality types (introvert, extrovert)
Reporter		Mythological types (muses, witches)
		Diagnostic types (schizophrenic, depressive)

DISTORTIONS. Typologies represent patterns abstracted from a multitude of events and are assumed to exist 24 hours a day regardless of circumstances. Emphasis upon the typology often obscures individualized features and blinds the doctor or the engineer to possible solutions based on detailed observation.

TYPOLOGIES OF FIELDS

Not only persons but also fields or situations can become the subject of typologies [20]. Such situations may gain their label through the time or place of occurrence, the kind of participants, the context, or any other distinctive feature (see Table 5-6).

TABLE 5-6
SCHEME OF THE FIELD TYPOLOGISTS

The Observer	The Observed Field	Observations
Physician	Festivities	Context (in terms of time, place)
Anthropologist	Catastrophes	Label of situation
Social psychologist	Business patterns	Rules governing situation
Psychiatrist	Recreational situations	Networks of communication
Criminologist		

DISTORTIONS. The personality features of the participants are neglected in that they are known only by their role or position—e.g., the bride, the victim, the deceased—which procedure describes people in organizational terms.

Communication as General Systems Theory

In a general systems theory of behavior, the characteristics of living entities have to be spelled out in such a way that cells, organs, organisms, organizations, or societies can be compared to each other. Table 5-7 describes the three types of observations necessary to evaluate a behavioral situation, irrespective of the entities involved:

TABLE 5-7
OBSERVATIONS FOR EVALUATING A BEHAVIORAL SITUATION

Characteristics of the Communicating Entity (cell, organ, organism, organization, machine)	Characteristics of the Connecting Processes (field)	Characteristics of the Outside Observer or Manager
1. Input functions (perception) 2. Central processes Data scanning (recognition) Data processing (thinking) Data storage (memory) 3. Output functions (expression, action)	1. Combination of functions of several small entities to form larger entity 2. Networks 3. Languages and codes 4. Content: referential property attributed to signals 5. Metacommunicative processes: instructions and interpretative devices 6. Feedback: reincorporation of information at the source	1. Scientific observation or measurement of ongoing signal exchange 2. Evaluation of exchange in terms of a theory or model 3. Assessment of effects produced 4. Management and organization: interference with the connecting processes 5. Treatment or repair, interference with the entities themselves

If we apply the above scheme to a two-unit situation, we have to introduce additional refinements. Two people or animals or machines who communicate with each other do not present the same state throughout the period of interaction. The most suitable model for the conceptualization of the variability of a communicating entity is the rotating globe. Each globe receives and emits messages or engages in actions. Inasmuch as these messages and actions have both quantity and direction, they can be conceived of as vectors. However, internal and external screens may prevent messages from being properly directed, quantified, or encoded. Only when sender and

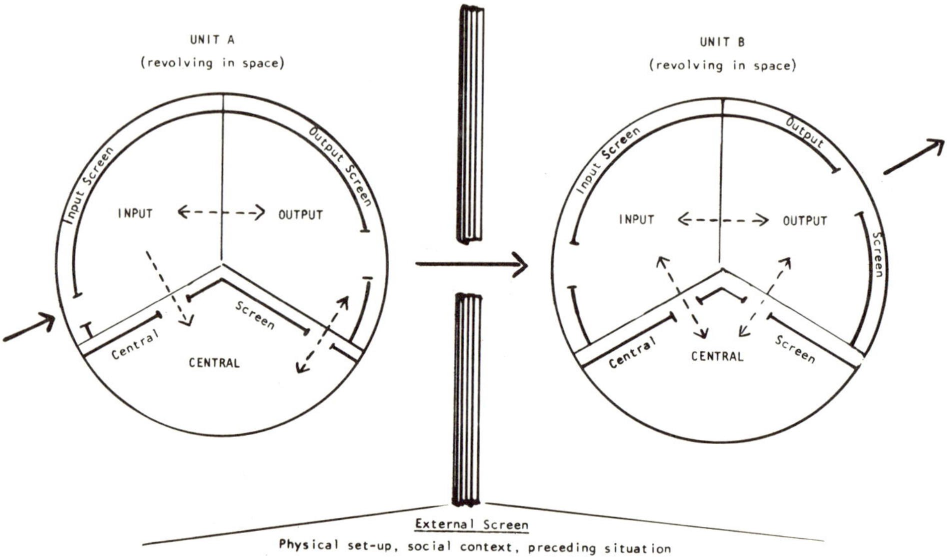

FIGURE 5-3. THE DIFFERENT FACES OF COMMUNICATING ENTITIES.

receiver are using the same code, when interference of internal and external screens is minimized, and when the signal is capable of passing through the selective screens and filters can communication be successful. Considering the obstacles presented by internal and external screens and the necessity for proper steering of messages and actions, it is a miracle that we communicate at all (see Fig. 5-3). The regulatory forces that influence the message exchange are shown in Table 5-8.

TABLE 5-8
REGULATORY FORCES IN MESSAGE EXCHANGE

Screens Located Inside the Communicating Entity	Screens Located in the Field
1. Input screens (control of perception)	1. Physical screens (distance, time, architecture, climate)
2. Central screens (control of storage, scanning, and evaluation of information)	2. Social screens (language barriers, rules, regulations)
3. Output screens (control of expression and action)	3. Organizational screens (lack of provisions for certain activities or message exchanges)

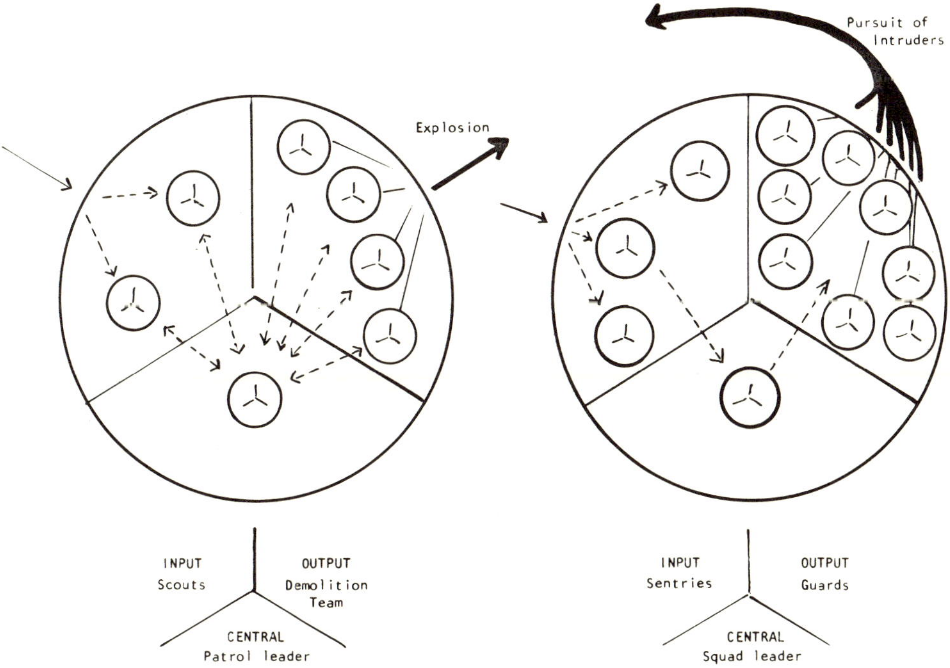

FIGURE 5-4. GROUP INTERACTION: DEMOLITION OF A BRIDGE IN WARTIME.

Group interaction can be analyzed the same way as two-person interaction. Group actions, for example, can be divided into input, central, and output tasks. The input task is in the hands of scouts, observers, intelligence officers, and other watchdogs; the central functions are handled by the group leaders and their assistants; and the output functions are performed by workers, writers, or speakers. The interaction between two groups is subject to limitations similar to those outlined for the two-person team (see Fig. 5-4).

At the societal level, a similar scheme occurs. The input functions are handled by public opinion experts, critics, reviewers, economic analysts, politicians, and technical and military observers. The central functions are held by the government executives, the justices, the parliamentarians, and the directors of business corporations, labor unions, professional organizations, and religious bodies (see Fig. 5-5). The output functions are carried out by workers, by the news media, and by the armed forces or the police.

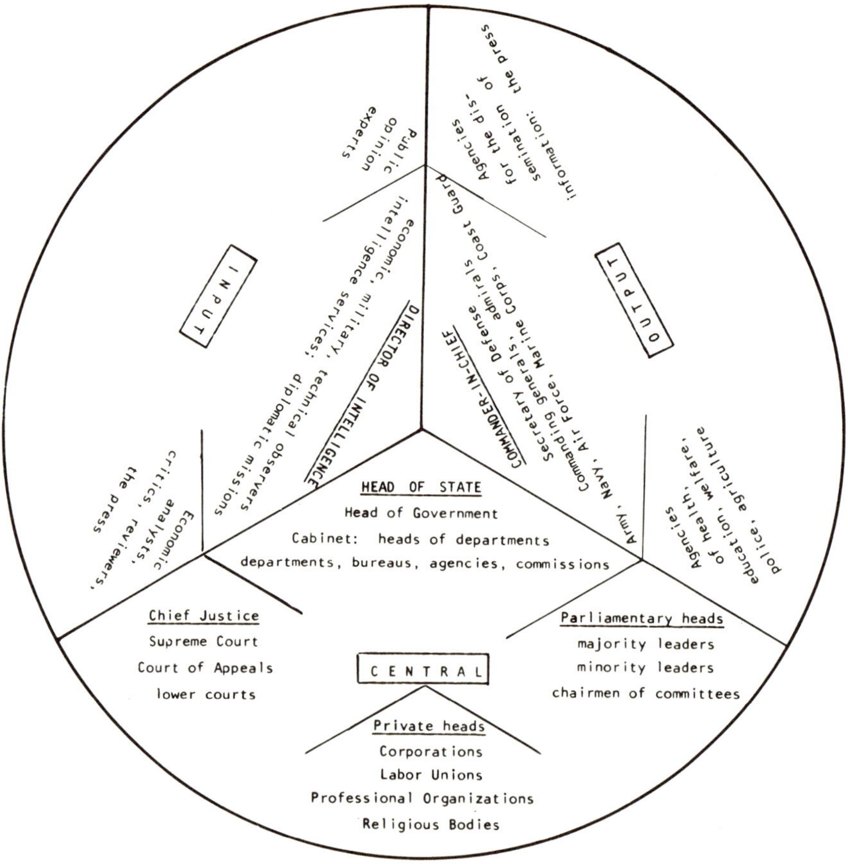

FIGURE 5-5. THE GOVERNMENT OF A NATION.

Minimal Requirements for a General Systems Theory of Human Behavior

In summary, then, the minimal requirements for a general systems theory of human behavior are:

1. A human observer (scientist), equipped with
   a. Technical extensions to observe events
   b. Language to record and codify events
2. Events (natural or man-made events)
   a. Action events
      (1) Physical entity that carries out action

(2) Physical field in which the action occurs

b. Symbolic events

(1) Symbolic entity that signals

(2) Social field that determines the rules of communication

3. Transformational devices (technical)
   a. To carry out action in terms of the available information
   b. To represent information in terms of action that took place
4. Feedback (to equate information and action as an alternating step function)
   a. To record at the original source of information the physical or social effects of any action that has been undertaken
   b. To translate into action the altered information to produce new effects

These four principal requirements will satisfy physical, biological, social, and behavioral scientists as well as humanists, philosophers, and theologians, who all believe that information controls action and that action influences information. And while action is assessed by the impact it has over time, a message is assessed by the reply it evokes in another communicating entity. Therefore a systems theory of behavior must always contain provisions for:

1. Entities that act or communicate.
2. Other entities that reply or react.
3. Connecting processes that regulate this interaction.

As we have moved from structure to process, from statics to dynamics, from single entities to multiple entities, and from univariate to multivariate approaches, we have taken a decisive step away from linear thinking and the observation of simple phenomenology toward circular or more complex thinking and the consideration of more general processes [31]. This is about where we now stand.

References

1. Allport, G. W. *Pattern and Growth in Personality.* New York: Holt, Rinehart and Winston, 1961.
2. Ashby, W. R. *Design for a Brain* (2d ed.). New York: Wiley, 1960.

3. Berelson, B., and Steiner, G. A. *Human Behavior*. New York: Harcourt, Brace & World, 1964.
4. Boguslaw, R. *The New Utopians*. Englewood Cliffs, N.J.: Prentice-Hall, 1965.
5. Caplan, G. *Principles of Preventive Psychiatry*. New York: Basic Books, 1964.
6. Cattell, R. B. (Ed.) *Handbook of Multivariate Experimental Psychology*. Chicago: Rand McNally, 1966.
7. Crick, F. H. C. *Of Molecules and Men*. Seattle: University of Washington Press, 1966.
8. Dubos, R. *Mirage of Health*. New York: Harper, 1959.
9. Eddington, A. S. *Stars and Atoms*. New Haven: Yale University Press, 1927.
10. Feigenbaum, E. A., and Feldman, J. (Eds.) *Computers and Thought*. New York: McGraw-Hill, 1963.
11. Gerard, R. W., and Duyff, J. W. (Eds.) *Symposium on Information Processing in the Nervous System*, Leiden, 1962. Amsterdam: Excerpta Medica Foundation, 1964.
12. Grinker, R. R. (Ed.) *Toward a Unified Theory of Human Behavior* (2d ed.). New York: Basic Books, 1967.
13. Hall, E. T. *The Silent Language*. Garden City, N.Y.: Doubleday, 1959.
14. Hollander, E. P., and Hunt, R. G. (Eds.) *Current Perspectives in Social Psychology*. New York: Oxford University Press, 1963.
15. Horrobin, D. F. *The Communication Systems of the Body*. New York: Basic Books, 1964.
16. Hoselitz, B. F. (Ed.) *A Reader's Guide to the Social Sciences*. New York: Free Press, 1959.
17. Lewin, K. *Principles of Topological Psychology*. New York: McGraw-Hill, 1936.
18. Litterer, J. A. *Organizations: Structure and Behavior*. New York: Wiley, 1963.
19. Marx, M. H. (Ed.) *Theories in Contemporary Psychology*. New York: Macmillan, 1963.
20. McLuhan, H. M. *Understanding Media*. New York: McGraw-Hill, 1964.
21. Miller, J. G. Living systems: Basic concepts; Living systems: Structure and process; Living systems: Cross-level hypotheses. *Behav. Sci.* 10:193, 337, 380, 1965.
22. Morris, C. *Signification and Significance*. Cambridge, Mass.: Massachusetts Institute of Technology Press, 1964.
23. Petrullo, L., and Bass, B. M. (Eds.) *Leadership and Interpersonal Behavior*. New York: Holt, Rinehart and Winston, 1961.
24. Pierce, J. R. *Symbols, Signals and Noise*. New York: Harper, 1961.

25. Ruesch, J. *Disturbed Communication.* New York: Norton, 1957.
26. Ruesch, J. *Therapeutic Communication.* New York: Norton, 1961.
27. Ruesch, J. Social process. *Arch. Gen. Psychiat.* (Chicago) 15: 577, 1966.
28. Ruesch, J. Epilogue to the Second Edition. In R. R. Grinker (Ed.), *Toward a Unified Theory of Human Behavior* (2d ed.). New York: Basic Books, 1967. Pp. 376–390.
29. Ruesch, J. Psychoanalysis Between Two Cultures. In J. Marmor (Ed.), *Modern Psychoanalysis.* New York: Basic Books, 1968.
30. Ruesch, J., and Bateson, G. Structure and process in social relations. *Psychiatry* 12:105, 1949.
31. Ruesch, J., and Bateson, G. *Communication, the Social Matrix of Psychiatry* (2d ed.). New York: Norton, 1967.
32. von Bertalanffy, L. The theory of open systems in physics and biology. *Science* 111:23, 1950.
33. von Bertalanffy, L. General Systems Theory and Psychiatry. In S. Arieti (Ed.), *American Handbook of Psychiatry.* New York: Basic Books, 1966. Vol. 3, pp. 705–721.

6

Systems and Psychosomatics*

ALBERT E. SCHEFLEN, M.D.

The paper "Psychosomatic Manifestations of Rapport in Psychotherapy" which is reprinted in another section of this book served as a stimulus for this paper and its general remarks. The reader may read these words either as an introduction to Charny's work or for their own value. Charny's paper (see Chapter 13 of this text) is not a psychosomatic study in the traditional sense. He does not assess psychological states or measure and correlate physiological variables as is usually done. Yet obviously there is some cogent reason for his work to have been published.

Nor can Charny's work be assigned to other classical traditions of medical or psychological research. Although he is a psychoanlayst by training he does not base his work on psychodynamic inference, and even though his observational field is a psychotherapy session, his method could not properly be called clinical research. He focuses upon behavior, but not in the behaviorist, neobehaviorist, or behavioral therapy traditions of psychological research. In fact, his method is not that of any experimental psychology, for he does not isolate variables. Rather, he describes units of individual behavior in a social context.

Charny studies social interaction, but not in order to define social structure per se. He uses the social relationship as a context to determine the meaning of individual behavior. It is obvious that he is relying upon some theory of the organism and some method of study that is basically different from the classical approaches of the psychological and clinical sciences. His work reflects a systems view of behavior and his methodology is a modern variant of the natural history approach that has evolved to synthesize elements and conceptualize systems as wholes.

In a systems view, classical divisions such as social, behavioral, and physiological have a different relationship whereby psyche and soma

* Reprinted with permission from *Psychosomatic Medicine* 28:297–304, 1966.

are not seen dichotomously. Since the behavior of the organism as a whole is seen in a social context, somatic processes come to have a different location in the conceptual schema. They are investigated at another stage of the research procedure and in quite another context. Charny does not decide not to assess traditional physiological and mental processes because they have no place in a systems view, but because they are approached at a level that Charny has not included in this particular study.

The important issue is not whether or not Charny's work is actually a psychosomatic study, but whether or not it gives us a fresh start on some old and vexing problems. It will be seen that it does. In summary, Charny's work depicts a structure of framework into which both psychic and somatic data will fit. His application of the systems view to the study of posture is a most welcome example in a new and most welcome trend.

Since I, too, have been active in this approach, I have been invited to write this introduction.

THERE IS NOTHING NEW in principle about a systems view. Rather, it is the newest form of a classical approach to the study of nature, one that evolved from ancient science to the naturalistic approach in Darwin's time, appeared as the holistic view in psychiatry, and was used with experimental verification by Einstein.

It is an oversimplification to reduce the complex history of science to this or that view. But I would, nevertheless, set in contrast two antithetical trends in order to build a crude perspective for the emergence of a systems view in the sciences of man.

Newton is often seen as the principal developer of the method in which some element of the whole is isolated for examination and measurement [6]. This approach became prevalent in the biological and psychological sciences about 1880 and eclipsed the naturalistic approach. In psychiatry, even though the naturalistic view has remained the backbone of *clinical* practice and observation, since World War II the isolation of variables, or American experimental methodology, has become ascendant in *research*. The naturalistic view has also been consistently advocated in the anthropological method and in psychology by men like Wertheimer [32], the gestalt theorists [13, 14], and Lewin [16]. It is a strange paradox that psychiatric research should shift to the Newtonian method at the same time that in other sciences a strong resurgence of naturalistic science was occurring.

In physiology, for instance, in the 1940's we stopped listing the endocrine glands and the Brodmann areas as emittors of discrete functions and we conceived of endocrine and neurosystems as integrations of simultaneous interactions. Many other sciences were formulated in systems terms. Cybernetics [2, 31], the social biologies [12, 17, 27] and Parsonian sociology [20] are clear examples. More recent developments have been (1) attempts to characterize the general nature of systems [1, 19, 21], (2) application of systems viewpoints to human behavior [18, 21, 22, 23], and (3) the formulation of research strategies for synthesizing elements and studying relations.

Material Systems

It is helpful to conceptualize living systems morphologically, that is, as material or concrete entities. This was one of the early steps in general systems theory [9, 19, 28]. It was seen that physically tangible systems are made up of tangible components: a group is made up of people, an organism of organs, an organ of cells, and so forth. Such material entities are seen as having specific relations and organization. Any given system is made up of particular components and these alone, which are arranged and organized only in particular, lawful, and standard ways. (A urea molecule consists always of given atoms in particular configuration; Congress consists of congressmen with specific roles and relationships.) Components of a system are themselves systems and are, in turn, made up of components that are systems. Hence, looking "upward" to complexity, a system is part of a larger system, and analyzing "downward" to elements, a system is made up of smaller systems. One relation between systems, then, is hierarchical: i.e., systems make up systems that make up still larger systems. For living systems, the hierarchy of levels is something as follows:

Group
Organism
Organ system (e.g., endocrine)
Organ (e.g., thyroid)
Cell
Molecule
Etc.

A different order of relation exists between two interacting systems at the same level (e.g., thyroid and adrenal cortex) than between a system and the larger system to which it belongs (say, thyroid and endocrine). It is also apparent that the relationship between a system and its components is not a simple Aristotelian matter of class and member, for the components of a system are interactive and the properties of the larger system depend upon this interaction. Recognition of relationships of level is crucial in old problems like the relation of psyche and soma and in the operations of the new research methodology.

Behavioral Systems

Such a description, though serving as a useful model for seeing structural and morphological characteristics, is not sufficient for depicting the dynamics of living systems. For it is not the spatial relations of organs alone that maintains the living organism. A cell is certainly more than a bundle of molecules and a group is more than a number of bodies. Dynamically, a system is made up of the *behavior* or outputs of material systems—in interaction or relation to each other. Mrs. Smith, Mr. Smith, Betty, and Johnny are not what make up a family. A family is made up of relationships, reliable reciprocals like father-son, mother-daughter, husband-wife, and so on, all maintained in an integration. These relationships are, in turn, made up of behaviors. Mr. Smith is not fathering when he is driving his car to work or engineering a business deal. Fathering is the behavioral unit that he contributes to the family as a family member.

Suppose for a working definition we call any state of a system, including the change that brought it into that state, a behavior. We can postulate that the equilibrium of a system depends upon there being some limits to the range of the behavior of its component systems. The behavior of a system must be lawful and determined, in ordered relation to the behavior of the other components of the larger system. If there should be a runaway in thyroxin excretion or a sharp change in quality of the hormone or a triggering of the excretion that is not related to endocrine regulation, a disruption of endocrine system, then organism disequilibrium would result. Similarly, a group may not survive or remain organized if one or more members runs amok, invents new sets of responses or behaves bizarrely. So it is clear that a systems concept depends upon a deterministic view (not a linear determinism, but a concept of multiple simultaneous interactions—feedback), and we can insist that communication and social organization is not possible unless human behavior occurs in standard forms lawfully arranged in reference to subsystems (physiological) and suprasystems (social and ecological).

The demonstration that human behavior meets these conditions came from structural linguistics. Language behavior occurs in standard units of articulation called phonemes (roughly equivalent to syllables), which are lawfully integrated into larger units called morphemes (roughly equivalent to words) that are, in turn, integrated into still larger units called syntactic sentences [10, 11, 21, 22]. In more recent years, it has been shown that many forms of human behavior occur in standard behavioral units integrated analogously [4, 5, 15, 21, 22, 25]. These units are culturally determined, evolved in the traditions of a people, and transmitted through some type of learning. The evidence for cultural determinism is that the morphology and arrangement of behavioral units is homogeneous among individuals within an ethnic, class, regional, and occupational category, but heterogeneous between such groupings.

Organic and Organismic Levels

The characteristics of the human individual accrue from the transmission of qualities via two large systems: (1) from the gene pool, a molecularly coded template guides the division and specialization of cells that form tissues, organs, and organ systems level by level, and (2) from birth on, this organism is raised in a social group and is thus exposed to a culturally transmitted coding for the patternment of changes and states that we call behavior.

Methodologically, then, we can examine the organism from two general directions. The choice of an option between two operational possibilities seems to explain a long-standing confusion in the sciences of man. Theoretically, any investigator beginning with the cell and synthesizing upward level by level should arrive at the same picture of the individual as one who begins with the group and analyzes downward to the individual. But historically this has not been the case. One investigator arrives at a concept of man entirely different from that of another. It is as though two men had approached each other from two sides of a checkerboard and had not met because one had moved on the black and the other on the white squares. The biologically oriented scientists (including the medical researcher, the organic psychiatrist, and often the psychologist) have ended up with the idea of an organism that largely ignored learning and behavior. On the other hand, the social scientist and the psychologist (who was wittingly or unwittingly socially oriented) arrived at a concept of the person or individual described only in terms of behavior—i.e., learning, psychic processes, roles, etc.

The biological thinker, arriving at the organism by way of material subsystems, stopped short in his synthesis without knowing it, and saw the organism as a collection of organ systems. Wedded to the Newtonian tradition in which the components and not their *organization* was central, the biological investigator was a level below where he thought he was, which is always the case with a reductive investigation. On the other hand, the social scientist moved downward from group to relationships to behavior, observing through time the outward changes and actions of the organism, but often failed to give serious attention to the structural and morphological subsystems in the hierarchy.

Psychosomatic research has been caught right in the middle, using both traditions without the difference having been clearly conceptualized. Soma, a construct that in theory can be considered organismic (and thus take into account culturally prescribed and learned behavior), is in research operations invariably defined *organically*, i.e., by measurement of some subsystem output like blood pressure, corticosteroid level, etc. Soma is not, therefore, operationally at the same level as psyche. As will be seen, psyche *must* be an *organismic* construct. The old arguments, then, about psyche and soma, either as interacting or as different aspects of the same thing, wash out when conceived in terms of the basic relations of systems in a hierarchy.

The concept of levels helps us to grasp how our predecessors got into their conceptual dilemma and gives us an opportunity to correct it. We have not in the past separated the behavior of the relevant systems level by level. We have so often used the old familiar operational definition and said behavior is what an animal does. This would include any activity from metabolizing to breathing, playing bridge, or mating. Now we can specify the level or range of levels of behavior that we are investigating.

SOMA AND ORGANISM

Behaviors of organ systems become integrated into a higher level of organization in the maturational process. At least in part, acculturation, socialization, and idiosyncratic experience create patterns of behavior which accompany somatic activities. Thus, one reference for these patterns is at the social level of organization: they are communicational, relational, and social insofar as their meaning, function, and formal determinants are concerned.

Consider some simple examples. The vibrating column of air that passes through the lips, pharynx, larynx, and trachea can be manipulated to produce vocalizations. The maturing child learns—in compliance with a cultural tradition—to form these sounds in definite quanta or units, phonemes, that are the basic building blocks of his language. At the same time he uses flush reactions, head and neck musculature, and other organ system activities to integrate these sounds in the enactment of the spoken statement. Another example is the formation of behavioral patterns in the process of ingestion. Such patterns involve not only the digestive system, but the articulatory system, posture, dress, cosmetics, and what not. Thus, eating may be genetic but, for the most part, when we eat, what we eat, and how we eat and act when we are eating are cultural and require that we see them also at a higher level of systems and behavioral integration.

PSYCHE AND ORGANISM

The difficulties we have had in the past in defining behavior and soma have been nothing compared to that in defining abstract constructs like psyche, mind, ego, and self. The matter is not straightened out by a systems model, but a few comments are in order. In common usage, three different kinds of definition have been customary for the term psyche:

1. *Behavioral*—psyche as the totality of behavior—e.g., in concepts like personality (by implication, a subject's psyche would have to be described by another person).

2. *Mentalistic*—psyche as a subject's conception of himself (described, of necessity, by the subject himself).

3. *Inferential*—psyche as an abstraction about mentalistic or neuroprocesses reached by inferences from behavior (using what we now call a "black-box" model).

Separating organic and organismic levels immediately allows a partial clarification of these various usages. Any black-box inference about neuroprocesses belongs at a lower level; it is not a statement about the organism as a whole. It is probable that affect belongs also to the organic levels, as an experience of organic system states (though recognition and conceptualization of affect are organismic behaviors).

There seems to be little value in retaining terms like psyche to represent directly observable behavior. The observable activity of the organism we can call behavior, and reserve the term psyche for a subject's impression of his own and of other behavior (i.e., "reality"). We can conclude immediately that most, if not all, of a subject's knowledge of himself and the world has been *learned*. Often, to be sure, it has been learned as "insight" from his own observations, but more often it is learned in his family and his culture.

The fact is that members of a cultural tradition learn not only what behavior to perform in a given situation, but what to *say* about it. What one learns to say becomes what one learns to think. It is, therefore, no longer tenable to base psychodynamic models on linear causation and say that thought *causes* behavior. For obviously the two are interconnected in multiple simultaneous causation, together with other aspects of organism state and social situation.

In any event, there seem to be two basic modes by which a subject can inform others (and himself) about his behavior: (1) linguistic and kinesic communication and (2) symbolic expression. Both are behavioral systems. So we are forced to classify the phenomenon of consciousness as a special subclass of organismic behavior.

The concept of the unconscious does not seem to change this self-interpretation [7, 8]. Some behaviors are observable and others are not. Operationally, the unconscious can be used to refer to the discrepancy between behavior (of self and all other systems) that is observable and reportable by the subject, and that which is not. (This discrepancy has to be extended to cover the difference between what significance a behavior is thought to have and what it can be observed to have.)

One can readily appreciate the disappointment of those researchers who have tried to make simple one-to-one correlations between psychic and somatic variables. The problem is that they have been studying them, whatever their theory, from different levels of organization.

Research Methodology

Assuming such an arrangement of matter in living systems as I have described and the occurrence of orderly changes at each level, an investigation in systems terms demands that we keep to a level and that we not confuse levels or jump unsystematically from one to the other.

The arrangement in a hierarchy is such that any behavior is mediated at the levels below and, in turn, sustains the order at higher levels. It is at these higher levels that the meaning or function of a behavior is assessed. This has led men like Simpson [26] to formulate rules something as follows: to describe a system we see it as a whole; to find out *how* it works we examine its component systems and their relations; to find out *why*, we examine its role in a larger system.

In the study of the human organism, we can describe what it is in terms of its behavior. The data about the organism are behavioral data, and these must be the focal point of organismic research. *Then* we can move to studying: (1) what people *think* and *feel* about behavior; (2) what the behavior means in a social and cultural context; and (3) what organic, cellular, and molecular systems maintain and mediate the behavior. It seems likely that states of human relationship are related to states of the organism that are related to states of all organismic subsystems. Changes at any level reverberate up and down the hierarchy until they are brought into equilibrium, so study at any level is vital. But behavior is the starting point and any change described must be related to the behavioral units that accompany it. The human being may eat when his blood sugar drops and when someone cooks dinner. When the focus of interest is in the individual, the behavior of eating should be studied at both levels.

An ultimate research program would include a team of specialists from the biological sciences and the behavioral and social sciences, men flexible enough to shift their frame of reference to those levels above or below their particular specialty. Short of such an ideal situation, Charny nevertheless embarks upon a relatively wide course of study. He begins by defining constellations of behavior. He has reason to believe that postural shifts will demarcate these units [24, 25], so he sketches the postural aspects and fills in some of the lexical behaviors. Then he relates these units to the therapist-patient relationship as a social context.

There is an important tactical gain in shifting from inference and subjectivistic data to direct observation of behavior in a social context. Certainly, the individual learns something of the function, significance, and value of his behavior in growing up and, theoretically, we can gather *some* information by asking him (directly or by projective and other inferential techniques). But we know how often he misconceives what he does and why he does it. We can circumvent this problem and gain direct and objective data on the meaning and value of behavior by observing systematically what happens in an actual interaction when given behavior does and does not occur. Thus in a sense we duplicate Bernard's [3] classical method at the social level and take the matter out of the realm of our notions or the subject's notions about behavior. I am not suggesting that we return to Watson's negativity [30] about subjective experience, but that we treat it in psychosomatic research as additional data. It seems short-sighted to use it alone to assess organismic states.

Another gain is made by shifting from the variable—often an a priori abstraction about behavior—to the behavioral unit. For the latter is presumed to be coded in the culture and to appear regularly in interaction, so that any investigator can repeat the observations.

An even more radical implication for research method in a systems view, one that pervades all of its operations, is based on the cardinal principle that the system is an organization. We are not, therefore, able to study a system cogently by means of the Newtonian methods, useful as they are to study sets of mechanical components that are not in significant interaction. The components of a system are specialized and a study of one does not tell us about the others. And qualities emerge by virtue of organization that are not in the components alone—e.g., H_2O is a liquid, a quality not predictable from the study of gaseous hydrogen and oxygen. Similarly, the study of the endocrine and cardiovascular systems will not tell us how an organism will behave as a whole; the study of personality will not predict how members will act in a group.

To study a system, we may break it down into parts as a step in the investigation, but we will have to then synthesize to examine organization. There are no independent variables in a system.

References

1. Ashby, W. R. *An Introduction to Cybernetics*. New York: Wiley, 1956.
2. Ashby, W. R. General systems theory as a new discipline. *Gen. Syst.* 3:1, 1958.
3. Bernard, C. *An Introduction to the Study of Experimental Medicine*. New York: Schuman, 1927.
4. Birdwhistell, R. L. Body Behavior and Communication. In D. L. Sills (Ed.), *International Encyclopedia of the Social Sciences*. New York: Macmillan, 1968.
5. Birdwhistell, R. L. In N. McQuown (Ed.), *The Natural History of an Interview*. New York: Grune & Stratton. (In press.)
6. Burtt, E. A. *The Metaphysical Foundation of Modern Science*. Garden City, N.Y.: Doubleday Anchor Books, 1954.
7. Fenichel, C. *The Psychoanalytic Theory of the Neuroses*. New York: Norton, 1945.
8. Freud, S. The Unconscious. In *Collected Papers*, Vol. IV. London: Hogarth, 1953.
9. Grobstein, C. Levels and ontogeny. *Amer. Sci.* 50:46, 1962.
10. Harris, Z. *Methods in Structural Linguistics*. Chicago: University of Chicago Press, 1951.
11. Hockett, C. F. *A Course in Modern Linguistics*. New York: Macmillan, 1958.
12. Klopfer, P. H. *Behavioral Aspects of Ecology*. Englewood Cliffs, N.J.: Prentice-Hall, 1962.
13. Köhler, W. Gestalten Probleme und Anfänge einer Gestalt und Theorie. *Jahresber. Ges. Physiol.* 3, 1925.
14. Koffka, K. *Principles of Gestalt Psychology*. New York: Harcourt, 1935.
15. Lawson, C. A. Language. Communication and biological organization. *Gen. Syst.* 8:107, 1963.
16. Lewin, K. In D. Cartwright (Ed.), *Field Theory in Social Science*. New York: Harper, 1951.
17. Lorenz, K. *King Solomon's Ring*. New York: Crowell, 1952.
18. Miller, G. A., Galanter, E., and Pribram, K. A. *Plans and the Structure of Behavior*. New York: Holt, 1960.

19. Miller, J. G. Living systems: Basic concepts. *Behav. Sci.* 10:193, 1965.
20. Parsons, T. *The Social System*. Glencoe, Ill.: Free Press, 1951.
21. Pike, K. L. *Language*, Part I. Summer Institute of Linguistics, Glendale, Calif., 1954.
22. Pike, K. L. Toward a theory of structure of human behavior. *Gen. Syst.* 2:135, 1957.
23. Sapir, E. In D. G. Mandelbaum (Ed.), *Culture, Language and Personality*. Berkeley: University of California Press, 1956.
24. Scheflen, A. E. The significance of posture in communication systems. *Psychiatry* 27:316, 1964.
25. Scheflen, A. E. *Stream and Structure of Communicational Behavior*. Philadelphia: Eastern Pennsylvania Psychiatric Institute, 1965.
26. Simpson, G. G. The status of the study of organisms. *Amer. Sci.* 50:36, 1962.
27. TinBergen, N. *Social Behavior in Animals*. London: Methuen, 1953.
28. von Bertalanffy, L. An outline of general systems theory. *Brit. J. Phil. Sci.* 1:134, 1950.
29. von Bertalanffy, L. *Problems of Life*. New York: Harper, 1960.
30. Watson, J. B. *Behaviorism*. New York: The People's Institute, 1925.
31. Wiener, N. *Cybernetics*. New York: Wiley, 1948.
32. Wertheimer, M. *Drei Abhandlungen zur Gestalt Theorie*. Erlangen: Philosophische Akademie, 1925.

7

Asymptotic Systems of Survival

JULES H. MASSERMAN, M.D.

THIS CHAPTER is essentially a succinct reply to a challenge by the editors of this volume to summarize whatever systematization of psychiatric thought and practice I may have derived from three score years of a full and perhaps only partly futile life spent in library, clinic, and laboratory. I cannot pretend that the theoretical residues of the triune experiences I shall review here are in any sense comparable to Einstein's tri-termed $E = mc^2$, except that my formulations as to the interrelationships of the causes, phenomena, and control of behavior also consist of but three interpenetrating concepts, to wit:

First, that we human beings have only these ultimate (Ur) sources of concern: (1) our physical health and longevity, (2) our interpersonal securities, and (3) our existential significance.

Second, that whenever these primary Ur-anxieties become manifest in excessive and disabling somatic and behavioral aberrations, we insistently seek, despite the ultimately Sisyphean futility of our efforts, a corresponding triad of therapies designed (1) to restore our well-being and manipulative skills, (2) to recultivate our social alliances, and (3) to reconstitute our faith in life's meaning and values.

Third, that these statements can be validated by recourse to the three epistemologic sources of all sciences: (1) the historical, (2) the comparative, and (3) the experimental. Let us explore each heuristically in the same order.

Historical Approaches to the Essentials of Therapy

PHYSICAL THERAPIES

Since paleolithic times, man has tried to mitigate his fears of his inimical universe through his sciences and technologies. In two of the most important of these disciplines—medicine and surgery—he has sought to find means to cure his ills and restore his skills by empiric medicaments ranging from herbs and minerals to the first specific use of quinine for malaria in 1820, and various procedures from the ancient binding of fractures to the modern embedding of plastic hearts. Psychiatry, too, as a branch of medicine, has always employed pharmacological, surgical, and other modes of relieving man's material fears and their somatic reflections and, fortunately, is once again intensely developing this important rubric of therapy.

SOCIAL

Scientific and technical advances, however, have historically proved inadequate for man's quest for security, inasmuch as:

1. Expanding knowledge revealed mysteries and challenges ever beyond man's puny powers, however much he learned and applied.
2. Each man realized that the tools his neighbors developed could be as lethal, or more so, than his own.

Ergo, man learned to seek ever more inclusive human allegiances ranging from parental and familial ties through the clan, tribe, city, and state to their present limits in highly uncertain international pacts. Within all these aggregations, specially selected men were given specific trusts and tasks toward enhancing group welfare: kings* and constables for internal government, soldiers for security (which of course included external conquest), and medicine men who, as protophysicians, treated illness, and as protopsychiatrists, corrected excessive deviations from locally accepted norms of conduct—a reintegration of medical and community function that, fortunately, is also being revived in modern psychiatry.

MYSTICAL

However, these irregular and sometimes inchoate strivings, even when combined with man's pyramiding technical knowledge and skills, have never yet been enough to allay a third and even broader

* Our term "king" (German *König*), head of state, is derived from Anglo-Saxon *cyning*, head of kin (Perry).

concern, since no group of human scientists, however large or learned, has ever been able to control the heavens, secure the future, explain the nature or meaning of life, or mitigate the inadmissibility of death. Man has therefore always erected transcendent cosmic systems, presumably within his ken, operated by supernatural beings whom he could conveniently control either directly or through special mediators here on earth. Designated as such were priests and medicine men who, in their temples and sanatoria, were thus wishfully charged with this third task: not only to minister to the physical and social needs of men, but also to influence in their favor the mystical goals and gods of the universe.

It would obviously require too long a series of ponderous tomes to cite in detail the triune treatment of these trepidations of man throughout history; but perhaps a brief recall of how, with both conscious and intuitive skill, each aspect of the therapeutic trinity was intuitively invoked during one enlightened period—for example, in the Asklepiad Sanatoria of ancient Greece—may serve to illustrate the general thesis. In brief, the methods employed in these classically effective mental health centers (named after the son of Apollo, the handsome, incorruptible god of music, mind, and medicine, and thereby the very image of every psychiatrist) mitigated man's Uranxieties as follows.*

First, by restoring physical well-being. After the patient had left his contentious home and traveled to one of the Sanatoria in the salubrious environs of Cos, Memphis, or Knidos, he was welcomed not by a clerk or social worker but by no less a parent-surrogate than the Head Priest or Priestess, who then further cheered and reassured him by conducting him past piles of discarded crutches and bronze plaques bearing testimonials from grateful ex-patients. Immediate attention was then concentrated on restoring the patient's physical well-being through rest in pleasant surroundings, nourishing and appetizing diets, relaxing baths and massages, and the carefully measured administration of nepenthics—drugs that resembled modern ataractics in that they apparently tranquilized both the patient and the doctor. Indeed, the Greek word *therapeien* itself meant *service,* just as later the Latin *curare, to care for,* gave rise to our term *cure.*

* Adapted from J. H. Masserman, Man's eternal anxieties and compensatory illusions. *Illinois Med. J.* 12:1–16, 1965.

Second, by recultivating human relationships. In accordance with this objective, equal effort was expended in counteracting the patient's Ur-anxiety of social isolation as follows.

1. Dyadic Confidence in the Physician. Then, as now, the patient was encouraged to relate to his therapist:

   a. As a kindly and protective parental figure who provided a source of security and comfort.

   b. As a learned and experienced teacher whose counsels for more restrained and balanced, and therefore healthier and happier, modes of life could be followed on rational and practical grounds (e.g., as in the various Stoic schools).

   c. As a more personal mentor uniquely interested in the suppliant's complaints ("present illness") and willing to explore their relationship to the patient's past experiences (psychiatric history), their meanings and values (symbolisms), and acquired patterns of goal-directed action (operational analysis), in order that the intellectual understandings ("verbal insight") so derived would lead to more satisfying, lasting, and useful (emotional, or better, operational) adaptations. Socrates required his students to work through their own perplexities under his mentorship, and Plato understood the unconscious significance of dreams and symbols. Aristophanes, in his delightful comedy "The Clouds," pictured the distraught Strepsiades lying on a couch and trying to acquire understanding through free-associations about controlling the phases of the moon so that his monthly bills would not fall due; less passively, Soranus records the cure of a case of "hysteria" in a virginal bride by a form of direct action (i.e., "lighting the torch of Hymen") that would shock a Freudian analyst.

   d. Finally, the needful dependence of the subject on the physician (transference) that was so cultivated also led to an avid acceptance by the patient of the efficacy of the physician's quasiscientific, quasimystical remedies, some of which were covertly punitive for, and thereby comfortingly expiative of, social and religious transgressions implicit in the patient's *illness*—a term related to Anglo-Saxon *yfel, evil.* Among these were not only a vast variety of unpleasant purgings, bleedings, coolings, roastings, and broilings, but even more direct physical and surgical interferences with cerebral function, such as the Egyptian practice of trephining the skull and incising the cortex or, as described by Pliny the Elder, subjecting the patient to convulsive therapy by discharging electric eels through the head. We read of Hippocrates' condemnations of the ignorance and superstition inherent in many of these "false remedies," but this alone proves how widely practiced, then as now, they must have been.

2. GROUP THERAPIES. But the patient's social rehabilitation was not left to such partial and intermediate transactions with the physician alone; concurrently, his essential group relationships and skills were recultivated through the following modalities.

a. *Music,* which provided aesthetic expression and encouraged group belongingness through feelings of conjoint rhythm and harmony.
b. *Calisthenics and dancing,* which afforded similar possibilities of reorientative interaction and communion.
c. *Competitive athletics,* not only for the joy of healthy action but for public recognition through nondestructive competition and reward.
d. *Dramatics.* Here the poetic psychiatrist (and there can be no other) may well ask: What writings better explore or epitomize basic human relationships than the plays of Euripides, Aeschylus, or Aristophanes? And what productions can offer the patient, either as witness or participant, more varied identifications, vivid experiences, or vicarious solutions of his own interpersonal problems? The Greeks cherished and utilized these tragedies and comedies for their deep human empathy and ageless significance, endlessly varied their themes, and were personally involved as actors, chorus, or affectively moved audience—and thus explored in essence the basic interactions utilized in modern forms of group therapy.
e. *Social rehabilitation.* This offered a transition between a passive dependence on the sanatorium to a growing recognition of the advantage of a return to the community and service for the common good.

3. METAPHYSICAL. Finally, the Greeks in their wisdom also recognized a third human Ur-necessity of a belief in some inscrutable but transcendent order—and, parenthetically, demanded that Socrates pay the ultimate penalty for threatening man's trust in the existence of beneficient celestial beings. To capitalize on this ultimate faith, the Asklepiad Sanatoria, like many hospitals today, were built and operated by one or another religious cult, which added the following powerful factors to therapy.

a. A "divinely revealed" doctrine in which all believers could feel an exclusively self-elevating bond of fellowship.
b. A reassuring ritual which, through its origin in human needs and through millennia of empiric refinement, included such exquisitely gratifying procedures as:
   (1) The symbolic eating and drinking of the parent-god's body in the forms of thaumaturgically potentiating food and wine (as exemplified in the ancient worship of Melitta and Mithra).
   (2) The temple hymns, sung and played in the simple, repetitive, hypnotic cadences of a mother's lullaby, and often resulting in "temple sleep." (Two thousand years later, Mesmer was to extol the mystic powers of "animal magnetism" and, through Charcot and Janet, initially mislead Freud and lead Bernheim to warn, "It is a wise hypnotist that knows who is hypnotizing whom.") Such escapist trances were then varied with food, drink, and Dionysian sexual indulgences to be triply enjoyed, since they also honored one's permissive and accommodating gods.
   (3) The "anointing" or "laying on of hands" to cure an injured bodily part—a direct reminiscence of the soothing parental stroking of an injured child, as ever since exemplified in the healing powers of the King's Touch, exploited in the seventeenth century by Greatrakes the Stroaker, and still sought by the emotionally immature and physically deprived avid of massage and chiropractic.
   (4) The ethereal, elevating emphasis on the spiritual—a concept as fundamental to life as is the neonate's first breath or *spiritus*. Every human is indeed variously inspired, ac-

quires an esprit de corps, becomes dispirited, and finally expires so that his immutable spirit can begin life anew. And here, too, the physician-priest functions in knowing the spiritual world, purveying professed contrition and remorse to the Spirits of our Fathers, and requiring only a gratifying small penance with which to avoid the horrors of eternal punishment. Meanwhile, the temple then and now furnished a divinely protected sanctuary from earthly stresses and problems, much as Lourdes is a haven of comfort and healing today.

(5) Finally, the priest also mediated the supreme promise of all religions—or, for that matter, of all "scientific" systems—the conquest, through life eternal, of man's most grim and implacable enemy: death itself.

These, then, were the ancient—and are the eternal—practices that embody what physicians have intuitively known for centuries: that although no man can ever be certain of his health, friends, or philosophy, the illusions of security in each of these spheres are essential to his welfare; and that all methods of medical-psychiatric therapy are effective only insofar as they restore physical well-being, foster more amicable interpersonal relationships, and help the patient amend his beliefs so as to render them more generally acceptable and useful.

The Comparative Approach to the Essentials of Therapy

To supplement this historical survey, it would require still another five-foot shelf to review the current and equally illuminating anthropological, ethnic, and transcultural studies of Gillin, Margolin, Lambo, or Kiev and Leighton, to cite Ziferstein's and my own comparisons of Soviet and American techniques, or to acknowledge the contributions of other sociopsychiatrists with sufficient breadth of interest and objectivity to investigate the therapeutic practices of other societies; moreover, it would take an integrative genius to demonstrate that all of these fall into the familiar rubrics of physical, social, and metaphysical readaptation. However, even if we do not go so far afield, perhaps a more parochial reexamination of the various modes of therapy in our own culture, from colonic lavage to Christian Science, may demonstrate a similar triune efficacy.

This, of course, requires an unusual willingness to concede that diverse concepts and cults are comparable to our own individually cherished one, a challenge few of us are willing to accept. Nevertheless, playing the role of a Socrates in a peripatetic dialogue with my psychiatric residents and analytic trainees, I have sometimes proposed some outrageous postulate such as: Resolved, that electroshock therapy and psychoanalysis are, as operational systems, essentially more alike than different in their basic therapeutic actions and effects. In the past, though not so much recently, this usually evoked a storm of protest that the two methods were manifestly, incomparably different, in that: electroshock therapy was "physical," "enforced," "impersonal," "stereotyped," "rigidly conducted," "suppressive," "antimnemonic," "intellectually impairing," and the like; whereas, in diametric contrast, psychoanalysis was "communicative," "psychologic," "voluntary," "exquisitely interpersonal," "flexible," "evocative," "restorative of memory," and "designed to develop to the full the patient's cognitive and adaptive capacities" through "insight."

At this juncture, I occasionally pointed out that the last assertion directly begged the question since, in the historical and comparative contexts in which I had posed the question, "insight" itself could be defined only as that transiently ecstatic state in which patient and therapist temporarily shared the same illusions as to the cause and cure of their mutual difficulties. With this additional goal, some of my residents began to explore subtler dimensions of therapeutically operative similarities between electroshock and analysis and came up with the following disconcerting symmetries.

PHYSICAL PARAMETERS

Both methods offer an escape from the mundane stresses of external reality onto a sensorially isolated bed or couch provided by society and protected by a parental surrogate for about the same average number of recumbent hours in hospital or office. Both methods serve to disorganize current patterns of deviant behavior: electroshock therapy by direct cerebral diaschises, analysis by semantic and symbolic alterations of concept and reaction.

INTERPERSONAL INFLUENCES

In both methods, the patient selects (whether "voluntarily" or by equivalent "external" pressure) the method of therapy and the therapist he regards as most suitable for his needs. In both methods, the therapist sympathetically accepts the patient as more or less helpless and "ill," is convinced of the special validity of his own therapeutic theories and the efficacy of his techniques—and thus rounds out an operationally effective folie à deux.

SOCIAL PARAMETERS

In both methods, there are covert but inevitable physical, economic, social, and other punishments for the persistence of abnormal conduct: in electroshock therapy, more incarceration, exclusion of visitors, postshock headaches, and other adverse sanctions; in analysis, more time, more expense, more patronizingly disillusioning "interpretations," plus the onus of having an "unanalyzable (i.e., incurable) character disorder." Conversely, there are desirable reacceptances and rewards for progressively more conforming behavior (the "patient compliance" of Ehrenwald): e.g., the expanding privileges, therapists' approval, renewed familial hegemony, and finally, membership in Recovery, Inc., after "successful" electroshock therapy; and after analysis—at least until recently—acceptance in the sophisticated elite of the "thoroughly analyzed" at cocktail parties or, in the case of institute trainees, admission to the local psychoanalytic society.

MYSTICAL

Finally, in both cases, patient and therapist join in an essentially worshipful belief that either Cerletti or Freud respectively brought providentially inspired salvation to ailing mortals, attainable through respectively prescribed rituals of suffering, expiation, enlightenment, and the reacquisition of metapsychological grace. Any agnostic who, at national meetings, has attended the Section on Electroconvulsive Therapy or, usually at another hotel at a discreetly noncommunicative distance, a Seminar of Psychoanalytic Theory and Therapy, will unmistakably have experienced the devotional aura as well as the scientific import of both proceedings.

Similar operational parallels can be constructed, say, between vitamin therapy and "nondirective" counseling or between dauerschlaff and psychodrama. Such comparisons should, of course, never obscure the specific attributes and special clinical applicabilities of each method, but outside the bounds of reductionist sophistry, objective analysis may help clarify the universal factors of therapy we have been considering, viz: to help the patient realize that his formerly cherished modes of conduct in the three spheres of physical, social, and philosophical adaptation are no longer either necessary or profitable, and to learn, by exploration and personal experience, that new ones will prove more personally pleasurable, socially advantageous, and existentially compatible.

Experimental Validations of the Essentials of Therapy

This brings us to our final heuristic resource: the experimental approach to the study of "normal" behavior and the treatment of its deviations. In the words of Adolf Meyer, every friend and patient can be studied as a ready-made "experiment of nature," but for more adequate controls and precise analysis, animal research becomes necessary. I have elsewhere rather extensively reviewed the rationale, ethological evidence, and experimental studies under this rubric; here, we can only in outline form summarize the premises tested and the methods of inducing and reversing aberrant behavior that emerged from such studies. The four "biodynamic principles" evolved, in briefest statement, are these:

BIODYNAMIC HYPOTHESES

PRINCIPLE 1: MOTIVATION. The behavior of all organisms is actuated by physiological needs, and therefore varies with their intensity, duration, and balance.

PRINCIPLE 2: PERCEPTION AND RESPONSE. Organisms conceive of, and interact with, their milieu not in terms of an absolute "external reality," but in accordance with their individual genetic capacities, rates of maturation, and unique experiences.

PRINCIPLE 3: RANGE OF NORMAL ADAPTATION. In higher organisms these factors make possible many techniques of adaptation, which in turn render the organism capable of meeting stress and frustration and maintaining an adequate level of satisfaction by (1) employing new methods of coping with difficulties when the ones formerly employed prove ineffective, or (2) by modifying the goals or substituting new ones when the old become unattainable.

PRINCIPLE 4: NEUROTIGENESIS. However, when physical inadequacies, environmental stresses, or motivational-adaptational conflicts exceed the organism's innate or acquired capacities, internal tension (anxiety) mounts, neurophysiological (psychosomatic) dysfunctions occur, and the organism develops overgeneralized patterns of avoidance (phobias), ritualized behavior (obsessions and compulsions), regressive, hyperactive, aggressive, or other deviant social transactions, or bizarrely "dereistic" (hallucinatory, delusional) responses corresponding to those in human neuroses and psychoses.

EXPERIMENTAL NEUROSES. As indicated in Principle 4, marked and persistent deviations of behavior were induced by stressing individual animals between mutually incompatible patterns of survival: as, for instance, requiring a monkey to secure food after a conditional signal from a box which might unexpectedly contain a toy snake—an object as symbolically dangerous to the monkey as a live one. In this connection we further amended Freudian doctrine by demonstrating that "fear" in the sense of threat of injury need not be involved at all, that is, equally serious and lasting neurotigenesis could be induced by facing the animal with difficult choices among mutually exclusive satisfactions—situations that parallel the disruptive effects of prolonged indecisions in human affairs.* Either form of conflict produced physiological and mimetic manifestations of anxiety, spreading inhibitions, generalizing phobias, stereotyped rituals, "psychosomatic" dysfunctions, impaired social interactions, and other persistent regressions and deviations of conduct.

CONSTITUTIONAL INFLUENCES. Animals closest to man showed symptoms most nearly resembling those in human neuroses and psychoses, but in each case the "neurotic" syndrome depended less on the "specificity" of the conflict (which could be held constant) than

* Applying this principle on an international scale, the economist M. F. Millikan asks: "How much freedom of choice can people not used to many degrees of freedom take without psychologic strain that threatens the stability of the system?" And answers, "My hunch . . . is that a good deal of the instability . . . in the underdeveloped world may result from the phenomenon [of indecision]."

on the constitutional predisposition of the animal. For example, under similar stresses spider monkeys reverted to infantile dependencies or catatonic immobility, cebus developed various "psychosomatic" disturbances including functional paralysis, whereas vervets became diffusely aggressive, persisted in bizarre sexual patterns, or preferred hallucinatory satisfactions such as chewing and swallowing purely imaginary meals while avoiding real food to the point of self-starvation.

Methods of Therapy. Since we induced experimental neuroses in animals not only to study their causes and variations but primarily to search for the principles of therapy, this portion of our work was assigned most time and effort. After the trial of scores of procedures, only nine general methods—significantly parallel to those used with human patients—proved to be most effective in ameliorating neurotic or psychotic symptoms. Again in summary, these techniques could be marshaled under "physical," "social," and—if the term be used in a broadly operation sense—"mystical" subheadings, as follows:

Physical Methods

1. The satisfaction of one or more of the opposed biological needs that induced the impasse by facilitating the achievement of one or another goal, such as hunger or thirst versus sex, or exploratory drives versus evasion of injury—conflicts of motivation that often produce adaptational difficulties in humans.
2. Removal from the laboratory to a less stressful environment, again, by human analogy, a better home, job, or climate—or even an occasional "vacation" from the cumulative strains of daily living.
3. The provision of opportunities for the reutilization of neurotically inhibited skills: in the cat, using another lever to obtain food; in the human, inducing a crashed, unhurt but temporarily phobic fighter pilot to fly another plane immediately, and thus dispel his mounting anxiety over his threatened loss of mastery over his milieu.
4. The preventive administration of various drugs such as alcohol that blunted perceptions of impending trauma or gave the animal later surcease by dampening excessively intense, complex, or prolonged residual responses to conflict. Some neurotic ani-

mals which experienced such effects from the ingestion of alcohol sought out the drug if made available and became avid alcoholic addicts, thereby extending the human parallel.

5. The use of electroconvulsive or other methods of producing sufficient cerebral anoxia and diaschisis to disrupt undesirable patterns of behavior, provided they were relatively recently induced and thereby more vulnerable than those more deeply established and ingrained—again, in direct analogy to clinical electroshock therapy.
6. The employment of surgical or electrocoagulative techniques to produce neurological lesions and consequent disorganizations of aberrant behavior similar to those produced by human lobotomies or thalamotomies.

Social Methods

1. Environmental. The resolution of the motivational conflict through progressive exposure to neurotically feared situations, always keeping the animals' reactions within their readaptive potentials—a method that may be employed clinically in the so-called "behavior therapy" of obsessive-phobic patients, provided they possess sufficient fortitude ("ego-strength").
2. Dyadic. The retraining of the animal by carefully individualized care and guidance, the experimenter acting as a "personal therapist" who helps the subject reexplore early experiential conflicts and dispel their neurotigenic effects; as in psychoanalysis, helping the patient recall early traumatic events and reevaluating their residues in the here and now by "corrective emotional experiences" vis-à-vis the therapist and elsewhere in a physically, socially, and philosophically reevaluated world.
3. The provision of a therapeutic association with an accepting coterie of well-adapted ("normal") conspecifics, in peer-pedagogic parallel to sending a "problem child" to a "good school," i.e., where children behave in the manner desired.

Mystical Connotations. Finally, if this term is extended to its full meaning of behavior based on faith in the unprovable, then a dog's unreasoning confidence in his master's magical prowess and devotion to its interests may also be extended to the therapeutic techniques (1) and (2) as described under "Social Methods" above.

Clinical Summary

If, then, all of the historical, comparative, and experimental data here indicated were to be integrated and interpreted, our three essential principles of therapy could be clinically formulated as follows.

To begin with, we must discard our cold armor of aloof "professional dignity" and accept each patient not as a diagnostic "challenge," or as a recipient of "specific therapy for the organic pathology" (a repulsive solecism)—and least of all as only another research datum—but as a troubled human being seeking comfort and guidance as well as mere relief from physical suffering. These larger requirements should be met in the following psychiatric aspects of general medical therapy.

Regardless of whether the patient's complaints are considered primarily "organic" or "functional," bodily discomfort and dysfunction are to be relieved by every medical and surgical means available including, when indicated, carefully prescribed sedatives and hypnotics temporarily useful to dull painful memories, relieve apprehension, and quiet agitation. In my own research studies and in my clinical experience, I have found that the barbiturates, bromides, aldehydes, and other well-tested drugs, when wisely used, are often preferable to many of the widely promoted but dubious "ataractics" and "tranquilizers"; however, in nearly all medical and surgical specialties we recognize that such surcease, usually of mixed suggestive and pharmacological origin, is merely the first stage of therapy. As soon as the patient's tensions and anxieties have abated sufficiently to make him more accessible and cooperative, we must strive to reevoke his initiative, restore his lost skills, and encourage him to regain the confidence and self-respect that can come only from useful accomplishment.

Second, since no man is an "iland, intire of itself," the physician, whatever his specialty, has a broader task: to recognize that his patient may be deeply concerned about sexual, marital, occupational, and other problems that may also seriously affect his physical and social well-being. This involves an exploration, varying in depth and duration but always discerning and tactful, of the attitudes and values the patient has derived from his past experiences, his present goals and tribulations, his successful (normal), socially ineffective (neurotic), or bizarrely unrealistic (psychotic) conduct, the ways in which these patterns relieve or exacerbate his current difficulties, and whether they are accessible to various other methods of medicopsychiatric therapy.

It is customary at this juncture for the psychiatrist to warn his colleagues in other fields off his supposedly esoteric preserves; instead, let me confess that it has been my gratifying experience that in most cases any intelligent, sensitive physician can, in time easily available to him, conduct the essential psychotherapy required. In essence this will consist of using gentle reasoning, personal guidance, and progressive social explorations to help the patient correct his past misconceptions and prejudices, abandon infantile or childlike patterns of behavior that have long since lost their effectiveness, revise his goals and values, and adopt a more realistic, productive, and lastingly rewarding ("mature") style of life. In this skillfully directed reeducation (good psychotherapy, despite a recent fad to the contrary, is about as "nondirective" as good surgery) the enlightened cooperation of his family, friends, employer, or others may, with the patient's consent, be secured and utilized to the full. By such means the patient's second Ur-defense will be strengthened by renewed communal solidarity and security—a sine qua non of dyadic and group therapy.

Finally, and to mitigate the third, or existential Ur-anxiety, the patient's religious, philosophical, or other convictions, instead of being deprecated or undermined, should be respected or strengthened insofar as they furnish him with what each of us requires: a belief in life's purpose, meaning, and value. In this fundamental sense, medicine, being a humanitarian science, can never be in conflict with philosophy or religion—since all three seem to be designed by a beneficent providence to preserve, cheer, and comfort man—and may thereby complement a trinity to be respected by any physician deeply concerned with man's health and sanity.*

* Indeed, with respect to these latter terms, it is of historical-philological significance that the term *sanatos* implied to the ancients the indissolubility of physical and mental functions (*mens sanis in corpora sano*); so also, our more "modern" word *health* can be traced to the Anglo-Saxon *hāl* or *hōl*, from which are derived not only physical haleness and healing, but the greeting, "Hail, friend!" and the concepts wholeness and holiness. Once again, Greek, Roman, and Gaul have bequeathed to us, in the rich heritage of a syncretic language in which "reality" and "illusion" merge, their recognition of the indissoluble trinity of physical, social, and philosophical components of medical and psychiatric therapy.

PART II

SPECIFIC THEORETICAL CONSIDERATIONS

AN INDICATION of the increasing recognition of the importance of general systems theory and psychiatry is the presence in this section of a contribution by the President-Elect of the American Psychiatric Association, Raymond W. Waggoner, Sr. In Part III a paper by Howard P. Rome, a past president of the American Psychiatric Association, will be found, while another past president, Henry W. Brosin, has played a leading role in encouraging the organization of programs on general systems theory and psychiatry in the annual meetings of the American Psychiatric Association.

In Part II our focus is on the developments in specific theory that emerge from the transaction of general systems theory and psychiatry. Such theories are the necessary intermediaries between the general and global on the one hand and the practical and actual on the other. Specific theories are fruitful in both directions, not only providing guidelines for general systems theory in action, as described in Part III, but also in contributing to the further growth and development of the more general theories dealt with in Part I. As Roy R. Grinker, Sr., stresses in his introduction to the second edition of *Toward a Unified Theory of Human Behavior*, such smaller and more specific theories have been exceedingly productive in stimulating empirical and experimental investigations on which the further development of general and global theories has depended.

In the eight papers of this section the basic principles of general systems theory are applied to specific theoretical problems in fields of considerable diversity, ranging from the study of cognition to problems in the time domain, from studies of growth and development and of the therapeutic process to specific theories of industry, the community, and the nature of mental disorder. In each, general systems theory is elaborated and tied in with the specific fabric of the field studied. Our authors, somewhat younger as a group than those of Part I, are all in the forefront of the new generation in American psychiatry as teachers, researchers, and theoreticians.

Our first author, Silvano E. Arieti, has long been an advocate of general systems theory and psychiatry. He is the well-known editor of the *American Handbook of Psychiatry*, one of the recognized basic textbooks in the field. A paper by Ludwig von Bertalanffy in Volume III was the first discussion on general systems theory and psychiatry to appear in a psychiatric textbook. Arieti's fundamental contributions to our understanding of schizophrenia add to his eminence as teacher, researcher, psychoanalyst, and editor. In particular he has pioneered in the previously poorly understood area of disturbances of cognitive functioning in schizophrenia and in the understanding of schizophrenic phenomenology as a part of the process of system dedifferentiation and disorganization. The wide range of Dr. Arieti's interests and abilities is well illustrated by the paper presented here.

Albert E. Scheflen, author of the second paper in Part II, has been able to transcend the purely verbal or purely behavioral analysis of human communication. Starting with the natural history method in the analysis of psychotherapy, he progressed to the use of contextual analysis and then to the use of a systems approach. To Scheflen, psychotherapy is a specialized variant of communications systems in general. His paper on behavioral systems in human communication was presented as part of the initial program on general systems theory and psychiatry at the 1966 annual meeting of the American Psychiatric Association. It enables the reader to become acquainted with one of the revolutionary advances in our understanding of the psychiatric interview, and as such is must reading for all modern psychiatrists.

Our third author, Warren M. Brodey, is not only a very original psychiatrist but also poet, psychoanalyst, cybernetician, innovator, family therapist, and environmental ecologist. His most recent work is *Changing the Family* [1]. As one of the early creative pioneers, he is responsible for helping to establish family therapy as a valid part of psychiatric treatment. After intensively studying problems of communication with the blind, Brodey turned to the intricacies of the interfaces in large man-machine systems in the four years he spent at the Massachusetts Institute of Technology. During this phase of his career, Dr. Brodey excelled in the beauty and accuracy with which he captured the essence of the time dimension, without which there is no process and no life. While most of us speak of time in terms of static cross-sections to which we add the magical symbol *dt*, and then paradoxically label it dynamic, Brodey has evolved a vital language that brings ongoing time into words and phrases themselves. Strange as his language may sound to the beginner, it is the language that will free us from Newtonian or clock time in the same way that Einstein freed physics from its prison. Time-driving, time-graining, and time-loading exemplify the dynamic language of the future which will communicate the excitement that occurs when each family or society encounters its next moment, and when what was not even speculation yesterday becomes reality.

Our fourth author, John MacIver, is the main representative of general systems theory and psychiatry in the world of occupational and industrial psychiatry. In his paper MacIver differentiates community and industry in terms of the far more tightly coupled systems of industry, and sees in industry the origin of the increasingly complex technological systems that are perhaps the most characteristic feature of modern times. He portrays with clarity and forcefulness the fundamental implications of these developments, and makes a beginning in the building of specific theoretical formulations that will link general systems theory to the problems and promises encompassed in the wide fields of occupational and industrial psychiatry. Basically a humanist in the world of technology, MacIver, while recognizing our increasing dependence on complex and multiple monitors and regulators as rigorous as we are able to make them, emphasizes the necessity that they be embedded in a human ecology that will preserve and enhance individuality and creativity to the maximum.

We are indebted to Raymond W. Waggoner, Sr., President-Elect of the American Psychiatric Association, for his most pertinent and profound comments on Dr. MacIver's paper. Dr. Waggoner's position for many years as Director and Chairman of the Department of Psychiatry at the Neuropsychiatric Institute of the University of Michigan Medical Center has brought him into intimate contact with the extensive research on general systems theory carried out by James G. Miller and his colleagues.

The interests, concerns, and work of our fifth author, Montague Ullman, also cover a wide domain ranging from teaching, psychoanalysis, and community psychiatry to the utilization of modern dream theory and technique in the scientific study of extrasensory perception. Dr. Ullman has become an acknowledged leader in the field of community mental health. His paper marks the beginning of attempts to develop specific theoretical constructs of a systems type which will be applicable to the new problems and areas of possible intervention that result from application of the community mental health model.

E. Joseph Charny, our sixth author, coauthored with Edward J. Carroll the first paper on application of general systems concepts to psychoanalytic case material [2]. Teacher, researcher, psychoanalyst, and leader in the community mental health field, Dr. Charny is distinguished by the originality and freshness of his approach. His paper on "Psychosomatic Manifestations of Rapport in Psychotherapy" is not a psychosomatic study in the traditional sense; rather it is a highly original use of a systems view of behavior, a modern variant of the natural history approach that has evolved to synthesize elements and conceptualize systems as wholes.

Our seventh group of authors, led by Nicholas D. Rizzo, has contributed the first paper, as far as we are aware, in which principles derived from general systems theory and psychiatry are applied to problems of growth and development as they occur in the setting of educational institutions. It marks the beginning of an attempt to develop specific theoretical formulations of the nature of successful educational institutions through the use of general systems theory as a metatheoretical grammar of integrated behavior.

The final contribution to Part II is a remarkably prophetic paper, written nearly two decades ago, by two of the leading members of the psychiatric and psychoanalytic professions, Judd Marmor and Eugene Pumpian-Mindlin. This paper marks the beginning of specific theoretical developments based on an open systems approach to the whole problem of mental health, and is a heartening illustration of the strong lines of congruence between psychiatric theorists and general systems theorists, both in the openness of their approaches and in the humanism of their values.

N. D. R.

References

1. Brodey, W. M. *Changing the Family*. New York: Potter, 1968.
2. Charny, E. J., and Carroll, E. J. General systems theory and psychoanalysis. I. Theoretical considerations. *Psychoanal. Quart.* 35: 377, 1966.

8

Toward a Unifying Theory of Cognition *

SILVANO E. ARIETI, M.D.

THE AIM OF THIS PAPER is to illustrate structural similarities or isomorphisms in the different forms of cognition.

Can general principles be found which apply to such different levels as perception, recognition, memory, learning, simple ideation, language, conceptual thinking, arithmetic, etc.? Moreover, if such principles are found, will they be so generic that nothing will be added to our common knowledge?

The author's answer to these questions is that such inquiry is worth being pursued, first, because what appears simple may imply much more; second, because principles found at a certain order of generality do not exclude that each subsumed level has an organization of its own. For instance, thinking may have some properties which are not inherent in learning, and learning may have some characteristics which are not inherent in perception.

In other words, following the method of von Bertalanffy [18], we shall pursue a general system theory, although one which can be applied only to cognition, not to the whole universe. However, inasmuch as cognition plays such an important role in the understanding of the universe, any presentation of this kind has implications that transcend cognition itself.

In our attempt we cannot follow a neurophysiological approach, as we do not yet have the necessary knowledge for it, in spite of the fact that great progress has been made (for instance, through the work

* Reprinted with permission from *General Systems* 10:109–115, 1965. (Parts of this paper appear in S. E. Arieti [10].)

of Eccles [13, 14]) and daring working hypotheses have been advanced (for instance, by Hebb [15]). The author will adhere to a psychological approach. In what follows a few new ideas and a considerable number of old ones will be presented. The old ones, however, will be considered in a new and larger context. Actually what we are going to adopt is the inductive method. From several instances, occurring at different levels of cognition, we shall infer general rules.

Sensory Organization and Perception

Sensory-perceptual data undergo some kind of organization as soon as they are experienced. We shall take into consideration here those senso-perceptions which do not remain predominantly experiences of the inner status of the organism (like pain, hunger, thirst, etc.), but those which become parts of the field of cognition [4]. These psychological phenomena seem to us to follow three types or modes of organization.

The first mode of organization is *contiguity*. Sense-data experienced together tend to be reexperienced together, if they produced *one* effect in the organism by the fact of being together. The effect in its turn connects more firmly together the sense-data.

Let us assume that the four components of Figure 8-1, which we shall call a, b, c, d, produce an effect E in the perceiving organism. The effect E binds together a, b, c, d. Obviously the qualities which a, b, c, d must have in order to produce E vary with each case, but what is important is that a, b, c, d must be experienced together. Only those elements will be perceived together which are retained together by the feed-mechanism of the effect. When we state that the elements should be experienced together we do not mean it in an absolute sense. We mean that the perception of them is either simultaneous or overlapping or contiguous in time or space. The organismic effect, important in segregating perceptual wholes from the infinity and indefiniteness of the universe, varies with the level of organization and evolution. At first it may be a not-felt chemical change; later a not-conscious nervous change; still later a felt change, etc. With the evolution of cognition, however, the immediate effect loses importance and what counts more and more is the externalization of the perception, more or less independently of the effect. In Figure 8-1 the

FIGURE 8-1. SENSORY ORGANIZATION BY THE MODE OF CONTIGUITY.

four elements by being contiguous constitute a form, reminiscent of a human face; but sensory-data may be experienced together even when they do not form a definite whole, or a definite Gestalt, but just a group.

In Figure 8-2 the nonsensical lines are seen as forming one "group" of lines.

FIGURE 8-2. SENSORY ORGANIZATION BY THE MODE OF CONTIGUITY.

The mode of operation by *contiguity*, which seems at first so simple, is something which instead requires the solution, through evolutionary mechanisms, of a difficult problem which we have discussed elsewhere: response to a part versus response to a whole [6, 10]. We shall not repeat here this important subject. In normal conditions, whole-perception wins out and the mode of contiguity generally applies to the contiguity of the various parts which form differentiated wholes.

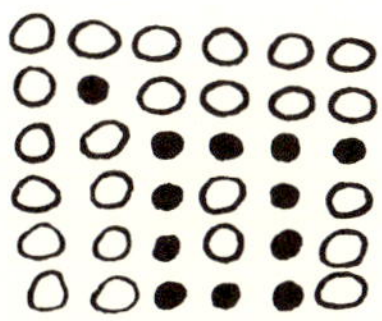

FIGURE 8-3. SENSORY ORGANIZATION BY THE MODES OF SIMILARITY AND PARS PRO TOTO.

The second mode of operation is the mode of *similarity*. Figure 8-3 contains dots of different shapes and colors. Alike dots tend to form a separate group or to be perceived together. Here similarity and contiguity reinforce one another. Elements which are similar or identical associate very well.*

The mode of operation by similarity applies to both part-perception and whole-perception.

The third mode of operation is pars pro toto. The perception of a part has upon the organism an effect which is equivalent to that of the perception of the whole.

If we look at Figure 8-4 we see a triangle, in spite of the fact that

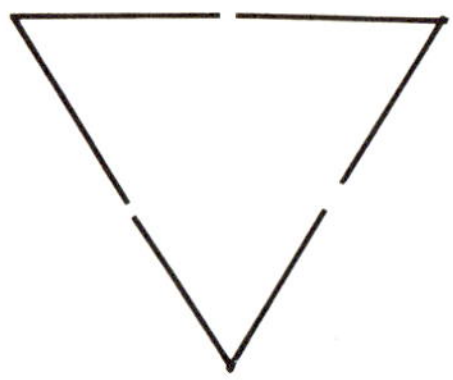

FIGURE 8-4. SENSORY ORGANIZATION BY THE MODE OF PARS PRO TOTO.

there are three gaps in the picture, which technically therefore is not a triangle.

The followers of the Gestalt school call this way of perceiving "closure." According to them the closure, as a principle of organization, permits the perception of the whole; the small gaps are filled in. This tendency to close a gap is considered by them the expression of a fundamental principle of brain functioning. Tension is supposed to be built up on both sides of the gap, like the tendency of an electric current to jump a small gap in the electric circuit. This tendency eventually would close the gap.

Let us give another look at Figure 8-4. The gaps are *never* closed. We continue to see the interruptions; but *in spite of* the gaps we perceive a triangle. The triangle of Figure 8-4 is not a whole triangle, but *stands for* a whole triangle. The Gestaltists speak of closure because they have especially taken into consideration experimentally devised stimuli with small gaps. Actually very little in nature is perceived totally. More often than it seems at first impression, we perceive parts which stand for wholes. For instance, I see only a crescent

* Advertisers know very well the effect of this reinforcement. Operators of newsstands often display the cover of the same magazine many times. The increase-effect is not merely due to the repetition of the advertisement, but to the fact that when similar elements are together they reinforce one another.

in the sky and I know I see the moon; I see only a side of the table and I know I see the table; I see only the facade of a cathedral and I know I see the cathedral. This phenomenon is not the simple phenomenon of closure of the Gestaltists, but a much more general faculty of the psyche, which permits a part to stand for the whole. As we shall see later in this paper, it is on this third mode of organization that the whole phenomenon of symbolism is based. Let us remember, however, that except for a few optic illusions purposely arranged, the whole is not experienced at a sensory level. The whole is filled in by our responses (inasmuch as we react as to wholes), or by our memories (of parts being generally associated with their respective wholes), or by symbolic processes.

At this point we must recognize that we have hit on important, apparently paradoxical facts. In other writings, we have determined how important the biological struggle is between part and whole, and how pathological or anti-evolutionary it is for the organism to confuse the part with the whole [1–7]. Now on the contrary, we come to the realization that the third mode of operation on which (as we shall see later) the highest levels of our cognition are based, actually consists of the ability of a part to stand for the whole. The whole seems again broken up and a part or a few parts together seem to resume importance. Often what is perceived is a *clue* or a *sign*, which stands for a whole. It could be that this facility with which a part can stand for the whole is somehow based on that ancient property of the organism of responding to parts rather than to wholes [5, 6, 10].

Learning

We shall consider now a different level of cognition, learning, and its relation to the same three modes of operation which we have outlined in reference to perception. By learning is generally meant a change in behavior as a result of individual experience.

There is no doubt that there is a strong relation between the mode of contiguity, or first mode of operation, and learning. Some learning theories, like those of Thorndike, Guthrie, and Pavlov, regard contiguity the most important factor in learning. According to these theories, an association (that is, a relationship based on temporal or spatial contiguity) occurs between stimulus and response. For instance, in conditioning, the dog learns to salivate in response to the ringing of a bell. But during the learning period the bell is rung shortly before or at the same time that food is seen. There is thus a temporal contiguity between food (unconditioned stimulus) and ringing of the bell (conditioned stimulus). In this type of learning a congenital response is extended to a stimulus which ordinarily would not elicit that congenital reaction. This type of learning is relatively simple. Nothing new has been added to the repertory of the dog's responses. The animal learns only to extend to a different context what its organism was already capable of doing.

It is beyond the purpose of this paper to evaluate all the ways in which the mode of contiguity operates in learning. The numerous psychological books on learning deal with this subject. It is beyond the purpose of this paper also to evaluate why, out of all possible associations, only some actually occur. This evaluation would require a study of the law of effect, reinforcement, etc. [10, Chap. 4].

Although the mode of contiguity plays a tremendous role in learning, and is probably the mode on which the other two are founded, it is difficult to subsume all types of learning under that mechanism. What psychologists call "transfer" may be viewed as an application at the level of learning of the second mode of operation, or mode of *similarity*. An acquired response is extended to many or all similar situations. For instance, if I have learned to avoid certain insects, like wasps, I may extend my avoidance reaction to similar insects, like bees. Learning by transfer is based on the fact that similar situations have identical elements. The learning is based on the identity or equivalency of certain elements [10, Chap. 7]. Learning by the mode of similarity requires a power of abstraction, just as does perceptual organization by the mode of similarity. Although generally useful, learning by similarity may lead to errors when the identity of a few elements do not warrant the same response [1, 2, 3, 6, 7, 8].

The third mode of operation, pars pro toto, operates in all types of learning, even the simplest. For instance, in conditioning, when the dog secretes gastric juice in response to the buzzing of the bell, the buzzing of the bell may be seen as a part which stands for a whole (buzzing of the bell plus sight of food).

At a neurophysiological level too, the whole process of learning can be seen as an application of the third mode of operation. The activation of certain neuronal elements through the stimulus brings about the arousal of the whole neuronal circuit or patterns involved in that given learning situation [11]. The phenomena which Gestalt psychologists attribute to insight may be viewed as based on the concomitant application of the second and third modes of operation. The "insight" is due to the fact that the organism responds as it would to a previous situation because of the presence of identical elements in the old and new situation. The *identical elements* stand for the total situations.

Since these three modes of operation are mixed together in learning, the possibility which one may prospect is that even these three modes may be synthesized or unified into just one mode. In learning, we could be tempted to reduce the whole process to the fact that reactivation of a part leads to the reactivation of the whole. However, we have to assume that prior to this reactivation the modes of contiguity and similarity have already been in operation.

Memory

We shall take into consideration only two mechanisms of memory: recognition and recall; and only in relation to the three general modes of operation.

Recognition is the simplest of the two and exists, at a more or less elaborate degree, in most animals. It means that a perception A^1, occurring now, is compared automatically or unconsciously with a past experience A, and found to be eliciting the same response as A. A^1 is then considered similar or identical to A. The first mode of operation, contiguity, differentiates perception A from the manifold of experiences. A will remain as a differentiated memory trace. The first mode of operation will later differentiate A^1 too. The mnemonic trace of A and the perception of A^1 are connected by the second mode of operation, similarity.

How this is possible at a neurophysiological level must remain the object of speculation. Perhaps the perception of A^1 sends several "echoes" throughout the nervous system, but the only echo which is absorbed is the one received by the engram of A. The engram A thus becomes associated with and reinforces the perception of A^1 and also confers the feeling of recognition. The perception of A^1, however, does not bring about the reactivation of A only, but also of many things formerly associated with A. We have thus again a pars pro toto mechanism.

Recall is a much more complicated mechanism. It differs from recognition inasmuch as it is not brought about by an external perception, but by a voluntary effort to reproduce stored images. However, images once recalled, organize also according to the three modes of operation that we have described.

SIMPLE IDEATION OR ASSOCIATIONS OF IDEAS

It is difficult to individualize all the cognitive forms which enter into the process of thinking. Simple ideation is a form of thinking which is organized by the simple laws of association. Simple ideation, however, although possible through sensory images, generally consists of verbal symbols. Nevertheless, for expository reasons, we shall postpone to the next section of this paper the discussion of language in relation to the three modes of operation.

The study of associations of ideas goes back to Aristotle. In the past fifty years this study has been discredited by people who have repeatedly pointed out that the laws of association are too simple, mechanical, and cannot explain the whole field of cognition. Of course, any attempt to explain the whole field of cognition by the laws of association is indicative of an extreme reductionistic approach. On the other hand, the phenomenon of association of ideas must be acknowledged and recognized as *one* of the foundations of the high levels of cognition. To discredit this phenomenon because it is insufficient to explain the whole human psyche is an excessive and inappropriate attitude.

The first law of associations of ideas, the law of contiguity, is another expression of the mode of contiguity. This law states that when two mental processes have been active together or in immediate succession, one of them on recurring tends to elicit the recurrence of the other. For instance, if I think of my grandmother I may also think of the house where she lived when I visited her in my childhood. The mode of contiguity which existed even in low animal forms now acquires an evocative and representational status.

The second law (or law of similarity) is another expression of the second mode of operation. It states that if two mental representations resemble each other, that is, if they have one or more characteristics in common, the occurrence of one of them tends to elicit the occurrence of the other. For instance, I visualize the Eiffel Tower and I may think of the Empire State building, because they are both very tall constructions. If I think of Beethoven, I may start to think of Brahms and Mozart, because the three were great composers.

A third law, mentioned in old books of psychology, is the law of contrast. For instance, the idea of white brings about the idea of black, its opposite. In normal mentation, actually, contrast plays a secondary role and may be subsumed under the second law (for instance, black and white are both similar, inasmuch as they are both "colors"). However, in primitive and schizophrenic thinking, association of opposites plays an important role, for reasons which cannot be discussed in this paper [10, 16].

Our third law, which is not mentioned in any book of psychology, corresponds to our third mode of operation: pars pro toto. This third law is actually inherent in the concept of association. The few ideas which are associated by contiguity and similarity stand for a whole constellation of ideas and tend to bring about the whole constellation. For instance, the idea of my grandmother may stand for my whole childhood and bring about many memories of my childhood.

This is also, in a certain way, *thinking by cue*. A small cue may arouse a complex pattern. A fragment of a situation may evoke the total situation; a member of a series may evoke the whole series.

Higher Levels of Cognition

High levels of cognition cannot be easily distinguished from one another. Some functions which appear different are based on the same fundamental processes. We shall try to avoid repetitions as much as we can.

In language, a sound (the word) becomes connected with an object or meaning by the mode of contiguity. It becomes applied to all the members of the class by the mode of similarity and may stand for the denotation or connotation of one or more members of the class or for the whole class (pars pro toto).

We can repeat approximately the same things for concept formation. The mode of contiguity is used to collect and connect the data which form the concept. The mode of similarity will make us extend the concept to all the members of a class. Finally, by virtue of the mode of pars pro toto, the concept will stand for the whole class.

In induction, the mode of contiguity makes us associate A and B because we have observed that B has followed A many times. The mode of similarity will make us associate all A's with all B's. The mode pars pro toto will make us extend to the whole series of A's and B's what we have observed in just a segment of that series. This apparently unwarranted generalization has worried philosophers since Hume. Just because we have noticed that B follows A a certain number of times, we are not entitled to infer that B will always follow A. Although the philosophers may be right in an absolute sense, life can exist only on such presuppositions as the one mentioned, that B will always follow A. Life is possible as long as the observed uniformities of sequences will recur. Let us examine again the conditioned reflex. If the sight of food is preceded by the ringing of a bell, the ringing of a bell (A) after a few exposures will make the experimental dog secrete gastric juice (B). The organism of the dog follows *induction.* The only difference between us, as conscious beings, and the organism of the dog (or even our organism), is that in us induction is accompanied by awareness, whereas in the organism of the dog it is an automatic mechanism. Most probably the same basic mechanism operates in us and in the conditioned dog; but in us, who are endowed with self-consciousness (or reflected activity), the mechanism has become subjectivized to a degree which includes a consciousness of it. The philosophers are right in stating that it does not follow that just because B has followed A several times, B should always follow A; but without such assumption life mechanisms cannot be understood. When we say assumption we actually anthropomorphize; we mean expectation which, although it is a human faculty, perhaps can be conceived as a property of life, a certain type of memory and preparation for similar events, which may exist in every protoplasm.

Life would not be possible without the living organisms functioning in an inductive way. Only the organism which can perform inductively and transmit genetically this inductive form of functioning can survive. Although induction may be based on something which cannot be proved philosophically, life depends on it.

In *deduction* the whole membership of the series is included in the big premise, for instance: All men are mortal. The big premise is made possible by the fact that the mode of contiguity (which connects "men" and "being mortal") is used together with the mode of similarity, which together extend the concept to the whole series (all men). The third mode (pars pro toto) is also applied in the big premise inasmuch as the symbol "all men" stands for the whole class (all past, present, and future men). The second premise (Socrates is a man), by following the mode of similarity, makes us attribute to Socrates membership in the series of men. The inference which we shall be able to make, using again the mode of similarity, will apply to Socrates the characteristics of the members of the class of men.

About deterministic causality we could more or less repeat what we have said about induction: an association is made between B and A (mode of contiguity). The association is extended to all the sequences A → B. Inasmuch as this law will be applied to all similar data in the cosmos, it will follow the pars pro toto mode.

The three modes apply, of course, to arithmetical thinking also. Although all numbers imply the three modes, we could say that in number one the mode of contiguity predominates. In fact, by the contiguity of parts, a unity is formed which is separated from the rest of the world and will be later recognized as being one [10]. Number two uses predominantly the mode of similarity. In fact, unless objects are at the same time seen as distinct and similar, no concept of two or of any subsequent plurality is possible. The properties of 1 and 2 and of 1 and 2 together are then applied by pars pro toto to the whole series of positive integer numbers.

Unifying Hypotheses

At this point we can perhaps conclude that the three basic modes of operation determine and structure our knowledge of the world.

The first mode determines *what is and what is not*. What is, is a unity.

The second mode *identifies*, by discovering similarity, or identity, and permitting class formation.

The third mode *infers* the not given from the given.

These three modes could really be considered three basic pre-experiential categories. We can, however, represent the general process of cognition with a different formulation and say that it is based on two fundamental characteristics: progressive abstraction and progressive symbolization.

The first mode abstracts unities and groups from the manifold of the universe. The second mode abstracts the similarity between different unities. The third mode abstracts (that is, infers) the not given from the given.

Abstraction, as important as it is, does not comprehend the whole cognitive process. Symbolization has to be added.

If we include both signs and symbols under the general process of symbolization, we can say that even a perception is a symbol, inasmuch as it stands for the perceived thing. Images, endocepts, paleosymbols, verbal symbols, preconceptual and conceptual, are different forms of progressive symbolization [9].

The ideal symbol is the symbol which, although different from the thing it represents, has the same properties as the represented thing. The symbol which comes closer to this ideal is not the onomatopoetic symbol or an iconic symbol, but the number. Although the number is not the thing that it represents, it has the numerical properties of the represented things. For instance, the symbol 15, although a symbol, has all the numerical properties of the group of 15 integer objects. Although 15 is an abstract symbol it retains its concrete numerical qualities. At the same time it is independent of its concrete embodiments and other concrete qualities, so that it permits the best possible identity with any of its embodiments.*

We could therefore say that abstraction includes all the modes of operation, whereas symbolization is a different name for the third. Although there may be some merit in visualizing cognition in this way, this writer's preference is for retaining the formulation of the three modes of operation. It seems to come closer to a unifying theory of cognition.

The question may be asked whether these three modes of operation rule the other important areas of the psyche, for instance, the experiences of inner status (sensations, feelings, emotions). The answer is that if we separate feelings from their cognitive associations it will be very difficult to apply to them the three basic modes. Unity of feeling as feeling is at most hazy. Moreover, if for instance, a feeling like a pain is recognized as similar to a previous feeling, this act

* We could approximately repeat the same things for algebraic symbols; *n* is independent of any embodiment and yet retains all its algebraic qualities.

of recognition is a cognitive act. Feelings do not have the finiteness, nor the potential symbolic infinity of acts of cognition, unless associated with acts of cognition. They have, however, the power to transform or determine cognitive processes.

The Biological Origin of Knowledge and the Mesocosmic Reality

It would seem thus that the basic principles of cognitive organization are responsible for the way we interpret reality. That is tantamount to advocating a biological origin of our knowledge.

Let us reexamine briefly some of our basic concepts. In other writings, we have shown how the possibility of perceiving wholes and parts eventually led to the distinction of subjects and predicates. The ability to recognize, or to respond in the same way to similar or/and identical things, led to the formation of classes and, incidentally, to the laws of Aristotelian logic. The excluded middle, or the third law of thought of Aristotle, represents the victory of identifying by wholes rather than by parts. Our process of induction is nothing else but the subjectivization of the inducting capacity of the animal organism.

We must add something to which we have not yet made reference: space perception. Three-dimensional or Euclidian space is based not so much on our visual perceptions, as some people believe, but on our vestibular apparatus with its three semicircular canals. It is not due to chance that we perceive space as we do: we have no choice.

Thus, our whole interpretation of nature is imposed on us by the biological origin of our cognition. Moreover, it would seem that *the psyche brings to awareness what was already implied in the living matter*.

Does this mean that in our understanding of the world we use a priori categories? Are, for instance, the three basic modes of operation that we have described a priori categories? We think they are, but not in the Kantian sense, because we believe that the animal organism, through evolutionary mechanisms, has assimilated them from the external world. They are a priori as far as the individual is concerned. However, they are a posteriori if we consider evolution as a whole. Evolution has to "learn" them from the external environment and transmit them from generation to generation.

These *biologicogenic categories*, however, permit us to conceive the universe in a frame of reference which seems to reflect the Euclidian-Newtonian-Kantian universe. Thus, in a certain way, they seem antiquated in a world which is viewed today in Einsteinian-Heisenbergian terms. To give just two examples of fundamental importance: (1) we perceive space as three dimensional, but modern non-Euclidian geometry considers it more appropriate to consider space as non-three dimensional; (2) we have seen how not only our cognition but life itself is based on the reliability of induction. And yet we know that according to Heisenberg's principle of indeterminacy, the ideas of exceptionless repetitions, induction, and strict causality have to be abandoned.

Reichenbach [17] and Capek [12] made it clear that modern physics does not deny the validity of the Euclidian-Newtonian-Kantian world. It only restricts it to the mesocosm, a world of middle dimensions. For the microcosm (a world of subatomic dimensions) or the macrocosm (a world larger than solar systems) the Einsteinian-Heisenbergian physics applies.

But life exists in the mesocosm. In order to originate and evolve it had to incorporate mesocosmic laws. Without such incorporation life could not exist. On the other hand, the fact that life exists proves that there is some order in the mesocosm. Perhaps it is an order which applies only to the mesocosm, but in its restricted dimensions it exists and has to be taken into account. If a man were deprived of his three semicircular canals and consequently of equilibrium and space-perception, he could not survive for more than a few days unless helped by others. If the principle of induction did not apply to life, the living world would be transformed into chaos.

The order which we, as animal forms, have incorporated is then reexternalized or projected into the external world from which it originated. What Euclid, Newton, and Kant did was to increase our awareness of mesocosmic categories and to enlarge our understanding of the mesocosmic reality of which we are the products. These categories represent what we have acquired from mesocosmic reality through a process of evolutionary adjustment. It would have been unnatural for Euclid to describe a non-three dimensional space.

We often read criticisms of Freud because he would have viewed the human psyche in the restricted Euclidian-Newtonian-Kantian world. Alas! This is not the error of Freud. The psyche itself, the whole life, as it is known to us, is based on the Euclidian-Newtonian-Kantian world. If some aspects of life follow mesocosmic categories, neither Freud nor anybody else could demonstrate the fact at this stage of our knowledge. The error Freud made was to apply to the psyche, especially in relation to the libido theory, concepts which pertain only to physics and economics, bypassing general biology and neurophysiology. Einsteinian-Heisenbergian notions, which are usually derived from the physical world, would be equally inappropriate if transplanted in the original form to the study of the psyche. The same thing would be repeated for more modern types of economics.

To our statement that the order of cognition derives from the order of the mesocosm we have to add important qualifications. At least one aspect of psychological life does not seem to derive from the mesocosmic environment: awareness, or subjectivity. As far as we know, only animal life is endowed with awareness. Automatic cognition can be seen as existing outside of the animal kingdom, and not exclusively in modern computers made by men. For instance, a solar system can be viewed as a machine which reflects a cosmic order—a machine, however, which in spite of its immensity is not, as far as we know, aware of its existence. Cognition acquires a human flavor when it is accompanied by awareness, and thus leads to the possibility of making choices and of *willing*.

Does the fact that our knowledge is mesocosmic in origin deny us access to the understanding of a macrocosmic and microcosmic reality or of an absolute or noumenal reality? Adventurous excursions into these other segments of reality are difficult and contrary to our present understanding of the biological nature of man. That is why they have occurred so late in history. Nevertheless, they are possible, as modern physics shows. My belief (or, perhaps, wish!) is that our mesocosmic knowledge (again applying the pars pro toto mode of operation) will permit men to understand not only the microcosm and the macrocosm but realms of reality now unimaginable and unthinkable. In fact, the pars pro toto mode implies that if a little order exists a big order also exists. But in believing so I have made an unwarranted generalization. I have applied to the unknown the inductive method which I know may be valid only in the mesocosm. My belief, thus, must remain a belief.

References

1. Arieti, S. E. Special logic of schizophrenic and other types of autistic thought. *Psychiatry* 11:325, 1948.
2. Arieti, S. E. *Interpretation of Schizophrenia*. New York: Brunner, 1955.
3. Arieti, S. E. Schizophrenic thought. *Amer. J. Psychother.* 13: 537, 1959.
4. Arieti, S. E. The Experiences of Inner Status. In B. Kaplan and S. Wapner (Eds.), *Perspectives in Psychological Theory*. New York: International Universities Press, 1960.
5. Arieti, S. E. The loss of reality. *Psychoanalysis* 48:3, 1961.
6. Arieti, S. E. The microgeny of thought and perception. *Arch. Gen. Psychiat.* (Chicago) 6:454, 1962.
7. Arieti, S. E. Studies of thought processes in contemporary psychiatry. *Amer. J. Psychiat.* 120: 58, 1963.
8. Arieti, S. E. The rise of creativity: From primary to tertiary process. *Contemp. Psychoanal.* 1:51, 1964.
9. Arieti, S. E. Contributions to cognition from psychoanalytic theory. In J. Masserman (Ed.), *Science and Psychoanalysis*, Vol. VIII. New York: Grune & Stratton, 1965.
10. Arieti, S. E. *The Intrapsychic Self: Feeling, Cognition, and Creativity in Health and Mental Illness*. New York: Basic Books, 1967.
11. Bugelski, R. *The Psychology of Learning*. New York: Holt, 1956.
12. Capek, M. The development of Reichenbach's epistemology. *Rev. Metaphys.* 9:42, 1957.
13. Eccles, J. C. *The Neurophysiological Basis of Mind*. Oxford, Eng.: Oxford University Press, 1953.
14. Eccles, J. C. *The Physiology of Nerve Cells*. Baltimore: Johns Hopkins Press, 1957.
15. Hebb, D. O. *The Organization of Behavior*. New York: Wiley, 1949.
16. Kaplan, B. On the phenomena of "opposite speech." *J. Abnorm. Soc. Psychol.* 55:389, 1957.
17. Reichenbach, H. *The Rise of Scientific Philosophy*. Berkeley and Los Angeles: University of California Press, 1951.
18. von Bertalanffy, L. General system theory. *Gen. Syst.* 1:1, 1956.

9

Behavioral Programs in Human Communication

ALBERT E. SCHEFLEN, M.D.

To communicate, participants must share a common coding system. In face-to-face communication without machinery this system is made up of patterns of behavior, recognizable because they occur in consistent patterns and are meaningful to all participants of common background. This is so because the patterns are conventional in a given tradition.

This paper describes these programs of behavior: their structure, how they appear to be performed and integrated in a group, and what significance their explication may have for behavioral science.

Observing Behavioral Structure

If you search human behavior in a certain way, there is every indication you will find reliable, determined order or pattern in any interaction you examine. The necessary strategies of search are these:

First, you must agree on the frame of reference. Focus on the *form* of behavior, resisting the temptation at this stage to abstract qualities, as you do in a personality study. Refrain from making black-box inferences about the mental or physiological processes that mediate behavior.

Second, your observations must be *first-hand* and not obtained by directly or indirectly asking the subject; he cannot adequately tell you what he is doing. What he will report are *feelings about* behavior or idiosyncratic or cultural myths about behavior. Such data are useful in making inferences or studying myth systems but not in determining behavioral pattern.

Third, you must not be satisfied to isolate bits of behavior and merely measure or count them. It is the *relations* of the elements or events, the configuration, the pattern we are after.

The first few times you observe the succession of events in a small group interaction you will see some recurring constellations. Now and again a participant will repeat the same gesture or light a cigarette in the same way, but the whole stream will not hang together in any apparent pattern. You will have to observe the same scene over and over—possibly fifty or a hundred times. This need for search and research, of course, requires that you have a recording, generally a motion picture or videotape of the interaction, for a tape recording of the voice alone or just a transcript of the content will not provide sufficient information about the behavior in communication [8, 9, 10, 39, 51]. You will also need to devise some kind of a coding system, for the complexity is enormous.

It is when you have watched the stream of behavior thirty, fifty, or more times that the pattern will begin to be evident. Dozens of microbehaviors will be seen to recur in the same sequence, and this sequence as a whole will be repeated. You will see sequnces in one person's acts, sequences *between* participants, progressions from one type of activity to another. There will not emerge a single pattern, but patterns of pattern. Your problem in recording will not be lack of order, but great complexity. You will not be able to manage so much data with simple description and narrative.

Integration of the Elements of Behavior

You recognize, of course, that I have so far described traditional natural history method. But now we come to a difference. We are not going to reduce this complexity by abstracting qualities or making a taxonomy of types. In recent years modern variants of naturalistic approaches have evolved for the analysis of behavior in systematic contexts [3, 11, 12, 18, 39, 41, 42, 46, 50, 51, 52]. In these approaches we *reduce complexity by determining the integration of the behavioral units level by level.*

The basic assumption here is that behavior appears in standard units in any culture because the members learn to perform to shape their behavior into these molds so it is mutually recognizable and predictable. Such units of behavior Pike calls "behavioremes" [41]. Others have called them structural or behavioral units [39, 50].* The units are found to form a hierarchy of levels of integration in a stream of behavior; that is, small units are lawfully put together with other units to form larger units, and these in turn form even larger units. The complexity is reduced to a unity when it has been shown how all these units come together in a single configuration, like baby bathing, a church service [41], a psychotherapy session [50], or a meeting.

Abstracting the Program of an Interaction

But most interactions show progression; that is, they have steps, and the configuration of behavior changes in stages. For example, in interactions in which a task is carried out, each step toward completion calls for a different type of activity and maybe for a different type of relation between the participants. So it is necessary to have a second manner of depicting the behavioral integration *with regard to time and steps in progression*. Technically speaking, this is done by defining a level of behavioral integration (when the structural analysis has made this possible) and showing stages or steps as units at this level. Fortunately, people signal, whether they know it or not, each terminus to a stage and each beginning of the next [26, 48] so we can cross-check the shift points in our structural analysis.

The properties of such a diagram are *programmatic*. They have implicit coded instruction, for those who know the interaction, equivalent to "After this is done, do that." And they have alternatives or branching possibilities, such as, "If so and so happens, shift to format so and so." If, for example, it begins to rain at a picnic there are traditional alternatives for continuing. If the business meeting is to be followed by a speaker, the short form of the meeting is used. After the baby has been dried, the mother shifts to activities for dressing him and, depending on whether the contingencies call for taking him out or putting him to bed, the dressing activities vary accord-

* These units are discovered roughly as follows: recurrent constellations of behavior are described in detail as tentative units. They are then tested to see (1) if they occur each time in the same configuration, (2) if they occur each time in the same context, and (3) if they have a differential effect when they appear that is contrastable to that of other forms. (See references above and also 23, 26.)

ingly. We can speak, then, of the interaction having a "program."*

If you wish to reconstruct the program for some traditional type of interaction, say, the Thanksgiving dinner or the business meeting of medical societies, you can observe many performances of this type of interaction and abstract the commonalities of units and steps. This is not unlike the way the musicologist tape records many renditions of a folk song and then reconstructs the score. Be certain, however, to have each example from the same ethnic, class, and regional background and for the same occasion, for each category has evolved its own programs and each is specific for certain contexts.

Here are some qualities of programs that are abstractable in this manner:

1. Programs provide for performance of standard, recognizable behavioral units, integrated hierarchically and programmed to be performed successively in steps.
2. Programs are specific to subcultural categories according to ethnic, class, regional, and institutional traditions.
3. Programs are context specific, that is, a given situation, a given task, and a given social organization evoke a given program performance.†
4. Programs have variants, or "branches," alternative units and steps to meet common contingencies that may arise.

Are All Interactions "Programmed"?

You will grant from common experience that such patternment occurs in interactions that we consider ritualized, for example, games, church services, funerals, weddings, college curriculums, and the

* Personality could be defined as an integration of all the parts an individual characteristically plays in programs and how he performs them. Culture, in this view, could be defined as a program of programs including all the statements about programs and the myths that are handed down with them.

† These elements of the context and the program are interdependent. They occur as a gestalten in perceptual terms. The social arrangements and situation may "trigger" the performance of the program, but the program will, in turn, determine the places of performance, the occasion, and the social organization of the performers. Thus, four people in a living room may decide to play bridge, or one who wants to play bridge invites three others and has them meet where a table and cards are available. Later I will have some comments on this "releasing" or evoking mechanism.

like. And it is also evident that the linguistic modalities of interactions are so patterned; standard units of vocalization called phonemes are integrated into morphemes that, in turn, are integrated into syntactic sentences [23, 26].* But are all types of interaction so tightly organized and structured?

It is my working hypothesis that all interactions and modalities of behavior appear to be programmed when they are examined by these methods. Not only is language structured in a hierarchy of levels, but Birdwhistell [8, 9, 11] has shown an analogous morphology for kinesic or bodily movement actions. Pike [41] has similarly demonstrated the structure of a family breakfast; we have described this configuration in the psychotherapy hour [50, 51] and, in unpublished works, for the party, courtship, and the bathing of babies.

Pattern has often been missed in psychological and social research for a series of reasons. Two of these seem related to cultural values: first, Western cultures accord a high value to individuality, indeterminacy, and nonconformity and, second, until recently, direct observation of behavioral form itself—people-watching, if you wish—has tended to be limited to watching people of non-Western societies. These factors probably influenced the focus and methods of research. But there is also no question that there are so many thousands of types of programs that any observation without strict control of subculture and context would give the appearance of probabilistic events and individual variation. I have already mentioned that each culture and subculture, each institution and each situational and social context has its own programs. In addition there are multiple roles in *each* program, multiple variations and multiple intercultural admixtures and changes due to cultural evolution and deviancy that contribute to the illusion of infinite individual differences. Finally, each program, as I will describe, has many statements about it, value judgments, descriptions, parodies, and so on, that proliferate, it seems, even more rapidly than types of programs themselves.

* The integration of elements of speech is referred to as syntax [17, 23, 47], and the abstracted rules of this integration as grammar. Many such rules have been abstracted for language behavior. Sometimes a great many "free" variants of a morphism can occur in a given spot in a program without preventing its continuance, and this fact adds to the complexity of programs. Nonetheless, "constraints" occur upon the possibilities. Jakobson and Halle [34] for instance distinguished syntagmatic constraints (that dictate the order of sequences) and paradigmatic constraints (that dictate the allowability of occurrence of a word in a particular slot in the program).

The point of the program concept is not to deny individual and social diversity, but to identify order. However many possibilities and alternatives occur, they can be seen to be determined and orderly if we can demonstrate their reliable dependence upon specific variations in context. There are even in use programs for the deliberate creation of innovative programs [40] and for facilitation of creativity and spontaneity (for instance, the blue sky tactic in industrial designing or the group psychotherapy experience that insists upon affective revelation and uncustomary action). Logically speaking, were it not that interactions were patterned, behavior would be unpredictable and unreliable, and it would be impossible to sustain, mediate, and form human relationships, complete coordinated tasks, and transmit a common culture. *Communication depends upon a common behavioral morphology of shared meaning.*

Communication: The Performance of Programs

Communication is a social-level process. It will require communicators in human communication, people who have (1) reasonably intact sensorimotor systems and (2) cognitive representation of pattern or coding *and* who are in social relations such that they can (a) see, hear, and maybe touch and smell each other and (b) learn or use a common system of meaning.

MEANING

To have meaning, the elements of the code—in face-to-face communication, the behaviors of the program—must be patterned and reliable or predictable. This patternment is accomplished by the hierarchical structure of behavior which provides integration of the bits [45] in information theory terms [16, 53], or recognizable gestalten in cognitive terms [40]. To *convey* meaning the participants must share the same coding system, a commonality accomplished if the participants have grown up in the same tradition. They can then relate the patterns of behavior to contexts: to common tasks, goals, events, punishments, rewards, and so on. So we can add another property of programs:

5. Programs have, for their experienced performers, common meanings—or function, or significance, or purpose—in reference to larger systems.

You see we do not have to have, though it may be useful in certain research, any concepts of outputs, transmission, or noise to deal with face-to-face human communication, as we do to describe telecommunication. *People behave in coded, patterned ways and others perceive and comprehend these patterns.*

We can also derive a systematic method of determining the meaning of behavioral units in the program. For language, Katz and Fodor [35] said that the meaning of a word emerges as we learn what words precede and follow it [2, 27, 37]. Others have written more exhaustively about semantics. Bateson [39] suggests a simple maxim: *meaning is relative; the less ambiguity about the contexts of a unit the more meaning can be ascribed to it.* Defining the meaning, in practice, becomes specifying the contexts level by level until a sufficient pattern has been specified to result in almost zero ambiguity. We could postulate a "unit of meaning" for such a size of pattern for a given participant in a given situation. Here is a simple example borrowed from Bateson, but modified:

> If I utter the phoneme, *p*, there are so many possible words and sentences it could refer to that there is no way to derive what I mean. If, however, I repeat the phoneme in the word *peter*, I could be referring to fewer general usages, but still too many to do any more than guess the context. If I tell you I "mean" *saltpeter*, then say, "They put saltpeter in the tea at X girl's school," there are but a very few possibilities. Either I am divulging a secret, I am joking, or I am paranoid. A further search of the context will probably settle the issue. If I am kidding I will reveal a metacommunicational indicator, for instance, a smile. If I am paranoid, it will come out in the larger expanses of my behavior. Pinning down the facts, of course, could take a much broader investigation.

By the same token I can see two men kissing and *by scanning the environment*, ascertain whether I have witnessed a homosexual act, a family parting, or a French military ceremony. On the football field I can tell if a touchdown was scored if I missed the sequence, for there will be a scoreboard indicator, a signal from the referee, a roar from the crowd, and formation for a field goal. Actually, redundancy in behavioral programs reduces ambiguity.

ON INFORMATION PROCESSING AND BEHAVIOR IN THE COMMUNICATOR

Programs, of course, are performed by people—who learn them, store them, apparently in neuronal circuits, and under appropriate conditions perform in them by providing the necessary behavioral components to complete the units. In learning the programs organisms come to be people of particular skills and social position, and in performing them people make social relations and perpetuate culture.

Modern concepts of ego psychology do not ignore learned patterns of culture [20], and cognitive theories allow for the recognition of complex representations of behavior and context as "maps" [55], configurations [14], gestalten [36], or images and plans [40]. In computer simulation models of brain functioning [13] complex servomechanisms are provided by integrations of neurons. Pribram [43] and Pribram and Melges [44] have put these ideas together in a theory of information processing and behavior. They see a dynamic equilibrium of two large central systems: one preparatory that inhibits input and maintains ongoing cognitive and behavioral plans (I would say programs), and one that participates, by permitting input and opening the activities to external feedback and control. We could conceive, then, of the continuous neuronal equilibrium as cognitive "rehearsal" of program imagery, inhibited from performance except in appropriate contexts. Psychologically these activities might be perceived as thought, conscious or not, and the continuous activity as providing for memory. Emotion Pribram sees as the experience of temporary discrepancies in the equilibrium, but the experience is retroactive so as to influence the programs and their representation.

SOCIAL ORGANIZATION IN PERFORMANCE IN INTERACTION

In some programs roles—the divisions of the behavior among the participants—are fixed by tradition. But more often the assignment of parts depends upon the skills, identity, status, gender, and social relationships among participants. Since the matter is at least in part traditional we can list another attribute of programs as follows:

6. Programs prescribe social organization and division of function in the performance. There are therefore "roles" and complementary, parallel, and other relations of individual action.

The arrangements between participants has been the focus of research in communication in the social psychologies and social psychiatry. The most common viewpoint here is the interaction concept that participants send messages to each other, take action vis-à-vis each other, and so on. The model is one of action and reaction. Since the program is internalized the actions of the various participants *is but one determinant* in the pattern of the interaction, playing a variously important role depending on the type of program. In the social view of communication the focus is upon the *people* and the *group*, sometimes to the point that the people are seen as originating the behavior. This focus distinguishes interactional views from cultural views, where the emphasis is on the program and the behavioral pattern.

The Palo Alto group [6, 30, 31, 58] has elaborated on types of interaction in communication, normal and pathological, and many studies have elucidated other qualities of relationships in communication in small groups [1, 7, 15, 19, 21, 32, 33] and many others. Social network views have attended less to the relationships of behavior and more to what members have access to and to whom they speak [28, 54].

A final social aspect of programs needs to be mentioned. Programs have to be integrated with other programs in a society, and they have to conform to and maintain the cohesion of larger groups and the values of the culture. For this reason the program is controlled at higher social levels. The mechanisms for such integration include retroactive loops in direct contiguity with systems outside the interaction (external feedback).

7. Programs have elements and sequences that represent the values, purposes, and actual activities of the larger social organization and cultural tradition, e.g., picture of heroes on the wall of the meeting room, "guests" from the home office, precepts verbalized during performance, and so on.

ORDERS OF BEHAVIORAL INTEGRATION IN COMMUNICATION

At *least* three orders of integration can be observed in an interaction.

1. FIRST-ORDER COMMUNICATION: SIMPLE COORDINATION OF ACTIVITIES. People who know, whether consciously or not, the same programs of action can perform their joint parts of the program without

words or special signals. They simply do what is expected, conjointly and in synchrony. Consciousness is not *necessary* to the integration. For example, lovers can sexually consummate without instructing each other or even being conscious of the mechanics of intercourse. Workers can put together a machine or harvest a crop while talking about something entirely different.

2. SECOND-ORDER COMMUNICATION: USE OF INTEGRATIONAL SIGNALS. But it is often the case that contingencies or ambiguities arise. Performers do not know precisely what to do or when to do it. Hence, special signals are introduced to modify or integrate the activities. These behaviors, which are also conventional and therefore mutually meaningful, can be classed as signals of integration and belong to a second order of communication. Common examples are the use of a conductor to synchronize parts in a concert or the use of commands, bugles, whistles, and other signals to coordinate a battle or a work party. I would classify the integrative signals in face-to-face communication as follows:

1. *Pacing* signals to regulate the speed of performance; e.g., in conversations head nods and facial expressions provide feedback of the comprehension of others listening.
2. *Identification* signals from the participants that indicate their roles and capacities in the performance. These are divisible into two major types:
   a. Indicators of categorical membership such as ethnic, class, occupational; of caste membership, that is, age and gender; and of status, skill, and other general qualities.
   b. Indicators of state that give information about the immediate readiness of a participant, e.g., his consciousness, attention and interest, health, mood, and so on.
3. *Social integration* signals that monitor deviance, inattention, improper types of relationship or territorial spacing, and so on, facilitating coordination of performance.
4. *References to contexts* that indicate external events that require modification in the program or that control the values, objectives or modes of performance; e.g., looking at the decor in a room may serve to remind the participants that they are in a religious service demanding somberness and so on.

Characteristics of second-order communication are: the signals are not always necessary to communication, so they appear irregularly in programmatic abstractions; they are not in themselves programmatic, having no meaning outside the basic program; their appearance does not change the program itself, but only the particular enactment. Methodologically, then, the significance of signals cannot be determined until the basic programs themselves have been explicated and mapped out.

3. THIRD-ORDER COMMUNICATION: METACOMMUNICATION. There are, however, other situations that arise in maintaining a society and its individuals. There are occasions when an interaction may be represented symbolically for someone who is not present, or a participant may learn his part better if the program can be explicated and discussed with him. There has evolved *communication about communication*—or metacommunication.* As a consequence we can talk about activities not then in progress or speak about them while we are enacting them.

An action that is metacommunicational in function can be as simple as a gesture or a smile. For example, men may smile to indicate that their pummeling of each other is to be regarded as friendly play. Or metacommunication may be as complicated as language. I would regard the whole iconic or symbolic representation and thus a language itself as one form of metacommunication in which events can be represented by words or codes and ascribed to something not present. Drama, music, and other artistic forms might be regarded as enactments of specialized metaprograms, and we can also think of contrived codes, such as the Morse code. Types of metacommunication have increased almost infinitely as representations of programs have become related to various contexts: scientific, artistic, moral, ethical, etiquette, and so on. Metacommunication can be elaborated about any part of a program, any participant, any relationship, any outcome.

* For this type of communication Bateson [4, 5] has used the term *metacommunication*. In other sciences the prefix *meta-* is used in analogous ways. Linguists and semanticists have used the term *metalanguage* to describe statements about language. Since I see language as already a metacommunication, I prefer to call language about language a *meta-metacommunication*, as does Bateson [5]. Some computer scientists [29] speak of the instructions in a program (e.g., "*do*" or "*if . . . then*" statements in Fortran IV) as being meta to the data.

In the enactment of a program the communicational and metacommunicational aspects can have at least the following relations:

1. There may be *no apparent relationship*. The interactants may be performing a task and talking about entirely different matters. For example, a mother may feed her baby while talking to a neighbor about a television show.

   Sometimes it seems that conversation must be "piped into" certain program performances because the participants, for various reasons, expect and demand it.
2. A metacommunication *commentary* may be conducted *about the program in progress*. For example, participants may verbalize what they are doing and evaluate it by some criteria or evaluate and make explicit what others are doing. This is likely to occur when novitiates are being trained or deviants corrected, as in psychotherapy.
3. Metacommunication *may distort, rationalize, disguise, or draw attention from the rest of the activity of the program*, either in conscious guile, unconscious delusion, or because in that culture the actual nature of the task has never been known or appropriately visualized.* For example, certain Africans designate their morning and nightly drumming as bringing up and putting down the sun.

Unlike simple integrational signals, metacommunication *can change the structure of behavior and programs themselves*. Representation, value judgment, and analysis are capable of resulting in relearning and in innovative change in traditional formats. In fact there are programs explicitly used for innovative or creative activities. It is equally obvious that metacommunicational activities can be

* When one is learning a program, the commentary may be learned with it and henceforth repeated with the formats without knowing why. In this way linguistic accompaniments that do not represent understanding or insight into the activities may be transmitted from generation to generation. It can be asserted that in some cases at least people not only learn how to perform in first order formats, but *learn how they are to refer to, think of, feel about programs*. This last point has a number of implications for psychological theory. We have traditionally assumed that feelings and thoughts precede and cause behavior; this view opens the possibility that behavior may precede or merely accompany behavior, rationalizing it, distorting it, or simply cluttering the field. A very crude example is the muttering of incantations while treating wounds or of endearments while feeding babies or making love.

used exploitatively to control, denigrate, defranchise, and otherwise destroy individuals and peoples. Both personality and organismic health depend upon programmed activity.

Often the metacommunication aspect comes to stand, in people's consciousness, for the whole program. The relationship between language channels and the whole remains unknown. Researchers who study only language in conversation as communication, therefore, rely indirectly on subjectivist approaches, cultural myths, then make up their data about communication. It often seems the case that the commentary is handed down in tradition without anyone knowing how it derived from or is related to the action as a whole.

MODALITIES AND CHANNELS OF COMMUNICATION

The participant in an interaction, then, has to deal with an enormous number of behaviors of different logical types, of different orders and at multiple levels. This amount of information could not be conveyed in any simple channel or performed in simple, linear sequences by the body as a whole. Multiple simultaneous units of behavior can be splitting the motor equipment into integrated clusters capable of performing different modes of behavior, e.g., language, eye movements, arm, leg, torso movements, changes in skin color, and so on.

An interactant thus uses (1) multiple modalities and (2) splitting of the body into regions such that one side of the body or the upper or lower portions of the body can engage in different subformats of activity in different relationships [49, 51]. Thus, a participant can flirt kinesically with one person, speak seriously to another, and have his arm around yet another. He can also convey different orders of behavior as to cultural significance in different channels simultaneously, for example, dressing in an upper middle-class fashion and using a lower-class dialect.

It is currently conventional to list these modalities in terms of the "channels" of communication through which their information is conveyed:

LANGUAGE MODALITIES

Vocal: Linguistic [47, 57]

Lexical

Stress, pitch, and junctures

Paralinguistic [56]
Nonlanguage sounds
Vocal modifiers

NONLANGUAGE MODALITIES
Kinesic and Postural, including voluntary and involuntary mediation behavior, facial expression, tonus, positioning, and so on [8, 9, 10, 39]
Tactile [22]
Odorific
Territorial [38, 59] or Proxemic [25]
Artifactual, including dress, cosmetic usage, props, decor, use of spaces, and so on

It would be convenient, if it is true that speech itself is metacommunicational, to posit a simple dichotomy and say that first- and second-order communication is nonlinguistic and third-order linguistic. *But this is not the case.* Some integrational signals are vocal; many kinesic behaviors act with language and appear to reduce ambiguities of speech; some kinesic actions are clearly metacommunicational. In general, there is considerable redundancy of channels. For example, if my child is too obstreperous I may regulate him simultaneously in, say, three channels, saying "Be quiet," touching his shoulder with my index finger, and frowning. Unfortunately, we do not know the distribution of function in relation to modalities.

A final abstraction about the nature of programs is possible:

8. Programs provide communicative integration through a number of channels, and each of these has subdivisions, crudely analogous to bands. Collectively, these possibilities convey behavior of different logical types, orders, and levels.

Conclusions and Implications

Views such as this—of behavior in communication and its reference to larger systems—alter our approach to communication theory, to behavioral research, and to the applied behavioral sciences. Some of the following implications are worthy of consideration.

IMPLICATIONS FOR COMMUNICATION THEORY

The behavior of participants in communication is not simply the expression of idiosyncratic traits or instinctive needs, nor is it simple action-reaction response to social cues or stimuli. Rather, both individual qualities and interactional negotiation must be brought to terms and both of these integrated with the necessity for performing standard units and maintaining a progression to completion. Otherwise, relationships would not be maintained, the tasks of society would not be done, and the individual would not be understood or serviced.

A comprehensive communication theory, then, will have to bring together theories about (1) communicators capable of handling complex learning and cognition, (2) social relations, and (3) culturally evolved systems of coding. Such a comprehensive theory will have to provide for face-to-face communication, for some model of multichanneled behavioral coding. For it is not enough to explain the process simply by extrapolating from information theory, for machine communication apparently involves but a single channel and a linear Markovian process.

IMPLICATIONS FOR THEORIES OF HUMAN DEVELOPMENT AND BEHAVIOR

An outline of what must be recognized, recalled, and performed in any interaction gives us a glimpse of how learning may take place and what must be learned. Each organism, to become a person, must learn multiple units and integrations, many sequences and variations, and numerous roles—each in relation to specific contexts. These matters he learns in a cultural tradition and in a social organization. Any theory of human development, then, must integrate genetic, idiosyncratic, social, and cultural determinants.

IMPLICATIONS FOR BASIC BEHAVIORAL RESEARCH

The multiple and special relationships of language and consciousness to behavior further reduce our reliance on subjectivist methods in studying behaviors that are not in the metacommunicational orders.

Since behavior is patterned in larger and larger levels and contexts, thereby deriving its meaning, the study of behavioral isolates out of context cannot take us to meaning, purpose, significance in the system, and explanation. When we study behavioral isolates, we may unwittingly rely upon some preconception or metacommunicational myth about their actual contexts to make inferences about their implications.

When, however, we do isolate elements of behavior as a step in their analysis, we can discover and identify naturally occurring structural units, rather than relying on abstracted qualities such as the usual variables of personality.

After the configuration of units and contexts is known, we can test more meaningfully the effects of experimentally introduced parameters on behavior and social organization.

IMPLICATIONS FOR APPLIED BEHAVIORAL SCIENCES

In education, deviancy correction, and social action programs, a generally talked-about problem is that of improving communication. Two aspects of such efforts will have to be explication of cross-subcultural differences in such dimensions of communicative behavior as spacing in territory, roles, and group structure, as well as metacommunicational signals, and general behavioral units and programming. Hopefully, these matters may be explicitly taught at some future date and taken into account in planning hospitals, public housing, conferences, preventive health programs, and so on.

A more precise knowledge of the nature and types of programs also offers a way of defining precisely and making conscious deviancy and miscommunication. It is possible, for example, using the model of the program to delineate aberrations such as (1) use of programs inappropriate to given contexts, (2) taking inappropriate roles, and (3) improper performance of units and sequences. Not only can persistent deviancy be defined by direct observation, but it can be distinguished from more simple failure to operate correctly in an unknown or foreign cultural program.

But I think these principles are already spreading in the sciences of man, though they have yet to be defined and described precisely. The study of systems was at first a view of material systems, that is, the group as a collection of people, the organism. But when the relationships of physical entities was more exhaustively examined, the emphasis shifted to the *behavior* of systems. In communication, for instance, it became increasingly evident that we need to know *what* is transmitted, and this point leads us to behavioral form and patternment. In this shift of interest to behavioral form and patternment, I believe, the behavioral science movement grew out of the social science era.

References

1. Bales, R. F. *Interactional Process Analysis*. Reading, Mass.: Addison-Wesley, 1950.
2. Bar-Hillel, Y. Logical syntax and semantics. *Language* 20:230, 1954.
3. Barker, R. G. *The Stream of Behavior*. New York: Appleton-Century-Crofts, 1963.
4. Bateson, G. The Message, "This Is Play." In B. Schaffner (Ed.), *Group Processes*, Vol. II. Madison, N.J.: Madison Printing Co., 1955.
5. Bateson, G. On the logical categories of learning and communication. (Mimeographed report, to be published.)
6. Bateson, G., Jackson, D. D., Haley, J., and Weakland, J. Toward a theory of schizophrenia. *Behav. Sci.* 1:251, 1956.
7. Bavelas, A. A Mathematic Model for Group Structures. In A. G. Smith (Ed.), *Communication and Culture*. New York: Holt, Rinehart and Winston, 1966.
8. Birdwhistell, R. L. *Introduction to Kinesics*. Louisville, Ky.: University of Louisville Press, 1952.
9. Birdwhistell, R. L. Contribution of Linguistic-Kinesic Studies to the Understanding of Schizophrenia. In A. Auerback (Ed.), *Schizophrenia*. New York: Ronald, 1959.
10. Birdwhistell, R. L. Some Relations Between American Kinesics and Spoken American English. In A. G. Smith (Ed.), *Communication and Culture*. New York: Holt, Rinehart and Winston, 1966.
11. Birdwhistell, R. L. Communication as a Multi-Channeled System. In D. L. Sills (Ed.), *International Encyclopedia of the Social Sciences*. New York: Macmillan, 1968.
12. Bock, P. B. *The Social Structure of a Canadian Indian Reserve*. Ph.D. Thesis, Harvard University, 1962.
13. Broadbent, D. E. Information processing in the nervous system. *Science* 150:457, 1965.
14. Bruner, J. S., Goodnow, J. J., and Austin, G. A. *A Study of Thinking*. New York: Wiley, 1957.
15. Chance E. Measuring the potential interplay of forces within the family during treatment. *Child Develop.* 26:241, 1955.

16. Cherry, C. *On Human Communication*. New York: Science Editions, 1961.
17. Chomsky, N. Three Models for the Description of Language. In A. G. Smith (Ed.), *Communication and Culture*. New York: Holt, Rinehart and Winston, 1966.
18. Condon, W. S. Process in Communication, 1967. (In press.)
19. Eisman, B. Some operational measures of cohesiveness and their interrelations. *Hum. Relations* 12:183, 1959.
20. Erikson, E. H. *Childhood and Society*. New York: Norton, 1950.
21. Festinger, L., Schachter, S., and Back, K. *Social Pressures in Informal Groups*. New York: Harper, 1950.
22. Frank, L. K. Tactile Communication. *Genet. Psychol. Monogr.* 56:209, 1957.
23. Gleason, H. A. *An Introduction to Descriptive Linguistics*. New York: Holt, Rinehart and Winston, 1955.
24. Goffman, E. *Encounters*. Indianapolis: Bobbs-Merrill, 1961.
25. Hall, E. T. *The Hidden Dimension*. Garden City, N.Y.: Doubleday, 1966.
26. Harris, Z. S. *Structural Linguistics*. Chicago: University of Chicago Press, 1951.
27. Hayakawa, S. I. *Language in Action*. New York: Harcourt, Brace, 1941.
28. Heise, G. A., and Miller, G. A. Problem solving by small groups using various communication nets. *J. Abnorm. Soc. Psychol.* 46:327, 1951.
29. Ingerman, P. Personal communication, 1965.
30. Jackson, D. D., and Weakland, J. H. Schizophrenic symptoms and family interaction. *Arch. Gen. Psychiat.* (Chicago) 1:618, 1959.
31. Jackson, D. D., Riskin, J., and Satir, V. A method of analysis of a family interview. *Arch. Gen. Psychiat.* (Chicago) 5:321, 1961.
32. Jackson, J. M. A space for conceptualizing person-group relationships. *Hum. Relations* 12:3, 1959.
33. Jaffe, J. Language of the dyad. *Psychiatry* 21:249, 1958.
34. Jakobson, R., and Halle, M. *Fundamentals of Language*. Hague: Mouton, 1956.
35. Katz, J., and Fodor, J. The Structure of Semantic Theory. In J. Fodor and J. Katz (Eds.), *The Structure of Language*. Englewood Cliffs, N.J.: Prentice-Hall, 1964.
36. Koffka, K. *Principles of Gestalt Psychology*. New York: Harcourt, 1935.
37. Korzybski, A. *Science and Sanity*. Lakeville, Conn.: International Non-Aristotelian Library, 1948.
38. McBride, G. *A General Theory of Social Organization and Be-*

haviour. St. Lucia, Australia: University of Queensland Press, 1964.

39. McQuown, N. A., Bateson, G., Birdwhistell, R. L., Brosin, H. W., and Hockett, G. F. *The Natural History of an Interview*. (In press.)
40. Miller, G. A., Galanter, E., and Pribram, K. H. *Plans and the Structure of Behavior*. New York: Holt, 1960.
41. Pike, K. L. *Language*, Part I. Summer Institute of Linguistics, Glendale, Calif., 1954.
42. Pike, K. L. Toward a theory of structure of human behavior. *Gen. Syst.* 2:135, 1957.
43. Pribram, K. H. Emotion: Steps Toward a Neuropsychological Theory. (In press.)
44. Pribram, K. H., and Melges, F. T. Emotion: The Search for Control. (In press.)
45. Rapoport, A. What Is Information? In A. G. Smith (Ed.), *Communication and Culture*. New York: Holt, Rinehart and Winston, 1966.
46. Rosenthal, R. On the social psychology of the psychological experiment. *Amer. Parent* 51:268, 1963.
47. Sapir, E. *Language*. New York: Harcourt, Brace, 1921.
48. Scheflen, A. E. The significance of posture in communication systems. *Psychiatry* 27:316, 1964.
49. Scheflen, A. E. Quasi-courting behavior in psychotherapy. *Psychiatry* 28:245, 1965.
50. Scheflen, A. E. Natural History Method in Psychotherapy: Communicational Research. In L. A. Gottschalk and A. H. Auerbach (Eds.), *Methods of Research in Psychotherapy*. New York: Appleton-Century-Crofts, 1966.
51. Scheflen, A. E. *Stream and Structure of Communicational Behavior*. Commonwealth of Pennsylvania: Behavioral Monograph No. 1, 1966.
52. Sebeok, T. A., Hayes, A. S., and Bateson, M. C. *Approaches to Semiotics*. London: Mouton, 1964.
53. Shannon, C. E., and Weaver, W. *The Mathematical Theory of Communication*. Urbana, Ill.: University of Illinois Press, 1949.
54. Shaw, M. E., Rothschild, G. H., and Strickland, J. F. Decision Processes in Communication Nets. In A. G. Smith (Ed.), *Communication and Culture*. New York: Holt, Rinehart and Winston, 1966.
55. Tolman, E. C. A Psychological Model. In T. Parsons and E. A. Shile (Eds.), *Toward a General Theory of Action*. Cambridge, Mass.: Harvard University Press, 1951.
56. Trager, G. L. Paralanguage: A first approximation. In W. M. Austin (Ed.), *Stud. Linguist.* 13: Nos. 1 & 2, 1958.

57. Trager, G. L., and Smith, H. L., Jr. An outline of English structure. In W. M. Austin (Ed.), *Stud. Linguist.* No. 3, 1956.
58. Watzlawick, P., Beavin, J. H., and Jackson, D. D. *Pragmatics of Human Communication.* New York: Norton, 1967.
59. Wynne-Edwards, J. C. *Animal Dispersion in Relation to Social Behaviour.* New York: Hafner, 1962.

10

Information Exchange in the Time Domain*

WARREN M. BRODEY, M.D.

THE EXPERIENCE on which this report is based is different from that of the average psychiatrist. I have had the enjoyable experience of reaching into the land of "the new technology," where tools are being built to use strategies that in man would be considered intelligent. I have been given the opportunity to think and work at the integration of man and machine, starting from the "man side" of the bridge and reaching toward those engineers and scientists whose technological skills are stronger on the machine side. This bridging is slowed by ignorance and by our own deficits as psychiatrists in finding ways of modeling our observations that would be more effective in communication. Our deficit, I believe, is due to the fear of science in which we psychiatrists lived—and I think justifiably so—in precomputer days, when there was no possibility of real time computation.

The Fear of Science

Why would the psychiatrist, the astute observer, the skillful clinician, the man who knows people, live in fear of science? The answer to the question is that he was preserving his science from the popular misinterpretation of science as a system for divorcing the facts from their relations and "truth" from its context. He had been a victim, often hurt by the logical use of his words by those who did not understand this need for protection.

* Presented at the 122nd Annual Meeting of the American Psychiatric Association, Atlantic City, N.J., May 9–13, 1966.

We have protected *our* science by coding it into handwaving and words which have meanings defined differently on each occasion. We have preserved our knowledge from the deadening oversimplification of experiments which imitated the rituals of scientific methodology without comprehending its substance. The experimenters heard *our* words as if they were simple in their way. One tried to explain: *That* word is meant *that* way only for *that* context at *that* time. These labels gained meaning as they provided context for each other. The physical scientist knew this relativism; the technologist imitator of the 1930's did not.

My own training in science was a doctor's training—a training in the ways of exploring, searching, or picking up curious relations—and of figuring out ways to check whether my observations were useful in terms of more or less rigorous testing. Rigor included striving for the level of accuracy meaningful for a purpose. Accuracy for its own sake, or objectivity, when it could only prove the truth or falseness of what I already knew, was not yet considered weighty enough to make a difference. Mostly I knew that the discipline I cared about did not significantly apply to the butchered pieces of phenomena that resulted when ordinary and simple measuring units were used. When complexity is denied, living things can no longer be studied with vigor and discernment: Procrustean "scientific methodology" produced valuable but nonetheless dead biology.

In this paper I will attempt to show that it is now time for us clinicians to come out of hiding and to leave behind the fear that science means to destroy our discipline in its complexity in order to make it measurable.

The Challenge

Join the ranks of scientists aware of the failure of the old scientific recipes. New tools allow us to reexamine the epistemological models that since Greek times were used to separate living phenomena into formal simplicities. These old traditions are hidden. They are built into our language of words, our grammar, our numbers, and in our out-of-awareness religions. Subject-object is a tradition of our culture —observer and observed are the same.

The prerelativism observer had the objectivity of God looking through a peephole from superspace. He was not affected by the system he observed in a way that meant the system changed by reason of being observed. All was true or false by some truthtable, statistical average.

As we approach observation systems in which the observer is less in the loop, the simplicities of the old hard sciences retain their usefulness unless we are seeking extreme rigor in our measuring. The astronomy enthusiast does not affect the moon by his observations.

The observer in the old "objective" position could not afford to get too intimate with what he was observing or what he chose to measure "objectively." For he could not easily measure his own participation without being entangled in the richer network of the myriad signals he was receiving and returning as he talked to the phenomena, even as he listened to the reply. This way of talking to the phenomena is the way scientists actually search. Later they research their territory into the scientific format for describing and proving to each other the observed phenomena. The research is a way of finally formalizing their discovery into a code for communication—the language of hard science. The hard science formalism structure has power when the observer and observed are not *inter*dependent.

Soft Science

In the soft sciences we seek to comprehend the phenomena we are observing, knowing from the outset that observer and observed are strongly engaged in their data exchange dialogue. This dialogue does not mold into the questions that can then be significantly researched in the old style. So we had search, but little research; we were full of hypotheses, post hoc explanations, and ongoing manipulations, all discussed as predictions. It could not be otherwise: we lacked both notation and measuring apparatus which could measure dialogue.

First let us approach the question of notation. We have as yet no way to symbolize this dialogue, no notation that can make this phenomenon translucent.

Now I am using the word *dialogue* with a very definite meaning. To illustrate, I refer to the network of systems which allows us, for example, to look into each other's eyes and to counterpoint each other's changing as we change, so that we evolve a sense of being known even as we together metabolize and unfold the next moment and its surprises.

Yes, a table of people, or a roomful, or a person, or a cell, or a city, metabolizes what was a moment ago merely "noise"—unpatterned background—into information. The surprises of the future becoming present grow as we grow with them. It is this growth into the unknown, and the rich power that systems have of organizing themselves and each other so as to metabolize the unknown, and finally to excrete the obsolescent, that is in the range of phenomena that applies to our problem and that needs modeling. This is the right arena.

In viewing man as metabolizing information we see him as a creature who has choices as to how he will pattern what he perceives and the way he will locate his perceptors—so as to adapt to the unknown and stay within his homeostatic limits.

As Ashby stated in the United States Air Force–sponsored Bionics Meeting held in Dayton in 1966, man does not operate within data processing limits that we know. In order to manage the millions of choices he can make at any moment, he clumps these into patterns, or programs. But remember, the question vital to this paper is not what these patterns are or where they come from but how man *grows them moment to moment.* How does he grow his capacity to metabolize into information what a moment ago he did not even recognize as meaningful? What units might be applicable to the description of the unfolding of a family? I do not know the answer. I know that it is not to be defined by timeless logical conundrums that legitimize the denial of the edge of change as it happens.

Having raised this question, let me remind you of the new developments which now direct us towards binding techniques of dealing with the level of complexity and interdependence familiar to the clinician. It is now becoming possible for man-machine communications systems to simulate complex man-man communication. In fact, in order to make full use of our artificial intelligence devices—the new generation of computers—we must tackle the dialogue problem directly. In work in which I am a participant at the computer laboratories at Massachusetts Institute of Technology and the National Aeronautics and Space Administration in Cambridge, Massachusetts, this problem is critical.

In the biocybernetics laboratory we are seeking to develop techniques for maintaining control of the relatively closed information ecology of several men isolated in a spaceship during a deep space mission. The first problem is to develop nontrivial questions which are relevant to the purpose of the search. Let us proceed. When observer and observed are in the same loop, each changing as a function of the other's changing, what can we specify as characteristic of the system? What can we specify as characteristic of that loop? What framework can we use to structure our notation system? What notation will allow us to tell of this framework so that we do not have to show the whole story in metaphor or handwaving each time we tell it? What we need is a structural framework for our model that will answer the following questions: What kind of model is appropriate for formalizing the behavior within the system that is relevant to our questions? What graining or texture shall we use to capture the features of the phenomenon we study so that we might at this beginning stage of study be able to reorganize and not go looking for a mountain with a microscope, or trying to measure significant movement of the human hand in a microsecond? We psychiatrists are experts in formalizing the dialogue in literary language. What can we say about information exchange characteristic of a dialogue?

The Dialogue

We realize first the power of a dialogue in driving its components. Two or more people speaking and contacting each other—en rapport—can become driven by their conversation so that they, the components, behave quite differently than if the loop of information exchange that generated between them were growing in a different way. It is in the nature of conversation that it would not develop were this not so. The spontaneous enthusiasm of two people close enough to the same "wavelength" may propagate to include a larger group, and the build of excitement may organize those who ordinarily do not vibrate to the "subject" context of the dialogue. For them the dialogue field created polarizes the neighboring noise into meaning, and context is engulfed and metabolized. What is the generalization of the above familiar situation? Complex fields of interdependent subsystems (humans or cells) do not vary randomly or continuously, but by reason of becoming interdependent they begin to lock in and out of step and organize. If each person in this audience were to stamp his boot now in time with his pulse so that the others could hear his rhythm, we would all soon succumb to a common beat or have to fight against rhythming together—being "entrained," as the engineers put it.

Behavior of interconnected objects or systems develops harmonies which allow the labeling of significant and meaningful groupings within the assemblage, or the interdependence breaks down and there is no dialogue. The dialogue of an interdependent system is maintained by growing and evolving, for simple repetition does not easily persist when there is adequate complexity.

But how can we conceptualize the dialogue so as to retain its feature of ongoing growing, of evolving not by stopping change but by playing against the kind of changing that is evolving. How can we do this? We psychiatrists are the experts. We join the dialogue—by playing the timing game. Can we formalize the way we time into the loop of an internal or interpersonal conversation? This is an important aspect of therapy; we change the dialogue's evolution by the way we time our changing as the patient changes. A mother who does not enter into a dialogue with her child cannot transmit the fine texture of information the child needs if he is to be skillful in evolving. The mother and child who can time to each other in a responsive way may each give to the other myriad data. But if each displays at the moment what the other cannot receive, the dialogue does not develop so as to facilitate the next message exchange. The predicament is self-reinforcing. Thus two children may, because of this timing problem, live in worlds which are different in information content by many orders of magnitude. This timing is an important aspect of the information exchange system of a mother and a child.

The loop as defined in cybernetics or general systems is never timeless. The simplicity we thought we lost in giving up (1) the historic, simple dependent-interdependent model of relations and (2) separation of the system into components as if the components contained all the information about their relations, we now trade off against data we can produce by grouping the interdependent phenomena of dialogue in terms of their timing.

Time Graining

But here we must begin, as you would expect, to introduce new words. First, let me introduce *time graining*. By this I refer to the kind of time lattice work, or texture, or grain (as on a camera film) with which one considers a system. This concept has been presented in a paper called "The Clock Manifesto" presented at the New York Academy of Sciences Conference on Time in 1966, and published in its *Annals* [1].

This paper deals with our unwillingness to consider clocks as existing for our use—and for our resourceful design. Clocks are considered to be divine, like the stars they imitate. You are all familiar with size graining and the need to distinguish the perspective of a mountain and a flea, for each perspective delivers a different system to our notice. This is equally true of time graining. History is commonly clocked into a lattice of important event time. The time distance between two important events, wars, for example, is considered the same. This time graining is not an equidistance in the astronomical sense, though it may be in the information sense—the wars occupying more space in the national memoirs than the poorly remembered periods of peace. This clock is marked by wars into equal amounts (a familiar story to the clinician). If this story still sounds strange to your ear let me remind you that it was not long ago that the clock used by common people was sunup to sundown as an equal span regardless of the season.

Time graining implies a necessary constraint in the discussion of complex information exchange. We must specify the kind of time regularity we will use in conceptualizing our loop phenomenon. No system which is complex and active will be without periods of relative stability and instability when it is more or less controllable. Phenomena are often quite naturally grouped by similarities in their time textures. Dialogue between systems of vastly different time textures is likely to be sparse, and thus these can more easily be modeled in the hard science mode. Complex systems have their own simplicities. Now you will see why we are not out to kill complexity but to use it for our study of the system.

The self-organizing power of our engagement with a system which already has some stability depends upon our playing jujitsu with it, by rolling with its changings just enough to find its moments of lessened stability, where a new piece of information may stream through many different levels—rippling throughout the system to produce a readiness to use the next ripple we create at just the right time in an even more significant way. You will recognize here the mechanics of innovation. When we can enter the dialogue we can examine each organism's way of evolving and evolve our own skill in adapting.

I am saying that each system has a clock and that if we wish to enter the dialogue this is to be measured in terms of the control periodicity natural to the system rather than mean Greenwich time. Only heroic change can produce innovation by a creation that is not evolutionary—and heroic change may kill the system's growth, slowing the adaptation that creates information out of the system's context. The mother crudely told not to spank her child may kick him; so it is, when change does not grow from inside. Modern control theory is quite different from the power control of the nineteenth century; it now conceptualizes control in terms of skillful use of information to make minor interventions count.

There are optimal time grainings for different purposes. This becomes important, for example, when one is trying to decide how responsive one should be in picking up the change in the system. Some people are set to pick up change in moments of very brief duration—they notice a flickering eyelid which is out of timing with their anticipation; others listening with longer moments will note how today's traffic movement is different from yesterday's. The choice of time grainings useful for a particular purpose with a particular system is limited. The time grain constraints are imposed by size and conduction delay in the communication network which integrates the complex system so that it can respond to the innovation. They are also a function of that different time delay for data processing, storing, and retrieving in memory. A psychiatrist may jump from responding to one time system, perhaps of years, speaking in words called *generalities* while still speaking with his hands in a time system of a few seconds, change being built together with response into these brief period dialogues. His jumping is itself a major communication—his jumping a message channel.

If one is too far outside the time graining of the system in one's choice of responding, communication is minimized and is said to be "simplified," for it is now without context. Consider a cheer leader who does not play against the timing of the crowd and only gains attention when she "works for it." Contrast her with a leader who gains the crowd's attention when it is just emerging from a hushed end run. The cheer leader moves at first in, then out of phase, and then to a faster pace, guiding her action in terms of the change in the total system (including the change within herself that occurs as she is changed by her involvement). She is active in creating the build—a nonlinear build—a build whose growth organizes context. She is able to transmit enormous amounts of information with significant effect. Similar time graining is important in making the decision as to when the display back to the user of a computer will have the most effective meaning in terms of enriching the man-computer dialogue so that it organizes what was peripheral context for the problems at hand. The possible effect of devices that can present data in terms of actively predicting user needs is an important area. Man-computer systems that will be able to use context as you and I do will require more knowledge of the timing which facilitates information exchange.

Time Driving

Another new word I would like to discuss is *time driving*. This is what the cheer leader did when she timed her action slightly ahead of where it was expected and called for. People easily lock step if they meet others who are sufficiently similar, and as I discussed earlier in the paper, the lock step may be one with continuous acceleration far beyond the usual threshold. The escalation in the system may time drive its components far past the point they wished to go. This happens most obviously in war when struggles intended to defend may go wild and destroy both attacker and defender without regard for their purpose.

The content or information exchange of the dialogue must not be separated from its time characteristic; these are different aspects of the same loop which act as context for each other.

The same dialogue may have profoundly different effects, depending upon its time driving characteristics—its timing or phasing or feedback.

Let us review. In discussing time graining we decided that for any phenomenon we intend to alter there is a set of particular periodicity ranges within which we must set the periodicity of our observing, if we are to influence that phenomenon by our return communication. Thus, Wiener points out that it is possible to influence the frequency of fireflies' lighting by flashing a light at about the fireflies' own periodicity—just then, blinking a little faster speeds up the fireflies' lighting. This illustration is like the cheer leader mentioned before who catches the tempo of the crowd, or the actor that of his audience, to gain a more significant communication position. But now let us move on from these homely examples to state the situation a little more formally. For each of these loops there are limited time ranges that will lend themselves to entering a dialogue with the phenomena. Flashing the blinker once a week or in each millionth of a second will not allow us to affect the fireflies' rhythm.

If we (1) choose one loop of a system of multiple connected loops, (2) seek out the limits within which it is unstable, insofar as it is reactive to an appropriately time driving input-output signal, usually translated by its environment in such a way that some part of it becomes an input used for control, and put it back a little sooner, giving it a slightly out of phase time feedback, and (3) perturb the system while engaging it in a dialogue by distorting the time function of its feedback, we then can find (a) the points at which the time driving fades out and at which it continues no longer listening, (b) when the system shifts to another mode, and (c) what loops are driven together. Thus, we begin to see how loops are interconnected in a control sense by the way they change together as a function of the altered periodicity, and at what point thresholds are reached so that the system becomes a different one in a discontinuous way.

You will think I am being highly technical, but I am describing in my more formal way what you must recognize a psychiatrist does quite intentionally when he is trying to explore what circularities a patient or family will be driven to, when one circular system is repeatedly primed into action by a suggestive comment at just the moment when the circle would not otherwise be repeated until after a rest period—but still is available to react if stimulated strongly. The same suggestive comment at a different time would not have the same meaning. The psychiatrist uses this device less intentionally when he places therapeutic comments just before the patient would have said it, the therapist predicting from context, or from his recollection of the beginning facial or lip expression being followed by this comment many times before. The patient gets the sense of rapport; they are timing together; he understands; more information can be exchanged, and more associations come that fulfill the prediction.

Such time driving, natural to humans in their dialogue exchanges, is frequently out of awareness, for our namings have not included such timing procedures.

When we consider the problems of education, time driving is particularly interesting. Actively unlearning the obsolescent is as important in education as presenting new data. In helping the child excrete informational organizations previously useful at earlier stages of his learning evolution, and taking care not to allow the newly presented information to crystallize too much that is unworked into words prematurely, one can present the obsolescent expectation just before the child is ready to present it *but* with an irrelevance associated, such as varied unexpected words from new contexts. Psychiatrists use this technique for helping people unlearn patterns they consider to be no longer useful.

Time driving is a powerful way of building and enhancing the exchange of information. A mother who presents her infant with food at a time when he has signaled that he does not quite want it, or after he is just a little too hungry, or just when he has signaled most clearly that he wants it and has the message out, is able to use the food not only for nourishing the child's tummy needs but also to build their communication. Being skillful in using timing is important to the earliest learning dialogue. The mother who knows the time graining which is appropriate to mapping her child's meaningful moments into her own can enter into a dialogue with him which, as it grows, organizes their communication so that just a glance between them is rich and carries with it a confirmation of their acknowledging together the meaning of a gesture just made.

But I have spoken to this point of time graining and time driving in the manner made necessary by our subject-predicate language, as if there is a "doer" of the action and "one done to"—a receiver. The loop concept has been subverted—if A and B are in a true dialogue their changings are as entwined as two lovers. Now there may be moments when the action is more unilateral, but the information exchange is the freshening of changings responsively timed to each other's changings. Who is father to the child fits the cause-effect model, but the evolution of the child and his family are a matter of dialogue.

An action spoken as a family builds a relation in which their shifting conversation carries their associations in an easy, timed flow is a million times richer than if mother's arm and infant's or father's do not time to each other's. Knowing the others' time graining and mapping it onto one's own is responding to how long a child's look sees when a child stares. Does he see very briefly a hand and a foot, or is he drinking in a whole scene? And this ability to teach complexity unintentionally by responding to moments which can be shared is a function of the dialogue which transcends the participants and evolves them. Each tests against the dialogue, each knows himself as a mix with another, and each knows what they can grow with timing to the music they evolve. This is a way of conceptualizing interdependent ecologies: families of similar cells or families of porpoises. I am using man as an example because it is easier to comprehend this idea in our own context of psychiatry.

Information Time

Having introduced the concepts *time graining* and *time driving,* now let me go one step further into the problem of *information time.* This is the periodicity of equivalence in information exchange quite unrelated to the familiar clock. Let me state that this concept is simple, but it is also commonly discussed only by handwaving or showing, as opposed to telling. It has no formal language that is available to the nonmathematician. Let me point out first the usefulness of this term in conceptualizing effective dialogue. But let me state, again, that formalization of the concept is just at its beginning. Let us begin as before with homely examples. Compare grandma's time to a 2½-year-old child's, and you will be aware that their times are very different. The child will move through his event space at a rate which far exceeds grandma's. His very biological changing, moment to moment, his fluctuation in heart rhythm and cell division, means he is processing more novelty simply by reason of changing his own receptor system. An hour of Johnny is equivalent in novelty to a day of grandma. Novelty has been used as a measure of information. Information in America is formally defined as "a measure of novelty." If equivalent periods of information processing were used as equivalences for building a clock (instead of rotations of the earth) we would perhaps have a more useful clock for measuring dialogue.

But to many this may seem an esoteric problem. The problem of giving up the hard science methodologies of the last century remains with the soft sciences. Physicists work with many clocks and build clocks to fit their purpose. That they be well defined is the only requisite. But let us return to the homely examples I mentioned. In communicating, Johnny and grandma both enter systems with different time grainings. If grandma is a warm person she can enter the child's time and the child will enter grandma's, but the communication in their dialogue will depend upon grandma's and Johnny's capacity to map each other's time.

To present yet another homely example, in an "emergency time" one has the sense that during the emergency, as when an accident is impending, 3 minutes are the equivalent of 3 hours—and one's ability to act in the 3 minutes may match 3 hours under ordinary circumstances.

When a creature who acts very quickly meets another who is very slow they will usually converge their pacings if they are to find a way of building a dialogue. The child who moves at a very fast pace and learns in a way that seems quick to one who moves slowly may be so surrounded by a pacing context of slow-moving objects, as in a traditional school, that he slows down to a pace which is no longer his natural pace. Such a child may not process ordinary information well enough to be a very good learner, or his "biological system" is not locked into his dialogue with those who teach him. Note here I am talking of the biological dialogue as the dialogue which includes the whole child and his whole manner, the dialogue of muscles sensing each others' movements as their "together movements," called the child, enable each next participation. If his behavior is not matched to his test time, the child is not synchronized and appears to be intrinsically awkward.

Information time means setting two time periods as equivalent when the same amount of data is processed. If our problem is the enhancement of dialogue, then timing as a function of information exchanged is one way of mapping two systems so that they may be synchronized in the sense of reaching critical change points in a way that allows optimal dialogue. It is the need to meet the requirements for control of man-machine dialogue that makes pursuit of the concepts of time graining, time driving, and information time important. The approach to complexity by using the time dimension as another dependent variable promises to bring the hard and soft sciences closer together.

Application

It would have been only recently that this discussion of dialogue would have been another discussion of the importance of the holistic approach I have always thought of as holy. This is no longer true. The modeling of information exchange in the time domain I have presented is a primitive start in thinking through practical problems. The bottleneck in making computer resources available to man is the need to interface their communication so that it will grow in the style of a real conversation—as exploratory conversation, as learning conversation, where the associations and experience of each will be available to the other. The innovative use of computers depends upon solving this dialogue problem. Man must adapt to the advantages bestowed on him by his inventions—fire, wheel, atomic energy, and computers, among many—or die.

The computer already has begun to alter man's science in a way that should please us who are used to complexity. But the dialogue of the soft sciences cannot be easily analyzed by the tools at our disposal, even yet, for we have no formal mathematics which can provide a calculus of relations. Our logic is as yet tautological and timeless. A statement of truth does not change because time passes. It does not decay or become obsolescent in a formal way though it does in the real world. Our power to develop the soft sciences as living sciences depends upon our learning to formalize. The computer's ability to match human complexity, to synthesize the dialogue, requires only that we prime it with our beginning analysis of the features of dialogue.

As we learn to build computer programs that have a little more ability than those we know so far, to enter into a rich and complex dialogue with man, we begin to learn a little better how to understand what we have hidden in our art and our practice—the science of information exchange. Having to formulate the skill in timing shown to us during our psychiatric apprenticeship may assist us in developing a science which encompasses the dialogue. We need not be afraid of such a science. Knowledge of the dialogue is needed if our machines are to engage in rich and complex dialogue with us and our children.

Reference

1. Brodey, W. M. The clock manifesto. *Ann. N.Y. Acad. Sci.* 138:895, 1967.

11

Implications of General Systems Theory in Industry and Community*

JOHN MAC IVER, M.D.

WE MAY BEGIN this consideration of the role of general systems theory in industry and in the community with the obvious, that industry is part of the community; but let us arbitrarily consider them as separate, albeit interacting. In fact, the human community existed in considerable complexity long before the rise of industry. Industry may be thought of then as one of the social institutions which have slowly risen in the course of social evolution. It tends to have, with fluctuations, "a life of its own." It emerges, like other social institutions—education, the military, and the like—to fill needs of the community otherwise impossible of achievement. We may note in passing that the political, juridical, and governmental apparatus can be conceptualized in a third category, with various unique functions, but in general providing cohesiveness for the community and its social institutions in their combined totality.

From a systems point of view the community and industry can be contrasted reasonably sharply. Interacting systems in the community can be characterized as loosely coupled; systems in industry are far more tightly coupled. Closeness of coupling is not itself an advantage or disadvantage; it may work either way. In industry various types of monitors or regulators, often quite sensitive, may permit exquisite control. Variety may be reduced, however, and the organization as a result be overly information tight. Market research and development is an example of a relatively new staff function which serves to bring

* Presented at the General Systems Session of the 123rd annual meeting of the American Psychiatric Association, Detroit, Mich., May 8–12, 1967.

in information and thereby forestall informational decay within the organization.

General Systems or Chaos

Why general systems theory for industry or community? One answer, perhaps the simplest and most direct, is that we have no choice at our present level of social complexity. Or rather, we have a choice between refining general systems and cybernetics approaches or descending into chaos, by which I mean a regression to a low order of human organization. For, despite the panning that the nineteenth century notion of inevitable progress has taken in the twentieth century, segments and pockets of human society *have* progressed in terms of living standards, invention, technology, and intellectual and artistic attainment. It is probably fair to say that we should not regard progress as inevitable, but I know of no conceptual approach other than general systems theory or cybernetics, dealing as they do with control and communication, that offers the likelihood of enhancing continuing social evolution. As Ashby has frequently pointed out, traditional Western science has been analytically, not synthetically, oriented and has studied highly selected types of systems, imposing rules of the game that eliminate vast areas of the first order of importance to the future of man. It has therefore been all too easy to forbid entrance to the traditional preserve of science to those whose areas of research require models different from those of the physicist and chemist.

To the present day, human systems beyond the family—if we assume the family has been to an extent biologically determined—have been added on to (or subtracted from) what already exists, in an empirical, cut-and-dried fashion, with little or no planning or indeed prevision of what the consequences to society would be. More accurately, before particular innovated systems come inventions—whether world-shaking, major, minor, or supplementary—and over the course of civilization myriad inventions. I here use *invention* in the widest kind of connotative fashion, whether conceptual, technological, psychological, philosophical, theological, or whatever other domain anyone might favor. The construction of systems, subsystems, and suprasystems follows, in the main, without much thought of consequences. This we can no longer afford. As Platt has said, the world is now too dangerous for anything but utopias. From now on we must know with reasonable surety where we are heading ecologically. As Ashby has pointed out [1], there are two major strands in general systems theory:

One, already well developed in the hands of von Bertalanffy and his co-workers, takes the world as we find it, examines the various systems that occur in it—zoological, physiological, and so on and then draws up statements about the regularities that have been observed to hold. This method is essentially empirical. The second method is to start at the other end. Instead of studying first one system, then a second, then a third, and so on, it goes to the other extreme, considers the set of "all conceivable systems" and then reduces the set to a more reasonable size. This is the method I have recently followed.

The basic science for Ashby's approach is cybernetics, and he has attempted to develop a science of systems as such. To quote him again [2]:

. . . an adequately developed logic of mechanism is essential [for his task]. Until recently, discussions of mechanism were carried on almost entirely in terms of some particular embodiment—the mechanical, the electronic, the neuronic, and so on. Those days are past. There now exists a well developed logic of pure mechanism, rigorous as geometry, and likely to play the same fundamental part in our understanding of the complex systems of biology, that geometry does in astronomy. Only by the development of this basic logic has the work in this book *Design for a Brain* been made possible.

As work proceeds in both areas of general systems, the theoretical and the empirical, and as more workers come to understand each other's viewpoints better, important new insights will very likely appear. Applications—that is, operational insights—are still hard to come by. This has afforded critics of general systems opportunities for cavil. There is a reality to be faced up to here, a serious difficulty, mainly the lack of availability to date of powerful general laws or criteria that would apply to human behavior. If highly generalizable laws were available, both general systems schools would likely be able to join forces with a consequence of greatly increased power and application.

Community Health

Community health, or public health, provides an example of a group of interrelated disciplines that has foundered in the past two decades after three-quarters of a century of brilliant achievement, achievement comparable to that of the nuclear physicists. The foundering was, and is, laden with overtones of great potential tragedy. The triumph was, of course, the triumph of bacteriology and its allied sciences of immunology and sanitary engineering. So powerful were concepts and applications of microbiology that the notion of the single causation of disease became preeminent and indeed persists in much of present-day medical culture. The ideology of a single causation persists despite the cogency of the attacks mounted against it, notably by René Dubos in his writings over the past fifteen years. Communicable diseases having essentially been conquered, public health, conceptualized as an applied biological science, found itself helpless in the wake of new demands on it as applied sociology. The infant mortality rate was universally accepted as the classical index of community health; one does not in these days of burgeoning populations find it vigorously defended. The words of John Dewey come to mind. "We have to include consequences impartially. It is willful folly to fasten upon some single end or consequence which is liked, and permit the view of that to blot from perception all other undesired or undesirable consequences."

Health indexes, social indicators, regulators, and monitors—I see these as a necessity for both industrial and community systems. Monitors and regulators will themselves be complex and multiple, and will have to be as rigorous as we can make them. Moreover, they must be embedded in a human ecology that will preserve individuality and creativity to the maximum extent possible.

Social indicators and theories of social evolution are closely related areas; the literature on these topics is often suggestive and provocative, but is on the whole quite unsatisfactory. What often appears to be brilliant and evocative nonetheless turns out on closer inspection to be clever retrospection, and it can be observed that prospection and operational statement are largely if not completely missing. One may recall Freud's statement that psychodynamics can throw light on the patient's past, but not on his future.

There are a number of objections, often repeated, that a social science is impossible and that a human ecological steady state is unattainable. Extremely able men, including Norbert Wiener, have taken this point of view. Among other things he says, "In the social sciences we have to deal with short statistical runs, nor can we be sure that a considerable part of what we observe is not an artifact of our own creation" [3]. This is an extraordinarily important issue, and while I would like to state that I do not agree with this point of view, I am not prepared at this time to offer extensive rebuttal. I would however point out that Ashby's writings suggest quite opposite implications from Wiener's on this point.

No paper of this sort is complete without a bow to the computer. It has the same importance for social evolution as the engine, the telescope, and the microscope—indeed, probably more. Little I have suggested in these remarks would be worth saying without it, at least in terms of operations and applications, simply for the reason that systems to monitor ecology require human and computer capabilities working in concert. The human being alone is just not sufficient.

In conclusion, I do not expect utopia, as I understand the meaning of this term. I do expect a high probability of achieving a human community which affords dependable and ever-increasing scope for social and individual psychosocial actualization.

References

1. Ashby, W. R. General systems theory as a new discipline. *Gen. Syst.* 3:1, 1958.
2. Ashby, W. R. Preface to *Design for a Brain.* New York: Wiley, 1960.
3. Wiener, N. *Cybernetics.* New York: Wiley, 1948.

DESIGNATED DISCUSSION

RAYMOND W. WAGGONER, M.D., SC.D.

THE GIFTS of science to man can be summarized in three words: prescience, control, and insight. The ability to predict regularly recurring events was the early result of systematic observation. This ability enabled man to adjust his activities to foreseeable conditions. Coupled with the experimental method and deductive logic, the ability to predict was extended to events never before observed as well as the ability to control events, and above all the ability to organize his environment and to adapt it to man's needs. Technology, as we know it, is the product of the power to control matter.

Insight is of a different order. Insight is an inner experience—the ability to see unity in diversity, to see deep relations among seemingly unrelated events. Reliance on insight has been traditionally the province of philosophy, religion, poetry, and the arts, rather than of science in the narrow sense of content-circumscribed investigations. Insight with a little mixture of serendipity has played a vital role in science as understood in its broadest sense. We can attribute to insight the formulation of atomic theory in antiquity by Democritus, of the self-recuperating powers of the body by Hippocrates, of the principle of natural selection by Darwin, of the role of the unconscious in human behavior by Freud, and of the unity of space and time by Einstein.

In and of itself, insight does not usually confer either prescience or control, as a scientific theory does, but it puts man, whole generations of men, into a frame of mind in which they are able to chart new courses for scientific advances.

General systems theory is not a scientific theory in the proper sense, that is, a clearly formulated set of axioms from which assertions with predictive content can be derived. General systems theory focuses rather upon new insights which have emerged from the scientific developments associated with the new system technology and those directions in biology which place homeostatic mechanism at the center of attention. These insights comprise three categories. The first revolves around the recognition that the idea of linear causality (that is, causality stated as an interaction between just two variables, which has been a standard paradigm of elementary scientific explanation) is not only insufficient but also frequently misleading when the scientist must deal with organized complexity. The second stems from the revival of the organismic analogy (as in the extension of the definition of "organism" to entities other than living individual organisms, for instance to colonies of organisms, communities, institutions, societies, subsystems of societies, and the like). The third insight stems from the discovery of formal mathematical homologies in the laws governing the behavior of systems with widely different constituent parts but similar organizational structures (as in the isomorphism between mechanical and electromagnetic systems, between a two-person zero-sum game and a linear programming problem, between a system of chemical reactions and a system of ecologically interacting populations, and so forth).

Dr. MacIver has singled out the first two of these approaches to general systems theory, which are, of course, intimately related. The notion of the functioning organism necessitates the abandoning of the notion of simple linear causality. Thus, the introduction of an industrial subsystem into a social system must and does have far-reaching repercussions in all the other subsystems, including family and social structure and the psychic life of the individuals who comprise the social system. The awareness of these profound interdependencies suggests ideas about man-in-society in a new light, as for example the component parts of industry or a community. It is no longer possible to pay attention to some one aspect of development, say, of the industrial complex, without giving serious thought to the repercussions which are bound to accompany it. Mutatis mutandis, it is impossible to understand some of the profound changes in man's psychic life without seeking its roots in the changes brought about in all of the subsystems of the social matrix in which the human individual is constantly immersed. This insight reveals the inadequacy of both the laissez-faire philosophy, according to which society is bound to move toward some optimal homeostatic equilibrium if all its components are only allowed to develop naturally, and the philosophy of total social planning with its tacit assumption that the consequences of controlled changes can all be foreseen and that the marginal utilities of all social values are self-evident.

In short, the outlook inherent in general systems theory reveals the inescapable necessity of fusing all available methods of science if the science of man is ever to keep pace with the science of the material world. It has become evident that the latter can be at best a mixed blessing and at worst a curse without the former, which is implicit in Dr. MacIver's paper.

12

A Unifying Concept Linking Therapeutic and Community Process*

MONTAGUE ULLMAN, M.D.

ONE OF THE IMPLICIT ASSUMPTIONS in the setting up of comprehensive community mental health centers is that psychiatrists can and should accept their share of the responsibility for meeting the mental health needs of an entire population living in a given catchment area. As exciting and as challenging as this idea may be to some, it has met with resistance from within the profession as well as from other interested professionals, ranging from a form of identity crisis to outright rejection in principle. In the first instance community psychiatry is seen as downgrading psychotherapeutic skills in favor of supportive and educational roles. In the second, questions are raised concerning the legitimacy of the endeavor [18] or its expected end-point, namely, the reduction of mental disturbances in the community [4].

The issues involved have been considered and answers suggested in the contributions of Duhl [3], Seeley [15], Linn [12], Hume [10], Riessman [14], among others. As we move into a broader social frame of reference and attempt to come to grips with an ecological point of view, we will need, as Duhl has pointed out, a new theoretical model of man, one which will link what we now think of as individual psychopathology to limitations in the social milieu, both directly and as expressed through significant figures. The maturational needs of the evolving human organism are social in nature and demand social solutions. It is in this sense that personality disorder may

* Presented at the 123rd annual meeting of the American Psychiatric Association, Detroit, Mich., May 8–12, 1967.

be considered as the endpoint of a series of enforced individual solutions to problems requiring appropriate social solutions.

The nature of a social solution is continuously defined by the potential of the immediate social context to so interact with the historically evolving personality as to exert an influence in the direction of healthy change. Put another way, a social solution is always the product of an interaction and not an arbitrarily imposed solution from above. Rituals, obsessions, phobias, psychosomatic symptoms, as well as the exploitation and misuse of one's biological potential for the enjoyment of sex, food, and drink are all individually contrived techniques for living arising under circumstances in which social solutions linked to adequate mothering, intact family structure, and educational and vocational opportunities were defective or lacking. The ecological point of view forces us to take a look at the actual way in which such social solutions are structured by the way people live in a given community.

In this presentation the focus will be on one particular problem which, if not attended to, will most certainly play into the resistances outlined. I refer to the lack of effective conceptual tools with which to bridge our involvement in clinical phenomena to our grasp of process and change in the extended social frame of reference. This is related to the communication problem referred to by Duhl [3], and the need for a more felicitous terminology proposed by Zwerling [19]. What I am suggesting is the need for concepts that are rich enough in meaning and application that they might well serve as the final common pathway linking clinical events to sociological data. Such concepts would provide the necessary leverage to open the door to a broader vista and to keep it open if their operational usefulness is established at both ends.

Interface Phenomena

A number of interface phenomena can be identified in the sense that they stand in a connecting zone between the clinical systems familiar to the psychiatrist—the dyadic relationship, the family, the group, the ward milieu—and the larger, ambient system extending outward to the community. Contributions from two sources structure this interface, but in different ways. The social scientist, viewing the various clinical groupings as special cases of group interaction and organization, tends to emphasize the structure and function of these larger groupings and to view individual behavior in its more macroscopic dimensions. Role theory and the concept of social role is a good example. Lasswell's [11] attempt to link psychological structure to political context and Fromm's [6] linkage of character structure to the laws of commodity relations might also be included. The second source of interface phenomena derives from the innovative efforts of psychiatrists themselves to enhance the availability and range of psychiatric services either through more extended treatment techniques, as in the case of family therapy, or through the evolution of reaching-out strategies as in the setting up of home services. The extra range and variety of therapeutic activity remains subject, however, to the traditional kind of microscopic analysis, meaning in this case the concern with intrapsychic mechanisms.

It is no wonder then that the psychiatrist peering across this interface into the enormously complex community that exists outside the confines of his familiar clinical arrangements is likely to ask himself three questions: Should anything be done? Assuming something is to be done, am I the one to do it? Assuming I'm the one to do it, how do I go about it? There are sufficient affirmative answers to the first two questions, as witnessed by the way in which the concept of community psychiatry has caught the imagination of psychiatrists and, at a pragmatic level, by the enormous investment of federal and local funds in community mental health centers. Our concern in this presentation will be in tackling one small part of the third question having to do with the search for unifying concepts which may be of help to the clinician as he begins to commute between therapeutic and community process.

The Concept of Power

The concept of power is one which might meet this need. It is a key concept, as yet not fully explored, in sociological theory. It also appears to be a concept which can be meaningfully translated into clinical terms. Community process is intimately linked to the deployment, accumulation, and operation of power along both formal and informal lines. Perhaps clinical symptomatology can be meaningfully extrapolated through the use of similar terms, and in a way that maintains a thread of identity between what happens in the specifics of a therapeutic engagement and what happens along more general lines when issues are identified and attacked at a community level.

When the marriage of anthropology and psychoanalysis took place, with psychoanalysis in the role of the virginal bride, the wedding was financed by the bride in the sense that it was the anthropologists who came to the analysts to learn and apply the new language of psychoanalysis to their own endeavors. With a marriage of another kind now on the horizon, namely, that of community psychiatry and sociology, the situation is likely to be reversed. If the marriage is to be consummated it is likely that its support in terms of conceptual flow will have to come from the groom's family, since in this particular instance they are the ones who possess a language capable of describing social events as social events rather than as psychological derivatives.

Whether or not the concept of power can serve as the unifying mode we are seeking will depend upon its strategic importance at a sociological level and whether or not clinical data can be meaningfully translated into states of power deficit and power operations. Let us consider these possibilities and begin by exploring problems of definition.

The first of a long list of dictionary definitions of the term *power* describes it as the ability to act so as to produce some change or bring about some event. This is then extended to meanings more heavily weighted to imply authority, strength, force, and coercion.* Interestingly enough, it is also defined somewhat negatively as the absence of such restraining influences, leaving power of volition to the subject. For our purposes we may accept Clark's [1] definition of power as energy available for work to bring about, sustain, or prevent change as the one coming closest to serving as a unifying concept linking the interpersonal to the more broadly social. Our task is to explore its relevance to clinical transactions and to examine some of the ways in which power operations occurring in the therapeutic context are congruent with power operations on a social scale. The concept of power as defined in an interpersonal setting has meaning only to the extent that it is isomorphic with the origin and exercise of power in

* *New Standard Dictionary of the English Language.* New York: Funk & Wagnalls, 1928.

more extended social systems. We will present a number of propositions which suggest that the concept may be used in this fashion.

Neurosis and Power Deficit

A number of reports attest to the higher incidence of psychiatric disorders occurring among patients in the lower socioeconomic brackets [5, 9, 17]. Hersch [8] notes the delay in establishing this relationship: "The hard evidence that poverty is a major epidemiological factor bearing on individual and family breakdown, and that there are clear associations between poverty and a number of specific psychopathological conditions, waited for the introduction of public health viewpoints and methodologies to be brought to the fore." Haggstrom [7], in an excellent discussion of the impact of poverty, notes that the consequences as expressed in the life style of the individual are directly related to a pervading sense of powerlessness and not to the absolute supply of money to the poor. The proposition we wish to consider is not that poverty causes mental illness, but rather that poverty is one among many circumstances of modern living yielding as a byproduct a feeling of powerlessness, and that it is the latter which sets in motion dysfunctional social behavior. We are suggesting that both poverty and mental illness are expressions of an existing inequity in power relations.

Poverty emerges as the concrete material precipitate of an underlying insistent imbalance in economic opportunity, while neurosis may be defined as the internalized reflection of existing inequities in power relations as manifested in the day-to-day lives of people. Put another way, we are saying that both the poor and the mentally ill are disfranchised, although in different ways and with different manifestations. Certain groups in our society tend to show a greater or lesser shift in the direction of second-class citizenship. This is so for females, for the elderly, for the Negro race, and for the poor. To the extent this shift takes place, civil, political, legal, and economic rights are increasingly precarious. What we are proposing is that there is another group that cuts across all these lines and is subject to the same institutional disfranchisement. Although not generally thought of in this way, the mentally ill do form an oppressed group, one for whom certain rightful social assists have been lacking and who as a consequence have had to develop individualistic solutions to problems demanding of social solutions. In the face of real deprivations in the area of effective mothering, effective family life, and effective educational and social opportunities, the socializing process is impeded, shunted, or grossly distorted.

What we call mental illness is, by and large, either due to or greatly augmented by defects in the socializing process. Neurotic symptoms are the surface excrescences of structures molded under less than optimal circumstances. It is in this sense that the mentally ill form an oppressed or a deprived group. The ultimate source of this deprivation would then be linked to institutionally sanctioned patterns of power operations. The specific nature of the oppression is an indirect result of a maintained state of power deficit. The language of neurosis seems to reflect the basic role of dysfunctional power relations. The initial balance between physical helplessness and omnipotence endures and is restated in different ways throughout life. Grandiosity is linked to feelings of insignificance, controlling needs to weakness, and power operations to feelings of powerlessness. In contrast to mutually enhancing or synergic* power relations which generate greater social power for both individuals, a state of asynergy exists in which another person is used as the object or source of power. In the first instance power facilitates the realization of social goals. In the latter, social goals are subverted to the establishment, maintenance, and enhancement of personal power. The techniques for coping with power loss tend to be individualistic rather than collaborative and tend to develop either in the direction of aggressive maneuvers directed outward (coercive, manipulative, competitive) or in the direction of withdrawal (denial, social constriction, and the schizoid maneuver) [16]. At a more socially organized acting-out level these diverse trends are expressed as delinquency on the one hand and in the evolution of various fringe groups on the other.

The thesis I wish to put forth is that neurotic suffering represents

* The two ways in which power can be expressed, to serve constructive or destructive ends, can be viewed in terms of synergistic relations described by Maslow [13]. He speaks of ". . . cultures with low synergy where the social structure provides for acts which are mutually opposed and counteractive, and of cultures with high synergy where it provides for acts which are mutually reinforcing. . . . I spoke of societies with high social synergy where their institutions insure mutual advantage from their undertakings, and societies with low social synergy where the advantage of one individual becomes a victory over another, and the majority who are not victorious must shift as they can."

a power deficit and that compensatory mechanisms take the form of seeking to fixate the environment through negative or destructive power operations. The essence of the neurotic state is helplessness, inadequacy, and aloneness, emerging in critical situations as the feeling of being trapped. In other words, a power deficit exists built up over the years, initially in the absence of available social solutions and later on despite the presence of social solutions. When we consider power as a compensatory effect, we have to examine it both objectively and subjectively. Its objective expression can be one-sided and destructive. This is so when it is exercised over another in an arbitrary way without taking into account the genuine self interest of the other, when it is used compulsively to maintain a structural status quo at any cost. In its most extreme form this amounts to a striving for power for the sake of power.

The strategies most commonly employed to achieve these ends involve control, manipulation, and exploitation. They represent the real manifestations of negative or destructive power. Omnipotence or grandiosity represents the subjective reflection or virtual image of the operation of negative power. Since the goal sought by the operation of negative power is impossible, namely, insulation against change, the strategies by which it is pursued must be endowed with magical and absolute qualities. Despite the difficulties experienced in trying to get neurotic operations to work and the fact that their success is increasingly contingent on an ever-narrowing constellation of special circumstances, even minor success reflected through the magical lens shaped by the driving necessities that beset the patient suffices to maintain the virtual image of omnipotence. This is why so much can go wrong and so little insight occur.

Power Operations and Therapeutic Process

In an effort to eschew the charge of superimposing our own values on patients, we lose sight of the fact that hopefully the therapist has more power to engineer a successful outcome in an interpersonal situation than does the patient to prevent one. What this means is that he can influence the situation along the lines that he wishes, with therapy coming through as the capacity of the therapist to wish for and evoke the greatest possible expression of autonomy by the patient. Every treatment situation is a struggle and in that sense a test of power. The thrust of the patient's power in the situation, if not deflected and transformed, would simply recreate another self-defeating experience. Although terms such as "power play," "control," and "the struggle for power" are often used in a pejorative sense, we do not ordinarily link therapeutic transactions to power in an explicit way.

In the one-to-one treatment situation the effort is made through the use of the therapist's power to force the patient to surrender his own neurotic power operations. The art of therapy is to bring this about through the establishment of socially congruent techniques of self-expression and self-assertion. The patient moves in the direction of enhanced self-esteem, a greater degree of psychological freedom, and some measure of effective control over his own destiny. In working with larger groups, such as the family unit, the task first involves the analysis of existing power relations, the way in which different members participate in the spurious use or misuse of power, the surrender of power, and the escape from power. Treatment involves a democratization and redistribution of existing power relations through the development of cooperative problem-solving techniques.

When we fail in treatment we do so by virtue of the fact that the weight of the patient's entrenched behavioral operations proves greater than the resources available to the therapist. Given this particular constellation of events, the therapist had less power and was the weaker of the two. Here we have an interesting example of what might be called negative power, that is, the power to resist change. The sick, paranoid patient may exhibit sufficient negative power to defeat the therapist and to structure the relationship along lines most comfortable to himself. He is unable to use his power for his own genuine self-interest. His enormous power to resist change endows this power with a negative valence rendering the patient invulnerable to change. Conversely, he is equally powerless to flexibly and adaptively meet the demands of daily living.

Treatment can be conceived of in terms of influencing the existing interplay of power operations in the clinical context in the direction of limiting the conditional base for the manifestation of one-sided, asynergic, or negative power and broadening the base for the emergence of positive or synergic power. Favoring the therapist is his knowledge of the fact that negative power exerts a social effect only in situations which facilitate, tolerate, or simply cannot prevent such an effect. Positive or social power, on the other hand, is defined as power capable of exerting a mutually enhancing effect on all parties concerned regardless of its source. The converse of social or positive power is failure in a particular context. The failure of negative power implies a sense of powerlessness in a general context.

Power Operations and Community Process

Effective decision making is always related to the availability of appropriate social solutions and the ability to use them. This, in turn, is based upon the existence of a certain equity in the deployment of power in interpersonal relations, buttressed by institutional pressure in the direction of equalizing power relations economically as well as politically. In the absence of this there remains a pervasive power deficit sanctioned by institutional support and so built into the fabric of our lives that it requires a determined effort to expose it and do something about it. Inequities in power relations with accompanying power deficit are both evoked and maintained by the social system. Poverty is the expression of these inequities economically; psychiatric disorder is their expression psychologically (see Fig. 12-1).

Power deficit refers to a failure of inner resources for appropriate decision making. For the social group this implies failure to provide or make available the necessary social resources for appropriate problem solving and decision making. In the realm of individual behavior, strategies of control or surrender bespeak an underlying power deficit. The frightened and helpless individual seeks to remedy the intolerable subjective consequences of a power deficit. He learns how to use relationships with others instrumentally to pursue power needs. By contrast the healthy individual engages in the instrumental use of power to further the natural and mutually beneficial ends inherent in the relationship. Such ends, whether they be sexual, educational, or therapeutic, come about in a setting of mutual respect, dignity, and growth. When the converse of this occurs, as in the first instance, we have the compulsive pursuit of power. This takes on many differing colorings based upon how one perceives power and the opportunity one has to learn how to use others instrumentally for one's own gain. The instrumental use of human beings in this way in the service

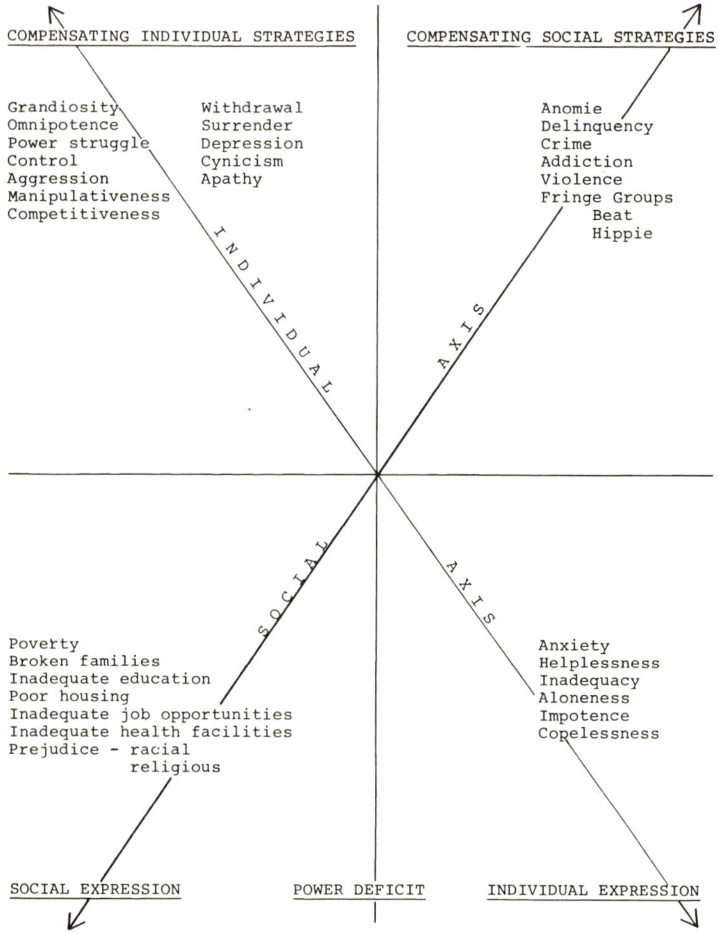

FIGURE 12-1. POWER OPERATIONS, COMMUNITY PROCESS, AND THE INDIVIDUAL.

of power operations is a reflection of as well as a contribution to the institutionalized sources of power imbalance. Individual strategies are amplified or dampened depending upon the prevailing strategy of power deployment at the social level.

In the approach to the community the first step is also the analysis of existing power relations and the way in which individualistic solutions have flourished at the expense of social solutions. At the social level, individuals sufficiently removed from decision-making processes on issues that vitally concern them turn to idiosyncratic or socially deviant solutions. Delinquency, crime, addiction, violence, and the like, may be regarded as compensatory social strategies of existing states of power deficit in the socioeconomic sphere, just as neurotic techniques of control or surrender are compensatory individual strategies related to the subjective residues of power deficit in the form of anxiety and helplessness. Certain analogies to the therapeutic process exist with regard to the ways in which community change is brought about. Just as in the case of the individual in whom some change is ego-syntonic, certain constructive changes can occur in the community within existing power relations. There are other situations, however, in which change can occur only through conflict resolution [2]. In the individual this brings about a shift away from compulsive power operations. In the community this involves a change in the existing power structure.

Summary

Qualitative changes in both theory and practice are likely to be an integral part of our currently expanding social program in psychiatry. In order to conceptualize the nature and direction of such changes, it becomes a matter of some importance to make explicit the links between community process fostering maladaptive behavioral patterns and therapeutic process attempting to cope with the tide of psychiatric casualties. If mental illness is related to or greatly augmented by defects in the socialization process, then it becomes very important to understand the nature of that process, its expression in a given community, and the way in which ideas about treatment have relevance to the social sources of pathogenicity.

The concept of a power deficit is developed as one such unifying concept linking therapeutic to community process. The mentally ill person is seen as a member of an oppressed group, a group deprived of adequate social solutions to the problem of individual growth and development. Disruptive behavior among groupings within a community is seen as also stemming from a power deficit resulting from the lack of adequate social solutions to the task of community growth and development.

At the individual level the power deficit is experienced as helplessness, inadequacy, aloneness, and impotence and is compensated for by grandiose and omnipotent strivings channeled through controlling, manipulative, and exploitative trends. Therapeutic effort is directed at replacing spurious power relations by a sense of conscious and effective personal power. At the community level the deficit is experienced as anomie, disfranchisement, cynicism, apathy, and powerlessness and is compensated for by strategies of delinquency, crime, violence, and cultism or just sheer ruthlessness. Programs aimed at community reform inevitably impinge upon existing power relations with both concordant and discordant effects.

References

1. Clark, K. B. Social Power and Social Change in Contemporary America. Department of State Publication 8125, Sept., 1966.
2. Dodson, D .W. The Community Organization Worker and the Urban Encounter. Presented at Convention of the Community Organizers Group, Jan., 1966.
3. Duhl, L. J. The changing face of mental health: Some ecological contributions. Scientific papers and discussions. *Amer. Psychiat. Ass. Dist. Branch. Publ.* 1, Feb., 1960.
4. Dunham, H. W. Community psychiatry. *Arch. Gen. Psychiat.* (Chicago) 12:303, 1965.
5. Dunham, H. W. Social class and schizophrenia. *Amer. J. Orthopsychiat.* 34:634, 1964.
6. Fromm, E. *Man For Himself*. New York: Rinehart, 1947.
7. Haggstrom, W. C. The Power of the Poor. In F. Riessman, J. Cohen, and A. Paul (Eds.), *Mental Health of the Poor*. New York: Macmillan, 1964.
8. Hersch, C. Mental health services and the poor. *Psychiatry* 29:236, 1966.
9. Hollingshead, A., and Redlich, F. *Social Class and Mental Illness*. New York: Wiley, 1958.
10. Hume, P. B. Community psychiatry, social psychiatry and community mental health work. *Amer. J. Psychiat.* 121:340, 1964.
11. Lasswell, H. D. *Power and Personality*. New York: Viking (Compass), 1962.
12. Linn, L. Letter to editor. *Science* 154:641, 1967.
13. Maslow, A. H. Synergy in the society and in the individual. *J. Individ. Psychol.* 20:153, 1964.
14. Riessman, F., and Miller, S. M. Social change versus the "psychiatric world view." *Amer. J. Orthopsychiat.* 34:29, 1964.

15. Seeley, J. R. Community psychiatry: The sociological specter. *Arch. Gen. Psychiat.* (Chicago) 13:289, 1965.
16. Silverberg, W. V. The schizoid maneuver. *Psychiatry* 10:383, 1947.
17. Srole, L., Langner, T. S., Michael, S. T., Opler, M. K., and Rennie, T. A. C. *Mental Health in the Metropolis: The Midtown Manhattan Study*, Vol. 1. New York: McGraw-Hill, 1962.
18. Szasz, T. S. Strategy of freedom. Community Leadership Project of Washington University, St. Louis. *Trans-action*, May-June, 1965.
19. Zwerling, I. Some Implications of Social Psychiatry for Psychiatric Treatment and Patient Care. Institute of Pennsylvania Hospital, Strecker Monograph Series No. 11, 1965.

13

Psychosomatic Manifestations of Rapport in Psychotherapy*

E. JOSEPH CHARNY, M.D.

A motion picture film of a psychotherapy session was analyzed for naturally occurring configurations of postures assumed by the patient and the therapist. The configurations were classified as congruent and noncongruent. The vocal behavior correlated with each of these types was analyzed structurally and thematically. It was found that the vocal correlates of congruent posture were significantly different from those correlated with noncongruent posture: they were consistently positive, interpersonal, specific, and bound to the therapeutic situation, whereas those occurring with noncongruent configurations were more self-oriented, negational, and nonspecific, and tended to be self-contradictory and nonreferenced. It was concluded that congruent postural configurations in vis-à-vis psychotherapy are behavioral indicators of rapport or relatedness. The results point up the value of using naturally occurring behavioral units in analysis of behavior.

IT IS THE GENERAL CONSENSUS among investigators of the process of psychotherapy that the patient-therapist relationship is the sine qua non for therapeutic effectiveness [1, 14, 15, 31, 32, 34] and that "the most important psychotherapeutic material with which patients and therapists must work stems from the vicissitudes of the doctor-patient relationship in its real and its distorted aspects" [15]. In other words, *what* a patient talks about in a therapy session and *how* he or she talks about it will be decisively determined by the state of the therapeutic relationship at the moment. Studies aimed at the elucidation of this state-of-relationship-at-the-moment (herein called "related-

* Reprinted with permission from *Psychosomatic Medicine* 28:305–315, 1966.

ness") have tended to focus on the analysis of artificially isolated predetermined units in the communication process. Language behavior has been examined principally by means of psycholinguistic content analysis of speech samples [11], or by the use of analysis of the forms of speech interaction [6, 7]; the area of "interpersonal physiology" has been investigated using autonomic nervous system parameters recorded simultaneously from both therapist and patient [10, 11, 19]. Studies of body motion behavior in communication have been limited heretofore to clinical studies of the significance of gestures as modes of affect expression [12, 13], and only incidentally have interactional factors been alluded to. However, recent developments in the science of human communication have made possible a systematic, detailed investigation of the communication process in psychotherapy which bears directly on the concept of relatedness and permits its explicit delineation in behavioral terms. These recent developments are: (1) the conceptualization of human behavior in communication as a unitary process involving at least four modalities of the communicational structure: lexical, paralinguistic, kinesic, and visceral;* (2) the development of the research cinema film, magnetic sound recording, and video tape recording which permits systematic repeated viewing of rapidly occurring phenomena [30, 33]; and (3) the development of specific systems for accurate, reliable, repeatable notation of speech behavior [28] and body motion behavior [2, 3]. Birdwhistell [3], McQuown et al. [20], and Scheflen [22–27] have applied these recent developments to evolve a specific method of *context analysis* of psychotherapy films which is multilevel and systematic, with no presumption of predominance or relative importance of any one channel of the communication process. This method is aimed at the discovery of naturally occurring units and determines their significance by analyzing their respective contexts.

In the search for the kinesic manifestations of relatedness, repeated study of unedited films of therapy sessions reveals multiple instances in which the postures assumed by the patient and the therapist are similar in arrangement of corresponding body parts and concurrent in time. Scheflen's studies have called attention to the occurrence of these "congruent" postural configurations in the naturally occurring

* The nomenclature here is in a state of flux. Here we follow Brosin [4]; Birdwhistell [3] includes peripheral autonomic manifestations in the kinesic mode. There is also a question as to the place of posture in this schema; however, this diagram appears adequate for purposes of this study.

body motion behavior of the patient-therapist dyad, and he has described and defined a practical system of classification of these interactional kinesic events [22, 24]. This system stresses the interpretation of these configurations as manifestations of the behavior of the dyad as a unit, rather than as individuals reacting or coacting with each other. As an extension of his work, this paper is a report of the results of an investigation of two concurrent modalities of patient-therapist dyad behavior recorded in a film of a psychotherapy session: (1) the program of postural configuration and (2) the concomitant program of lexical behavior as manifested in discourse.

Hypothesis

It is the hypothesis of this study that the patterns of postural relationships are specific behavioral indicators of the moment-to-moment relationship operating within the dyad. At the organismic level, any individual's postural behavior is the resultant of multiple simultaneous factors, including constitutional predispositions, learned patterns, cultural influences, states of health, occupation, and sexual identity. However, the postural configurations which actually occur are the behavior of the dyad itself; it is the immediate state of relatedness which is the *decisive* determinant of these configurations.* The other factors are viewed as "baseline" factors [3], important principally in limiting the repertory of postures from which specific ones are evoked by the dyadic communication process. It is most likely that in the sitting position this repertory is limited to a set of four to six fairly standard postures which have been learned as appropriate to given situations or contexts, especially those involving reciprocal role relationships (e.g., employer-employee, doctor-patient, therapist-client, etc.). As Scheflen has pointed out [22, 27], encounters in a psychotherapeutic context are programmed with an initial calibration period followed by a series of phases, each marked off by a postural shift or configurational rearrangement or both. It is the relationship of these shifts that leads to the congruences and non-congruences operative within the dyad.

* This is a "levels" problem. Thus in a dyad (A ⇄ B), the "state" of A, that is, the state of A's anatomy, physiology, health, self-attitude, etc., limits what A can do. But what the dyad does is a matter of analysis at the *social* level of organization, rather than at the organismic level [27].

In the course of the analysis of this film, it was necessary to assume that the space which patient and therapist occupy is bounded by what is visible and audible in the film. The possibility is not precluded that some of the postures assumed by the subjects may have occurred in relationship to others, either others immediate upon the scene, such as the cameraman or director, or the "whole host of others" referred to by Sullivan [32]; however, for the purposes of this investigation, these variables like the other mentioned above are seen as limiting ones rather than selective ones.*

Method

FILM

The sound film used in this study was made in the office of the therapist, an experienced psychologist, during a regularly scheduled psychotherapy session (the thirty-seventh) with a 27-year-old married white female who had agreed to the filming procedure about two weeks previously. The film is an unedited black-and-white 16-mm. sound film of 33 minutes, 17 seconds duration, photographed at 24 frames per second with the camera fixed in position and both participants "on stage" for the entire time except for a 34.1-second interval during the twenty-third minute.

The "action" in the film consists of therapist and patient, seated vis-à-vis, engaging in a therapy session. The discourse consists primarily of a set of utterances by the patient; she speaks first and also last. The therapist makes only six sentence-length utterances and multiple brief vocalizations.

EQUIPMENT AND TECHNIQUE

Two types of 16-mm. motion picture projectors were used. One was a Filmsound Projector,† Model 302, modified by the addition of remote control and by rewiring of the amplifier so that sound is available at 16 frames per second speed ("silent" speed) as well as at the conventional 24 frames per second. The other projector was a Time and Motion Analyzer,† Model 173 BD, with hand crank, frame counter, heat filter, and calibrated speed control which allows frame-by-frame viewing at full brightness.

* The problem of the interference with a naturally occurring event by the recording process is dealt with extensively by Brosin [5].

† Bell and Howell Company, Chicago, Ill.

Magnetic tapes reproduced directly from the sound track of the film run through the Filmsound projector at two-thirds speed are used for making the transcript; this speed facilitates tape listening and transcribing. The films used are frame-numbered so as to facilitate accurate recording, and are treated to increase wear resistance.

PROCEDURE

After multiple viewings of the film as a whole, the patient's initial utterance to the therapist was examined in detail. This utterance was found to be preceded by a shift by the patient into the vis-à-vis position with the therapist; at the onset of *her* vocalizations, *he* makes a postural shift which brings him into a postural configuration which is the mirror image of her posture; it is maintained for 4.3 seconds; as she terminates this utterance, she moves her head to her left, terminating the congruent postural configuration. Further repeated viewings with the concept of congruence in mind led to the discovery that "mirror posture" occurs with four of the five sentence-length utterances of the therapist while he and the patient are face to face; in each case the mirroring is initated by the therapist and terminated by the patient. Each occurrence is preceded by a statement from her requesting some verbal response. Prior to each of these statements, he takes his pipe from his mouth or moves his lips against each other, or both, moves his head anteriorly, and nods. During each of these four utterances he shifts his trunk and hips; he also in each instance has his legs crossed and makes a to-and-fro movement (foot sweep) with the elevated foot.

The fifth of his vis-à-vis utterances differs in many ways: it is not accompanied by mirror posture, it is not preceded by a request for response, and there is an absence of the movements which precede, and occur with, his speaking the other four utterances. The discovery of this regularity of interactive sequence suggested that postural configurations throughout the film might serve as indicators of some kind of "coupling" or linkage between the therapist and the patient, and that these linkages would be regularly associated with certain kinds of lexical items, or with certain content themes. These postural configurations, occurring at the initiation of the interview, during the utterances of the therapist, and also at the termination of the interview, appeared to have occurred at significant moments in the course of the therapy session; out of this initial set of observations came the hypothesis that these configurations may be manifestations of relatedness. In order to test this hypothesis, the following procedure was used:

FRAME-NUMBERING THE TRANSCRIPT. An accurate lexical transcript was made from the tape of the sound track run at two-thirds speed. This transcript was then frame-numbered; the numbers of the frames in which each word occurs were obtained with the modified projector running, with sound, at two-thirds speed. With the remote control and clutch, such can be recorded to the limits of ±3-frame accuracy (⅛ second).

CONFIGURATIONAL ANALYSIS. With the Time and Motion Analyzer, the film was studied frame-by-frame for three separate run-throughs. Each postural configuration in each frame was judged to be either "mirror" congruent, "identical" congruent, or "non"-congruent, separately for upper body and lower body.* These judgments were made according to the following criteria:

1. The mirror posture category ("mirroring") is composed of all configurations in which each person's posture behavior was the mirror image of that of the other; in the face-to-face position, laterality is reversed, so that one person's left side is equivalent to the other person's right.

2. The identical posture category ("matching") is composed of all configurations in which each person's posture matches, or is identical with, that of the other; here, right matches right and left, left.

3. To satisfy the criteria for upper body congruence, a given postural configuration must include an arrangement in which the head, shoulders, and trunk of each of the subjects was in a congruent relation to the corresponding body parts of the other, and held that way for at least 0.4 second (10 frames). A maximum angular deviation of 10° rotation from exact correspondence is allowable for each body part.

4. To satisfy the criteria for lower body congruence, a given postural configuration must include an arrangement of congruent posi-

* Scheflen [24] has found, with regard to postural interactions, that the body tends to be split either upper-lower or right-left, and that configurations involving half the body appear to have interactive significance; in the vis-à-vis, the split is usually upper-lower.

tioning within 10° maximum angular deviation of the hips and lower limbs of each of the subjects, and must be held in that position for at least 0.4 second (10 frames).

5. To meet the criteria for congruent movement, the body motion had to start from a congruent posture, be synchronous and congruent for its duration, and terminate in a congruent posture of the same kind. A maximum time lapse of 0.4 second between the start, or finish, of one person's move and that of the other was allowed.

The 10-frame or 0.4-second minimum duration required for an event to meet the criteria for postural congruence was originally arrived at empirically, for it was noted that any briefer congruent events were too difficult to distinguish from transient or momentary pauses in a movement (in a 24-frame-per-second film). The concept of posture involves the idea of some body position held over time; positions held for less than 0.4 second did not fit this idea of what posture is. Interactional events shorter than 0.4-second duration are better dealt with at a micro level [9], and body movements of less than 0.4-second duration appear to have greater significance at the intrapersonal level [17, 18].

Analysis of the Data. Three types of data analysis were performed. (1) The patterning of the postural configurations was charted so that the distribution of the congruent periods might be visualized throughout the film. (2) The lexical transcript was examined to determine pattern of content. (3) The lexical items concurrent with the various categories of postural configuration were extracted from the transcript and subjected to semological analysis.* Four semological categories were examined in detail and analyzed statistically; they are:

1. *Subjectives.* All members of the class of lexical items referring to subjects of verb forms. These were divided into first person pronoun (or its equivalent "you"); other person, object, or referenced pronoun; and nonreferenced nouns and pronouns.

* Semology is "the study of the structuring of discourse itself, the structure of multiple, systematically patterned, semological occurrences," so that semological analysis consists of determining "the patterning of distributions within the occurrence, and the patterned distribution of occurrences within discourse" [29]. The specific method of semological analysis is described by Sarles [21].

2. *Verbals.* All members of the class of lexical items referring to verbs. These were divided into action verb forms, reflexive verb forms, state-of-being verb forms, and negation forms of each.

3. *Locatives.* All members of the class of lexical items referring to place or location. These were divided into definite references and indefinite references.

4. *Temporals.* All members of the class of lexical items referring to time, either time of occurrence or time duration. These were divided into definite references and indefinite references.

Findings

DISTRIBUTION OF POSTURAL CONFIGURATIONS

The upper body identical congruent postures were examined as outlined above. They were few in number and of exceedingly brief duration (mean duration, 1.3 + 0.7 seconds), and do not appear to be of sufficient significance to detail. Therefore, as used below, the term "congruent" will refer to mirror congruent configurations exclusively.

There were significant differences in the durations of the individual configurational events and in the time taken up in each segment by the configurational events. As the film progresses, the duration of the individual upper body congruent events tends to increase; conversely

TABLE 13-1
MEAN DURATION OF POSTURAL EVENTS
(IN SECONDS)

Successive Time Periods[a]	Congruent Configuration	Noncongruent Configuration
I	2.4± 1.1	24.0±27.8
II	1.9± 1.5	12.3±18.9
III	2.7± 2.4	7.0± 8.0
IV	3.5± 4.1	10.3±12.4
V	6.8± 6.9	15.7±12.2
VI	5.3± 7.3	9.8±12.3
VII	8.1± 7.4	8.7±11.8
VIII	11.1±12.2	8.7±10.8
TOTAL	5.4± 6.7	11.0±14.6

$\chi^2 = 38.48$; $p < 0.01$.
[a] Roman numerals represent successive time periods of approximately 6000-frame (250-sec.) duration; slight deviations from this figure were necessary to keep individual postural events intact.

the duration of the noncongruent events tends to decrease; this trend is statistically significant (Table 13-1).

TABLE 13-2
TOTAL TIME OF POSTURAL EVENTS
(IN SECONDS)

Successive Time Periods[a]	Congruent Configuration	Noncongruent Configuration	Totals
I	14.3	244.7	259.0
II	22.3	202.5	224.8
III	56.4	205.6	263.0
IV	66.9	185.8	252.7
V	61.6	190.9	252.5
VI	84.0	165.6	249.6
VII	105.9	143.1	249.0
VIII	144.4	104.6	249.0
TOTAL	555.8	1442.8	1998.6

$\chi^2 = 304.95$; $p < 0.01$.
[a] For explanation of time periods, see Table 13-1 footnote.

Similarly, as the film progresses, more and more time of each successive segment is taken up in congruent posture, and less and less in noncongruent posture. The pertinent distributional differences between the congruent events and the noncongruent events are sum-

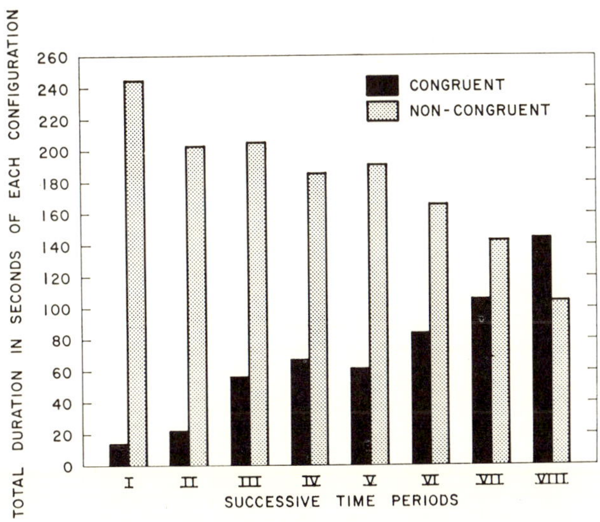

FIGURE 13-1. DISTRIBUTION OF POSTURAL CONFIGURATION EVENTS.

marized in Table 13–2 and shown in Figure 13–1. Illustrated is the significant trend toward an increasing amount of time spent by the participants in upper body congruent posture, and a decreasing amount of time spent in noncongruent posture, as the interview progresses ($p < 0.01$).

Determination of the lower body configurations proved to be unreliable. The table between the camera and the therapist blocked the view of his hips, lower trunk, and thighs sufficiently so that judgments about their position could not always be made with any degree of reliability. It is doubtful that the patient herself could see much, if any, of the therapist's lower body. For this reason, data pertaining to lower body postural configurations will not be included in this report.

SEMOLOGICAL ANALYSIS OF LEXICAL CORRELATES

Table 13–3 summarizes the numerical data pertaining to the distribution and frequency of the four semological classes of lexical items chosen for analysis of their correlation with the postural configurations. It can be seen that the congruent periods were correlated with lexical content referring to other persons as well as to the speaker; to positive statements of action, state of being, thought, and feeling; and to specific references to time and place. The noncongruent periods are significantly ($p < 0.05$) different: There was greater use of the first person and much less use of references to others; there was greater use of negation statements pertaining to action, thought, and feeling; and there was greater use of nonspecific temporal and locative references.

ANALYSIS OF CONTENT THEMES

The lexical content concurrent with the postural configurations was culled from the transcript and examined for thematic content. In the examples below, the figure before the decimal refers to the reel number; the figures after the decimal refer to the frame numbers in that reel. Material coincident with the noncongruent periods dealt principally with a description of the emotionally arousing events of the previous 24-hour period. Although numerous other persons entered into the scene during this period, the descriptions were highly self-oriented, often self-contradictory, and frequently negational or nonreferenced. As an example:

TABLE 13-3
SEMOLOGICAL ANALYSIS OF LEXICAL CORRELATES OF POSTURAL CONFIGURATIONS

References	Congruent Configuration	Noncongruent Configuration	Total	χ^2	p
		SUBJECTIVE			
First person	158	386	544	7.23	<0.01
Other person-object	61	75	136		
Nonreferenced	29	68	97		
Distribution				13.35	<0.01
TOTAL	248	529	777		
		VERBAL			
Action	43	88	131		
Negation of action	13	42	55		
Reflexive	34	91	125		
Negation of reflexive	17	70	87		
State of being	100	181	281		
Negation of state of being	27	55	82		
Distribution				11.27	<0.05
Total negation				4.29	<0.05
TOTAL	234	527	761		
		TEMPORAL-LOCATIVE			
Referenced temporals	29	47	76		
Nonreferenced temporals	5	20	25		
Referenced locatives	11	19	30		
Nonreferenced locatives	1	12	13		
Distribution[a]				6.58	<0.10
Referenced vs. nonreferenced				5.93	<0.02
TOTAL	46	98	144		

[a] Not significant.

1.2320–2890

I just I—I kept trying not to think because I didn't know what I would do if I was gonna think I just felt like I was—I didn't feel I could even safely—uh—cry because I didn't know where I—or what I could do n' Billie was there I was there all alone with 'em I just—I just didn't—uh—I was waiting for today I think of all the days not to get gypped out of about half of it—this is rough.

There was a relative paucity of references to others, a high degree of ambiguity, and infrequent references to therapy or to the therapeutic relationship.

By way of contrast, the content themes coincident with the upper body mirror congruent periods tended to be more focused, direct statements. Such descriptions of "day-residue" events as there were were clear, positively stated, and relatively comprehensible. As an example:

1.14710–14993

. . . [a neighbor woman] has been pretty nice, she came, she knew I was upset last night, she came over and stayed for a while un an' put together a jig saw puzzle 'n stuff with me . . .

These content themes were principally centered upon three areas: critical commentary on events of the previous day, statements of realization or awareness of the crucial tasks at hand, and statements engaging the therapist directly or indirectly in quest for relief of distress. There were specific references to others as well as to self, especially to the therapist and the relationship to him. Negational, nonspecific, or nonreferenced statements were infrequent.

Discussion

The material presented here represents a pilot effort to determine the pattern of motor sequences of interactive significance. Based on the assumption that communicational behavior is a continuous multichannel process which involves information-exchange transactions in a context of definitive relationship patterns, the search for naturally occurring interactive units becomes crucial to the discovery of the significance of any given piece of linguistic-kinesic activity.

One such pattern of motor (behavioral) sequences which is sufficiently recurrent and specific to be considered a naturally occurring interactive unit is the upper-body mirror congruent posture. These configurations were not randomly distributed in this filmed interview; as the interview progressed, more and more of the time was taken up in mirror congruent posture, and the events tended toward greater and greater duration. Concomitant with this trend are the patterns of discourse structure and content themes. As noted above, the subjective references tended to be specific references to other persons as well as to the speaker, that is, more object-oriented or interpersonal than at other times during the film; the verb forms are notably positively structured, and the time-space references notably specific. These findings, especially when contrasted with the more self-centered, negational, nonspecific qualities of the content during the noncongruent periods, would be in keeping with the ideas presented above concerning the concept of relatedness as a description of the moment-to-moment relationship within the patient-therapist dyad. As Cherry states, "Communication means a sharing of elements of behavior, or modes of life, by the existence of sets of rules" [8]; it seems probable that the behavioral manifestations of this sharing process indicate the state of relatedness of the communicants. The findings would appear to lend support to the hypothesis that upper-body mirror congruent posture functions as a signal system correlated with signal systems at the vocal level which serve as relatedness indicators. Such is not to exclude the possibility of other kinds or levels of relatedness indicators, but rather to point out that external, objectively recordable behavioral rapport systems are available for detailed study. It is clear at any rate that the semological structure of this particular psychotherapeutic interview is significantly related to the configurational structure of the postural relationship occurring between the patient and her therapist.

The thematic correlates of mirroring also point in this direction. The concomitant content is decidedly interpersonal; the comments on personal experiences are discussed in accordance with the awareness of a necessity to examine, with the therapist, what has happened and what reactions were produced by the happenings. It may be speculated that the physical fact of mirror image posture is an element in the total set which facilitates this self-examination in the presence of the therapist, a characteristic of successful therapy.

Further support for this connection is seen in an examination of the postural concomitants of the therapist's utterances throughout the film, which tend to occur in the context of a common patterning of motor behavior. The lexical-kinesic pattern can be readily specified in four of the utterances. A fifth utterance is uniquely in contrast, in that it lacks the preceding request for response, the accompanying mirror congruence, and the simultaneous body shift and foot sweep of the other four; with regard to content, it lacks either the interpretive quality or the supportive quality of the other four. It would appear that the other four statements are "packaged" in a way that the fifth is not, once again pointing out the importance of the relatedness function that mirror congruence plays in interactional events.

The hypothesis of this study was that the patterns of postural relationship as manifested in this psychotherapy film are specific behavioral indicators of the moment-to-moment relationship operating within the patient-therapist dyad. The findings tend to lend support to this hypothesis, specifically pointing to the function of naturally occurring interactive units, described above, as facilitating or potentiating therapeutic effectiveness. These data appear sufficiently encouraging to indicate that further, more complete configurational analysis of other films of psychotherapy might yield significant findings relative to the study of communicational interaction in psychotherapy. The data also point up the value of searching for the naturally occurring interactive unit and then determining its significance by analysis of the larger context in which it occurs.

Recent work on other films would indicate that noncongruent postural configurations can be divided further, into subtypes. Similarly, in this study, units of different size (i.e., time duration) and direction (i.e., open, closed, forward, backward) were put into only one of two categories. We now recognize that although the correlations with structural and thematic lexical classes were significant, a finer analysis of the subcategories of postural behavior is necessary for a richer understanding of the complexities involved in communicational behavior. These points are currently under active investigation and will be reported on at a later date.

Summary

Previous investigations of the psychotherapeutic process aimed at the elucidation of the state of the patient-therapist relationship moment-to-moment ("relatedness") have not paid sufficient attention to the body motion behavior of patient and therapist; specifically the postural patterns of the patient-therapist unit ("dyad") have not heretofore undergone specific, detailed investigation. With the development of a method of context analysis of psychotherapy films, it has become possible now to fill in this gap in knowledge of the communication process in psychotherapy.

It is hypothesized that the pattern of postural relationships as manifested in a psychotherapy film is a specific indicator of the relatedness of the participants. With the use of a 16-mm. sound-film projector with a remote control device, a 16-mm. time and motion analyzer, and a tape recorder, each of the postural configurations assumed by the patient and the therapist were recorded from a single 33-minute, 17-second psychotherapy film. These configurations could be classed as mirror congruent, identical congruent, and noncongruent, for upper and lower body separately. The vocal content of the interview was subjected to structural analysis ("semological" analysis) and analysis of content themes, and then correlated with the postural configurational contexts.

Significant findings were:

1. As the interview progressed, the trend was toward more and more time spent in upper-body mirror congruent posture.

2. The lexical content of these mirror congruent periods was notably interpersonally oriented, positive, and specific; in contrast, the noncongruent periods were marked by a greater frequency of self-centered, negational, nonspecific lexical references. The content themes coincident with mirroring were principally centered upon critical awareness of the significance of events in the patient's previous day and attempts to effect therapeutic gain by examining the patient's reactions to them. In contrast, the content themes of the noncongruent periods were highly self-oriented, self-contradictory, and frequently negational or nonspecific.

It was concluded on the basis of the differential distribution of the postural configurations, and their significant correlations with the various classes of lexical references and content themes, that upper-body mirror congruent posture in the vis-à-vis position in psychotherapy is a naturally occurring interactive unit indicative of a state of therapeutic rapport or relatedness. A second conclusion (although not secondary) is the importance of the search for the natural unit which finds significance only in relation to its context. Work on other films of psychotherapy for further testing of this hypothesis appears to be indicated.

Acknowledgments

The film was made available for study by the permission of the producer, Gregory Bateson.

The author wishes to express his appreciation to Henry W. Brosin, M.D., and Jack A. Wolford, M.D., for their support of this project and also to his colleagues Felix F. Loeb, M.D., William S. Condon, Ph.D., and Harvey B. Sarles, Ph.D.

References

1. Alexander, F., and French, T. M. *Psychoanalytic Therapy: Principles and Application.* New York: Ronald, 1946.
2. Birdwhistell, R. I. *Introduction to Kinesics.* Louisville, Ky.: University of Louisville, 1952.
3. Birdwhistell, R. L. Body Motion. In N. A. McQuown (Ed.), *The Natural History of an Interview.* New York: Grune & Stratton. (In press.)
4. Brosin, H. W. Studies in human communication in clinical settings using sound film and tape. *Wisconsin Med. J.* 63:503, 1964.
5. Brosin, H. W. Implications for Psychiatry. In N. A. McQuown (Ed.), *The Natural History of an Interview.* New York: Grune & Stratton. (In press.)
6. Chapple, E. D., Chapple, M. F., and Repp, J. A. Behavioral definitions of personality and temperament characteristics. *Hum. Organiz.* 13:34, 1954.
7. Chapple, E. D., and Erensburg, C. M. Measuring human relations; an introduction to the study of the interaction of individuals. *Genet. Psychol. Monogr.* 23:3, 1940.
8. Cherry, C. *On Human Communication.* New York: Science Editions, 1961.
9. Condon, W. S. Synchrony units in communication. Unpublished observations.
10. Demascio, A., Boyd, R. W., and Greenblatt, M. Physiological correlates of tension and antagonism during psychotherapy. A study of "interpersonal physiology." *Psychosom. Med.* 19:99, 1957.
11. Demascio, A., Boyd, R. W., Greenblatt, M., and Solomon, H. C. The psychiatric interview: A sociophysiological study. *Dis. Nerv. Syst.* 16:2, 1955.

12. Deutsch, F. Analytic posturology. *Psychoanal. Quart.* 21:196, 1952.
13. Feldman, S. S. *Mannerisms of Speech and Gesture in Everyday Life.* New York: International Universities Press, 1959.
14. Freud, S. On Psychotherapy. In *Collected Papers,* Vol. 1. London: Hogarth, 1946. P. 249.
15. Fromm-Reichmann, F. *Principles of Intensive Psychotherapy.* Chicago: University of Chicago Press, 1950.
16. Gottschalk, L. A. (Ed.) *Comparative Psycholinguistic Analysis of Two Psychotherapeutic Interviews.* New York: International Universities Press, 1961.
17. Loeb, F. F. Lexical-kinesic correlations. Unpublished observations.
18. Loeb, F. F. The fist. In preparation.
19. Malmo, R. B., Boag, T. J., and Smith, A. A. Physiological study of personal interaction. *Psychosom. Med.* 19:105, 1957.
20. McQuown, N. A. (Ed.) *The Natural History of an Interview.* New York: Grune & Stratton. (In press.)
21. Sarles, H. B. The question-response system in language. Unpublished observations.
22. Scheflen, A. E. Communication and regulation in psychotherapy. *Psychiatry* 26:126, 1963.
23. Scheflen, A. E. On the nature of human communication. Presented to a panel of the American Group Psychotherapy Association, January, 1964.
24. Scheflen, A. E. The significance of posture in communication systems. *Psychiatry* 27:316, 1964.
25. Scheflen, A. E. Natural History Method in Psychotherapy. In L. Gottschalk and A. Auerbach (Eds.), *Methods of Research in Psychotherapy.* New York: Appleton, 1966.
26. Scheflen, A. E. Quasi-courtship behavior in psychotherapy. *Psychiatry* 28:245, 1965.
27. Scheflen, A. E. *Stream and Structure in Communicational Behavior.* Philadelphia: Mental Health Research Foundation, 1965.
28. Smith, H. L., Jr., and Trager, G. L. *An Outline of English Structure.* Washington, D.C.: Amer. Council of Learned Societies, 1956.
29. Smith, H. L., Jr. Linguistics: A Modern View of Language. In L. Bryson (Ed.), *An Outline of Man's Knowledge of the Modern World.* New York: McGraw-Hill, 1960. P. 364.
30. Sorenson, E. R., and Goldusek, D. C. Investigation of nonrecurrent phenomena; the research cinema film. *Nature* (London) 200:112, 1963.
31. Stevenson, I. The Psychiatric Interview. In S. Arieti (Ed.), *American Handbook of Psychiatry.* New York: Basic Books, 1959. P. 200.

32. Sullivan, H. S. *The Psychiatric Interview*. New York: Norton, 1954.
33. van Vlack, J. D. The Research Cinematographer. In L. Gottschalk and A. Auerbach (Eds.), *Methods of Research in Psychotherapy*. New York: Appleton, 1966.
34. Wolberg, L. R. *The Technique of Psychotherapy*. New York: Grune & Stratton, 1954.

14

A General Systems Approach to Problems in Growth and Development*

NICHOLAS D. RIZZO, ED.D., M.D., WILLIAM GRAY, M.D., AND JULIAN S. KAISER, M.D.

IN AN AGE when the behavioral sciences are searching for theoretical models, general systems theory [4, 8] offers a new method of conceptualizing psychiatry in its broadest aspects. The method will enrich the working hypotheses from which scientific principles and corollaries are developed. General systems theory, originally proposed by von Bertalanffy [9] thirty years ago, has had a revolutionary impact by expanding the scientific frame of reference to include important areas of study in those behavioral and biological fields previously not amenable to such inquiry. The expansion of this frame of reference is illustrated by a number of central features of general systems theory.

Central Features

Foremost among these features is the concept of an open system, the model of which has required an expansion of the second law of thermodynamics, which had been previously applied only to closed energy systems. Thus it is seen that growth and development occur in open systems by the import of "negative entropy." In a similar fashion, information import must be negentropic in order for education to occur.

* Presented at the 123rd annual meeting of the American Psychiatric Association, Detroit, Mich., May 8–12, 1967.

Another central feature of general systems theory is the concept of equifinality as a fundamental property, meaning that equilibrial positions are not dependent upon initial states but are determined by the specific open system parameters, involving import, export, and transactions going on within the system itself. A third feature is the steady state concept which is a fundamental property of all open systems and implies that the equilibriums reached involve the maintenance of states of tension, in contrast to homeostasis which generally implies equilibriums in positions of rest. In addition, steady states are conceived to be multiple and to involve step functions. By way of clinical example, a neurosis could be defined as a seeking of instant homeostasis.

The concept of isomorphism is also fundamental in general systems theory because it makes possible advances toward the unification of science and allows for the conceptualization of how systems at different hierarchical levels can develop harmonious transactions. Another fundamental feature is the emphasis general systems theory places on spontaneous activity as a necessary property of all open systems, a concept which has great meaning for the behavioral sciences, particularly psychiatry, since it rejects the stimulus-response model as an inadequate explanation for living systems. Stimulus-response features are considered to be secondary, added onto the primarily spontaneously active open systems. The properties of self-regulation, self-direction, self-organization, and self-differentiation, previously considered as teleological and beyond the reach of science, are now seen as necessary properties of open systems, which develop as the degree of organization and complexity of the living system increases.

Finally we include organismic theory because it introduces notions that have not appeared in conventional physics and which are characteristic of organizations, whether of a living organism or of a society. Organismic theory introduces the concepts of wholeness, growth, differentiation, integration, hierarchical order, dominance, control, competition, centralization, leading part, equifinality, and feedback. A historical note of interest is a school of psychology founded on organismic theory which was pioneered in this country by R. H. Wheeler forty years ago [10–12]. He formulated laws or principles of human nature which removed the limiting effect of the mechanistic point of view and developed a more unitary theory appropriate to living organisms. Subsequent developments in general systems theory have repeatedly confirmed the validity of Wheeler's early work.

In education a general systems theory approach to growth and development recognizes decision procedures and attitudes that will maximize the growth and development not only of the particular system of interest but also of relevant ecological systems. It stresses concentration of attention on the "spontaneously active" nature of living systems, in addition to the concern for their reactivity. Finally, it develops an ability to see the common operational features in the various systems of relevance in terms of their capacity for adaptation, control, and change. As we apply this novel approach it has been necessary to study and observe the students experiencing difficulties in growth and development as well as the ecological systems of reference which include a highly successful school approaching its two-hundredth anniversary, a New England town founded more than three hundred years ago, and a medical department housed in an accredited fifty-bed infirmary-hospital composed of a full-time medical director and a part-time staff including a psychiatrist, a psychologist, several counselors, and twenty medical and surgical consultants. Other ecological systems of reference include the department of physical training, the school administration, the teaching faculty, housemasters, extracurricular activities, and the family systems of the students. Discussion of a general systems approach will be concerned with the dilemma of the underachiever, the problem of the disruptive, maladapted adolescent, and the relatively rare problem of the suicidal patient.

Phillips Academy and Its Student Body

The observations on which the present report is based were made at Phillips Academy [2, 3], a preparatory school for boys located in Andover, 25 miles north of Boston; the campus consists of many buildings, classrooms, laboratories, studios, and athletic fields. Phillips Academy has been highly successful, in a continuing way, in being able to change and yet remain stable in the face of the many radical changes that have occurred in the society and country in which it exists. By these observations it is implied that the academy must be understood from the viewpoint of organismic theory as a healthy institution which has grown, developed, and evolved successfully in its many years of existence. It is an important observation that its buildings, classrooms, laboratories, and athletic fields form an integrated structure and are meaningfully interrelated to one another.

The academy campus is located within the boundaries of a town whose population is 25,000, a setting which provides almost ideal boundary conditions for the development of an integrated and vital community with a real sense of identity. The students come from families representing all walks of life, diverse religious creeds and races, and, in a small percentage of the student body, remote nationalities. Selection procedures are such as to provide for a high probability that the students will be able to identify with and form a meaningful engagement with the Phillips Academy subculture, with appropriate allowance for necessary diversity. During the past twelve years, from 1955 to 1967, the enrollment grew from 800 to over 850 boys. Most of the students live in dormitories supervised by teachers whose homes are in the same building. Faculty selection and tenure are also consistent with the fundamental organismic and open-system properties of the academy.

THE SPECIAL COUNSELING SERVICE

During these twelve years a special counseling service [9] has been developed whose chief functions are the evaluation, diagnosis, management, and treatment of emotional problems arising in the student body. In addition to the medical service function, there has gradually emerged a pattern of working with all the interacting systems, keeping in mind the health, growth, and development needs of all these systems and of their integration into Phillips Academy. An advisory function was formed which is devoted to assisting the school authorities and faculty in redefining the academy's role in maintaining and expanding its unique position of excellence among secondary schools in America.

The student body in general can be described as bright, healthy, active, vigorous, imaginative, and always interesting. During these twelve years, two hundred boys were referred for consultation to the school psychiatrist. Perhaps six times as many were interviewed by the school psychologist and the medical director in the same interval of years for relatively minor problems. Certain basic findings have been summarized from these interviews.

TABLE 14-1
MEAN S.A.T. SCORES OF SENIOR CLASSES (PHILLIPS ACADEMY)

Class of	Verbal	Mathematical
1955	610	627
1956	587	633
1957	597	649
1958	603	661
1959	615	651
1960	639	674
1961	621	670
1962	632	680
1963	652	677
1964	639	690
1965	651	719
1966	646	672
1967	650	672
	Mean of standardizing population:	500
	Standard deviation:	100

Table 14–1 shows the mean values of the verbal and mathematical standard scores earned by thirteen successive senior classes. The Scholastic Aptitude Test is the most widely used index of its type in contemporary American colleges and universities. The test undoubtedly measures something important, but it measures neither what its ardent proponents claim nor as well as some of its adherents insist.

Tables 14–2 to 14–5 deal only with the boys whom the consulting psychiatrist interviewed.

TABLE 14-2
FREQUENCY OF INTERVIEWS OF 200 BOYS

Number of Visits	Number of Boys
1– 5	119
6–10	32
11–15	21
16–20	11
21–25	5
26–30	4
31–35	4
36–40	0
41–45	2
46–50	0
50 and over	2
Total:	200

A cutoff point, at two hundred case studies, was chosen for our immediate purpose. It can be seen that approximately 60 percent of the boys referred to the psychiatrist were seen in five or fewer interviews. An additional 16 percent were seen in twenty or fewer interviews, and approximately 24 percent were seen more than twenty times. Conclusions concerning psychological dynamics are therefore not as revealing as other aspects of our study.

TABLE 14-3
SOURCE OF REFERRAL OF 200 BOYS

Source	Number of Boys
Faculty and administration	140
Self	25
Family	17
Medical director	37
Other students	3
Multiple sources	22

Table 14–3 represents an attempt to specify how the boy was brought to the attention of the school psychiatrist. In about 70 percent of the group under review, the teachers, coaches, and advisors who come in daily, routine, and repeated contact with the boys suggested and initiated the referrals. Many boys referred to the Special Counseling Service, the number at present being well over one thousand since 1954, have been successfully dealt with or "screened" and were therefore not sent along to the school psychiatrist.

TABLE 14-4
REASONS FOR REFERRAL

Reason	Number of Boys
1. Poor marks	94
2. Disciplinary problems	56
3. Disorganized living habits	26
4. Withdrawn behavior	19
5. Social immaturity	17
6. Disruptive, destructive, impulsive behavior	16
7. Psychosomatic disorders	14
8. Sexual problems	14
9. Suicidal thoughts, gestures	13
10. Dislike of school, strong wish to leave school	5
11. Philosophic-religious problems	5

The reasons given by the referring person—the advisor, teacher, housemaster, or the boy himself—are listed in Table 14–4. In other words, Table 14-4 is a tabulation of the chief presenting complaints. We have tried to note the most important single reason for the referral, but this was not always possible. Maladaption has many manifestations, and during adolescent years changes in attitudes occur dramatically. Boys must produce acceptable academic work in order to remain in school. Poor marks or underachievement were the stated reasons in 47 percent of the referrals, and disciplinary problems accounted for an additional 28 percent.

TABLE 14-5
SEVERE EMOTIONAL DISTURBANCES

Disturbance	Number of Boys
1. Major disciplinary problems	10
2. Delusional thinking	9
3. Suicidal preoccupation, gestures	8
4. Sexual problems	8
5. Psychosomatic disorders	7
6. Depressive reaction	4
7. Chronic drinking	2
8. Impulsive, destructive behavior	2

Table 14–5 presents descriptive phases applying to the most seriously disturbed group of boys examined and evaluated during the twelve-year period. It was possible to help these youngsters sufficiently that all of them earned their high school diplomas at Phillips Academy except for two, who were graduated from other secondary schools.

Comments and Discussion

The underachiever and his difficulties have been recently discussed with thoroughness and competence by Benjamin Fine [1]. Simply stated, underachievement is a "learning gap," observed in a student whose potential is higher than his output. The terms are relative and the causes of underachievement varied, at times signaling the presence of deep-rooted organismic insufficiency. What has become increasingly apparent is that underachievement does not arise in a strongly deterministic fashion from the "initial state," that is, from very early childhood experiences, but more likely represents examples of equifinal and steady state phenomena. As such, underachievement is dependent upon the parameters of function of the significant systems involved.

The outlook, viewed from this approach, may be more optimistic, although the complexities of the systems involved remain immense. Some of our clinical experience has convinced us that frequently, by concentrating upon one or two factors, small changes can be amplified into large therapeutic progress. It should be kept in mind that one goal of the educational process is to help all learners along, through the unfamiliar trip among new and difficult intellectual and emotional experiences. It is well to remember that every generation starts at a zero level of educational attainment, with an open system full of its own potential for growth, but needing an experiential and informational environment in which it can, by negentropic processes, grow and develop.

By way of significant contrast, newborn organisms are equipped with metabolic and physiological models which change and become differentiated with minimal interference from the ecological systems of reference under ordinary circumstances. Underachievers lack educational and emotional models. If the learning materials have seemed too easy, the development of usable models may be unfortunately postponed. T. S. Eliot once was heard to remark, on a visit to Phillips Academy, that any boy who deals only in learning tasks which result in instant mastery never truly becomes educated.

Our contention is that the models acquired in physical training, drown-proofing, competitive sports, and more recently in the Outward Bound movement serve a function of incalculable benefit in maximizing total growth. Whereas true achievement in education can at times seem remote to the conscientious student, because he invariably finds out there is a lot more to learn, following physical training the results may appear rather suddenly. Some cogent observations on physical training and athletics have been made by Frederick Harrison [7], Director of Athletics at Phillips Academy for more than fifteen years. We are in agreement that physical activity provides release from daily tensions, a change of pace from the difficult, persistent demands of scholarship, and provides a different and friendly atmosphere characterized by less inhibited behavior.

A boy often discovers that he must learn to develop, coordinate, and control his body. In order to achieve maximal organismic growth one must also develop feeling tones and know experientially the meaning and impact of joy, sorrow, success, and failure, as well as the satisfaction which comes from the mastery of skills. A purely intellectual approach becomes too distant. There must be more total involvement, a more global participation of all organ systems, and it is here that physical training makes its greatest contribution.

Among the boys who have been unusually successful in athletic endeavors only an insignificant proportion developed either severe emotional disorders or chronic, persistent disruptive behavior requiring disciplinary action by the academy authorities.

Disciplinary problems, social immaturity, disorganized habits of living, and chronic disruptive behavior are encountered with relative frequency among the boys who are referred to the Special Counseling Service, but it must be kept in mind that the total number of boys referred has rarely exceeded 3 percent of the student body. In other words, the procedures by which students have been selected over the years have been remarkably effective.

What can be said of the boys whose behavior is unacceptable? Are they suffering from severe emotional disturbances? We have chosen to make a positive psychiatric diagnosis only in the groups described in Table 14–5 as showing "delusional thinking" and "psychosomatic disorders." In the remaining six categories listed in Table 14–5 we feel that the diagnosis of Adjustment Reaction of Adolescence [6], though applicable, needs further clarification. The boys whose behavior led to their inclusion in Table 14–5 were maladaptive, they "did not get the message," or they were deficient in conformally patterning to the demands of the total academy environment, which apparently is unable to reach or influence every one of its students. Stated in another way, Phillips Academy is not to be viewed as a high school with national geographic representation. A great deal more is expected of each boy admitted, but there is an enormous amount of help available if the boy is willing to accept it and be influenced by it. Stated in general systems terms the individual boy needs to develop the capacity for relatively big negentropic experiences. We view all the examples of problems under review as incident to growth and development, and therefore deal with them accordingly. A study of maladaptive behavior may also give some valuable clues as to the effectiveness of the various systems in their operations.

Suicidal gestures, attempts, and preoccupation, depressive reactions, and chronic withdrawn behavior have been observed in only 5 or 6 percent of the group of boys evaluated by the special counseling service. Two boys committed suicide after leaving the academy, one while attending college, the other after being granted a medical leave of absence in order to be near his family while undergoing more intensive medical treatment, which the academy was not able to provide. We view suicidal preoccupation as being symptomatic of maladaptation, which all the collaborating systems attempt to remedy through the guidance of growth and development. Naturally, all members of the special counseling service are alert to the ominous possibilities, but we make no special provisions to avoid or prevent suicide. In view of the high level of performance expected of each boy, it is inevitable that some youngsters will experience the same feelings of futility as their elders and may attempt to take their own lives. One boy, for example, had been the focus of attention, special favors, and continual praise from the moment of his birth. Everything had come easily to him all his life, including the number one position in his class, but at the academy he discovered that forty other boys had been the number one students in their respective classes. A series of systems breakdowns occurred in the boy, culminating in his swallowing fifty aspirin tablets. We feel there is no special contribution we can make at this time to unearthing hidden or unusual causes of suicidal behavior.

We do, however, wish to repeat a highly significant finding from one area under discussion. In all cases of what we refer to as "severe emotional disturbances" in Table 14–5, completion of secondary schooling was accomplished, either at Phillips Academy or elsewhere. In other words we do not have any high school dropouts in our series of two hundred boys, despite the presence of rather severe emotional disturbance.

Summary

In summary this is an initial report of our continuing effort to understand the functioning of a successful educational institution, through the use of general systems theory, as a metatheoretical grammar of integrated behavior. It is our conviction that this will reveal pertinent and far-reaching insights into the nature of growth and development. A parallel to the method we have employed here is seen in the application of linguistics to the study of language. It appears that many can use language without any knowledge of underlying principles. In a similar way, many can contribute appropriately and well to youth in the process of growth and development, but are unable to describe clearly what it is they do. Our future research will be focused upon the development of a clearer understanding of tradition as an open system, and of mapping out the leading parts in the educational process.

References

1. Fine, B. *The Underachiever.* New York: Dutton, 1967.
2. Fuess, C. M. *An Old New England School.* Boston: Houghton Mifflin, 1917.
3. Fuess, C. M. *Men of Andover.* New Haven: Yale University Press, 1928.
4. Gray, W., and Rizzo, N. D. History and Development of General Systems Theory. Chapter 1, this book.
5. Harrison, F. Personal communication, 1967.
6. *Mental Disorders, Diagnostic and Statistical Manual.* Washington, D.C.: American Psychiatric Association, 1952.
7. Rizzo, N. D., Clark, D. M., and Roebrig, A. K. Emotional Problems of Adolescent Boys. A Review of Nine Years' Experience of the Special Counseling Service at Phillips Academy, Andover, Mass. Presented at the Annual Convention of the American Personnel and Guidance Association, Boston, April 9, 1963.
8. von Bertalanffy, L. *Comments in Discussions on Child Development,* Vol. 4. New York: International Universities Press, 1960.
9. von Bertalanffy, L. General Systems Theory and Psychiatry. In S. Arieti (Ed.), *American Handbook of Psychiatry,* Vol. 3. New York: Basic Books, 1966.
10. Wheeler, R. H. *The Laws of Human Nature.* New York: Century, 1932.
11. Wheeler, R. H. *The Science of Psychology.* New York: Crowell, 1940.
12. Wheeler, R. H., and Perkins, F. T. *Principles of Mental Development.* New York: Crowell, 1932.

15

Toward an Integrative Conception of Mental Disorder*

JUDD MARMOR, M.D., AND EUGENE PUMPIAN-MINDLIN, M.D.

This article was conceived and written in 1946–1948, immediately following World War II; thus it reflects, in many of the clinical examples it cites, the recent military experience of its authors. It was, however, one of the earliest papers in American psychiatry to reflect an open-system approach to the whole problem of mental health, neurosis, psychopathy, and psychosis, conceiving of them as different points on a continuous spectrum of interaction between the individual and his "field." It recognized, moreover, that the "field" was not merely a spatial one but also a temporal one; and that qualitative as well as quantitative factors were involved.

It is, I believe, a tribute to its relatively advanced thesis that, if this article were being rewritten today, its essential conception would not have to be fundamentally altered. The discussions concerning schizophrenia and manic-depressive psychosis would have to take into consideration the newer knowledge concerning genetic and neurohumoral factors in these disorders; discussions of social and economic stresses would include more reference to the problems of race, cold war, urbanization, technological change, overpopulation, and the "culture of poverty"; discussion of psychological stress would take note also of concepts of developmental crises and problems of individuation and identity-formation; and, unhappily, the roster of potential physiological stresses would have to include the effects of nuclear radiation. Finally, of course, the conceptual framework *within which the thesis is presented would today incorporate specific references to information theory and systems theory.*

Judd Marmor

* Reprinted with permission from *The Journal of Nervous and Mental Disease* 3:19–29, 1950.

IN THE PAST TWO OR THREE DECADES there has been a significant, far-reaching change in the fundamental philosophy underlying medical thinking. Virchow's static pathology of individual cells and organs has given way to the study of the pathophysiology of the living organism and the interrelationship of organ systems. The purely mechanistic idea of disease being "caused" solely by a noxious agent has been abandoned. It has been replaced by the more dynamic concept of a functional interrelationship between the individual and a so-called pathogenic agent. Thus an organism which under certain conditions lives in a state of harmless equilibrium with its host may, under changed conditions, become lethal to him. The resistance of the host, the virulence of the organism, the portal of entry and the size of the invasion at any point in time constitute some of the variable factors which may determine whether or not a pathological process will develop. An important consequence of this conception of disease etiology has been a diminution of the tendency to "explain" diseases in terms of inheritance or constitution. The genetic factor is now recognized as but one element in a parallelogram of forces. Thus while we may recognize a constitutional factor in such conditions as diabetes, hypertension, or rheumatoid arthritis, we do not consider it per se as causative.

One of the most lucid presentations of this modern viewpoint on etiology has been that of Halliday [9], who states: "Illness is regarded not as a fault in the parts, but as a *reaction* or mode of behavior, or vital expression of a living unit in response to those forces which he encounters as he moves and grows in time. Cause is therefore *twofold* and is to be found in the nature of the individual and the nature of his environment at a particular point in time."

Implications for Psychiatry

What are the implications of this approach to the nature of illness when applied to the field of psychiatry? The whole conception of personality in a static sense has changed to one of a constant dynamic interplay between the individual and his environment. Events and situations may at certain times be traumatic, at others relatively innocuous. The receptivity of the organism for the stimulus has become as important as the stimulus itself. Stated broadly, this concept implies that all human behavior is the expression, at a psychophysiological level, of a dynamic interrelationship between the individual and his environment—a relationship which is in a constant state of motion and change. *Mental health, psychopathy, and psychosis can then be understood as expressions of varying quantitative aspects of this interrelationship, which at certain crucial levels result in qualitative changes. Moreover, this concept implies that these changes are potentially reversible.*

This thesis may be broadly schematized in the following diagram (Fig. 15–1), which endeavors to present in highly schematized form the interrelationship between the individual and his environment in the production of mental disease. The vertical line represents the nature of the individual. This is a dynamic complex made up of the dialectical interrelationship of the inherited biological constitution with the individual's life experiences up to that point in time (Table 15–1). Together they form a unit which constitutes the individual's personality structure at a given point in time. Depending upon the nature of these forces, this structure may have a high resistance to environmental stress (in which event we call it stable), or a low resistance to environmental stress (in which event we call it unstable). The horizontal line, on the other hand, represents the nature of the

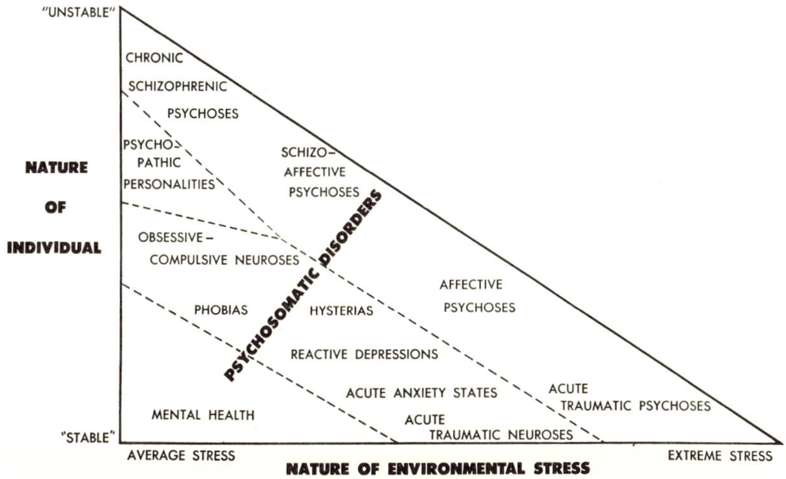

FIGURE 15-1. NATURE OF THE INDIVIDUAL: NATURE OF ENVIRONMENTAL STRESS.

TABLE 15-1
NATURE OF THE INDIVIDUAL ("PERSONALITY STRUCTURE")

Biological factors ("constitution")

1. Physical appearance
2. Endocrine balance
3. Intellectual endowment
4. Natural immunity
5. Sex
6. Age

Life experiences

1. Family relationships
2. Interpersonal relationships
3. Education
4. Social status
   a. Religion
   b. Race
   c. Class position
   d. National origin
5. Sexual development
6. Cultural factors

environment at any particular point in time. This may be broken down into component elements (Table 15–2), which together form

TABLE 15-2
NATURE OF THE ENVIRONMENTAL STRESS

1. Psychological stress; e.g., bereavement, loss of love
2. Physiological stress:
   a. Physical; e.g., heat, cold, fatigue, surgical procedures, bodily injury
   b. Chemical; e.g., drugs, poisons, alcohol, dietary deficiencies
   c. Bacteriological; e.g., infectious diseases
3. Sociological stress; e.g., cultural restrictions and taboos
4. Economic stress; e.g., unemployment, poverty

a unit constituting the degree of stress to which the individual is being subjected. This stress must be evaluated not only quantitatively, in terms of severity, but also temporally, in terms of duration, and qualitatively in terms of its special significance for the particular individual involved. Thus, for a soldier in combat, the quantitative factor is represented by the immediate severity of the battle conditions; the temporal factor, not only by the duration of the immediate combat situation, but also by the duration of previous exposure to combat conditions; and the qualitative factor by the specific conscious and unconscious significance of the total war situation to the particular soldier.

RELATIVE FACTORS

The relative nature of these factors is of the greatest importance. What is considered psychologically healthy in one era or culture may not be considered so in another. For example, the normal sexual behavior of an adolescent girl among the Marquesans or Trobrianders [12] would in our society be set down as constituting nymphomania or "psychopathic personality." Again, while we label homosexuality as pathological in our Western culture, among the Tanalans and Japanese [4, 12] it is accepted as a normal behavior variant and meets with no opprobrium. In our own Western culture, within the span of the last five hundred years, individuals who insist that they see visions, hear voices, and communicate directly with God have at varying times been revered as prophets, burned as witches, or hospitalized as psychotic. Moreover, even within the same broad cultural group there may be significant variations in standards of accepted behavior [14].

Again, what is traumatic for a single individual may not be traumatic in a group situation. The experience of the London blitz is a case in point. The number of neurotic casualties was far less than was anticipated owing to the fact that danger was experienced in a group setting. Further, a situation which is traumatic for one individual may be a source of gratification for another. It was not too infrequent for psychiatrists in World War II to see individuals who were previously maladjusted in civilian life make a good adjustment in the Army.

Examining Figure 15–1 we now see that mental health can be hypothecated as that state in the interrelationship of the individual and his environment in which the personality structure is relatively stable and the environmental stresses are within its absorptive capacity. Ideally, this implies socially adequate behavior in an individual who is consciously and unconsciously well integrated. If the level of the environmental stresses are increased beyond a certain point, however, neurotic symptoms will develop in the most stable individual, and in cases of extreme stress even psychotic symptoms may develop. Thus the first important point which is implicit in this conception is that there is no such thing as *absolute* immunity to mental disease. "Every man has his breaking point." This is borne out clinically in numerous ways. Mild anxiety states, manifesting themselves in insomnia, digestive disturbances, or headaches are universal under stress. Experiences with normal selected males under conditions of inordinate stress, such as those undergone by American soldiers at Guadalcanal, revealed that entire companies developed transitory acute traumatic syndromes of striking similarity [22]. Other psychiatrists have dramatically described the transitory psychotic states which previously well-adjusted soldiers may develop in the face of extreme danger or stress [8].

The one characteristic which all of these mental disorders developing in previously stable personalities have in common is their benignness. Rest, quiet, and relaxation—in other words, removal of the environmental stress—often result in rapid return to normal even in the absence of any direct psychotherapy [22].

SCALE OF DISORDER

When, on the other hand, the individual personality structure is unstable owing to unfavorable hereditary or early environmental influences, the amount of environmental stress necessary to produce abnormal symptoms is correspondingly less. Thus at one end of our scale we have our theoretically perfectly integrated individual whom only realistic threats of greatest severity and duration can succeed in unsettling; while at the other extreme we have individuals whose personality structure is so poorly equilibrated, so fraught with internal tensions and contradictions, that the simplest routine of everyday living is too much for it, and it gradually disintegrates in the form of an insidious schizophrenia. Between these two extremes there are infinite gradations. At the "stable" end of the scale the clinical picture is more likely to be colored by the "traumatic situation," as in the acute reactive anxiety states, reactive depressions, and the traumatic syndrome. At the "unstable" end of the scale, on the other hand, the traumatic situation itself tends to leave little mark upon the clinical picture, which is dominated by the individual's preexisting conflicts and personality structure. Between these two extremes one finds infinite variations and admixtures.

This conception has considerable usefulness in prognostication. One could anticipate, for example, in a group of soldiers, all manifesting similar neurotic or psychotic symptoms, that those whose past history reveals an unstable personality structure will have the poorest prognosis. Similarly, the more acute the onset of a syndrome and the greater the environmental stress with which it is associated, the better the prognosis is apt to be. These criteria have been found to be equally useful in prognosticating the long-term results of shock therapy in psychotics.

Another implication of this conception, which follows from our diagram (Fig. 15–1), is that an unbroken line of continuity exists from normal behavior through neurotic to psychotic behavior. This corresponds to the general principle in modern scientific thinking that "no boundary in nature is fixed and no category airtight," and that " 'mesoforms' are found at the transition point of one level of organization to the next" [18]. In the field of mental disorders such mesoforms are encountered in the frequent borderline cases in which one is hard put to make a definite commitment as to whether an individual is normal, neurotic, or psychotic. This does not mean, however, that there are no qualitative differences between these categories. As in other fields of matter, the quantitative differences at certain levels result in real qualitative changes—just as raising the temperature of water to a certain level results in its qualitative transformation into steam, or lowering its temperature transforms it to ice.

This in turn leads to another significant implication, namely, that it is possible for the same individual, under varying circumstances, to pass from normal to psychotic behavior and back again. Neurosis and psychosis should not be regarded as static and fixed entities. *They are dynamic and changeable states of behavior which are potentially reversible, the borders of which are often indistinct.* One of the residuals of mechanistic thinking in the field of psychiatry is the still commonly held conviction that a neurotic individual can never become psychotic, and vice versa. When the facts tend to disprove this theory, they are sometimes tailored to fit it by post hoc reasoning. For example, if a neurotic individual becomes psychotic, it is stated that he really had been a "latent psychotic" all along. Conversely, the psychotic who ends up with a picture of neurosis is said not to have really had a genuine psychosis, or else is considered as still a "latent psychotic." Used in this post hoc manner, latent psychosis obviously means nothing more than the possibility that an individual may become psychotic under certain conditions. The term thus actually becomes meaningless, for, as we have indicated, given the appropriate environmental conditions, any person may become psychotic. There are numerous examples in the literature of typical neuroses which under observation have changed into psychoses. Myerson [17], Miller [16], Caldwell [5], Zilboorg [26], and Jung [11], among others, have described such cases. In this regard it is also of interest to note that Rorschach records of psychotic patients often show the intermingled presence of neurotic elements [24].

INTERACTION IN HUMAN BEHAVIOR

Still another implication of this point of view is that, strictly speaking, there is no such thing as an "endogenous" mental illness. Human behavior, whether normal or abnormal, is always the resultant of an *interaction* between the nature of the individual and the nature of his environment. The fact that not infrequently a neurosis, psychopathy, or psychosis develops insidiously without any *apparent* outward contributory factor has led to a widespread conception that such conditions are of "endogenous" origin—that is, that they are either inherited or "constitutional," or else the result of certain changes which take place within the individual and which have no relationship whatever to environmental factors. Occasionally one encounters this belief as a result of a faulty understanding of the psychoanalytic formulation that mental illness arises out of conflict between the ego and the id, or between the ego and the superego. This is due to the mistaken conception that these designated aspects of personality are intrinsic within the individual and bear no relationship to the environment. Actually, of course, only the id represents the reservoir of biological impulses, and even the strength of id impulses can be affected by such environmental influences as stimulation, fatigue, malnutrition, disease, and physical castration. But the ego and superego are *always* resultants of interaction with environmental influences. Thus, no conflict which includes one or the other of them (as every conflict must) can be said to be independent of environmental influences. The term *cryptogenic,* as used in general medicine, would seem a more appropriate designation than *endogenous* because it implies that external factors, although present, are at the moment unknown.

Every psychiatrist has had the experience of having patients come to him with stories of neurotic or psychotic illnesses "coming on for no reason at all," only to discover on careful study of the patient's past history considerable material pointing to prolonged and severe conflicts with various factors in the environment. Zilboorg [26] and Schilder [21], among others, have described patients with schizophrenias of apparently insidious and nontraumatic origins in whom psychoanalytic study disclosed a wealth of environmental factors. Evidence is accumulating as a result of such studies that schizophrenias of insidious onset are the resultants of the exposure of susceptible individuals to severe and painfully crushing experiences very early in their ego development, probably in the first three or four years of their lives, with consequent serious crippling of their capacity to adapt to other people.* The schizophrenic withdraws from life because he has been deeply hurt—not because he is unable to relate to people but because he is *afraid* to. The therapeutic problem, as Fromm-Reichmann [7] has shown, is that of overcoming an overwhelming and overpowering distrust.

Similarly David Levy [15], Lauretta Bender [3], and others have demonstrated that the study of the early life of the so-called psychopathic personality reveals most often that these patients have suffered a deprivation of warmth and affection from maternal or maternal-surrogate sources in the first two or three years of their lives.† The result has been a distortion of the affective life of these patients.

With our increasing understanding of the dynamics of these conditions, they are gradually becoming accessible to various psychotherapeutic approaches [17, 25]. This indicates their essential reversibility and their accessibility to environmental influences.

Role of Constitutional Factors

Today it is fairly generally conceded that neurotic disorders arise in connection with various environmental stresses or frustrations in the course of individual development. But psychopathic personality, schizophrenia, and manic-depressive psychosis are still widely regarded as conditions which result primarily from hereditary or constitutional factors. Consequently, they are considered therapeutically more or less hopeless and preventable mainly by eugenic measures. In our

* The work of Spitz [23] is also suggestive in this regard.

† Another group of psychopathic personalities develops as the result of excessive parental indulgence unaccompanied by any form of discipline [15].

opinion this viewpoint is untenable, both on theoretical and clinical grounds. This is not to deny that constitutional (genetic) factors play a role in schizophrenia, manic-depressive psychosis, and psychopathic personality. Constitutional factors are operative in *all* human behavior, whether normal or abnormal. One has only to go through a ward of newborn infants to see that babies differ from one another biologically. But from the moment of birth on, all human behavior becomes an increasingly complex resultant of the interaction of these genetic factors with environmental forces. Abnormal behavior must always be evaluated in terms of this twofold aspect of causality. Constitutional factors do not operate directly in the production of psychological phenomena, but function as a dynamic substrate, being molded by, and in turn molding, the environmental impact. It is doubtful whether heredity ever contributes more than a greater or lesser propensity to mental disease, a sort of psychological threshold which is analogous to levels of so-called natural immunity in the field of general medicine. But like natural immunity, it is never absolute, and its relative importance depends on the quality, quantity, and timing of the environmental stress.

It will be noted further from Figure 15–1 that no attempt has been made sharply to "compartmentalize" or segregate the different descriptive types of neurotic and psychotic reactions. The more minutely one examines the mental reactions of neurotics, the more one finds that there are no sharp lines of demarcation between the various types. It is most rare, for example, to find a "pure" hysteria or "pure" obsessional neurosis. Most neurotics when studied intensively are found to be mixed types, with hysterical, obsessional, anxiety, and psychosomatic manifestations. Even the psychopathic personality is not always sharply demarcated from the neurotic. Mixed clinical pictures in which neurotic and psychopathic traits are intermingled are not unusual.

This is not intended to imply that there are no differences between the various types of neuroses and psychopathies. Obviously there are; the quantitative predominance of one or another group of symptoms at certain crucial levels leads to genuine qualitative differences. It is not that there are no differences, but rather that there are no sharp mechanical lines of differentiation between any of these varied mental disorders.

Nowhere are these interwoven dynamics more clearly exemplified than in the whole group of so-called psychosomatic disorders. These represent the exaggerated physiological response of the organism to prolonged psychological stress or tension. As such they may occur in all human beings, normal, neurotic, psychopathic, or psychotic. Most of the classical psychosomatic disorders, such as asthma, migraine, hypertension, ulcer, and colitis, are found on analysis to be accompanied by psychoneuroses and to have similar dynamic background and structure as the psychoneuroses. Why the same amount of anxiety or tension tends to provoke so much greater somatic reaction in some individuals than in others is still one of the important unsolved questions of medicine. It is likely that further elucidation of the physiology of the autonomic nervous system and the individual differences which exist in this regard among human beings may help to clarify this point. There is evidence that in some cases the choice of organ symptomatology may be psychologically determined [2].

In the field of the psychoses also, it may be worth considering whether the sharp differentiation hitherto made between schizophrenia and manic-depressive psychosis is warranted. Anyone who has ever worked in a state hospital will recall the countless hours spent in clinical conferences discussing whether a particular case should be placed in one or the other category.* Psychiatric literature is replete with investigations endeavoring to establish the differences between them. For example, Rachlin [19, 20] and Hoch [10], in a series of careful papers, demonstrate conclusively that many cases originally diagnosed as manic-depressive psychosis eventually end up as clear-cut schizophrenia. They do not, however, conclude from this that one condition may merge into the other, or that our descriptive categories are inadequate, but instead assume that wherever this has happened, the original diagnosis was incorrect. This kind of post hoc reasoning requires careful reevaluation. Is it not possible that the difficulty in differentiating the two conditions, and the occasional merging of one into the other, simply indicates that they are not the sharply demarcated nosological entities which they have always been assumed to be? This is further suggested by the existence of intermediate forms which the French have labeled "schizomanie," and which Kasanin [13] and Campbell [6] have reported as "schizo-affective psychosis."

* This is particularly true at the inception of psychotic illness. The sharply demarcated Kraepelinian categories are most often seen in the later stages of chronic psychoses.

Summary

The thesis of this paper is that human behavior is the expression of a dynamic interrelationship between the individual and his environment at a particular point in time and space; that mental health and the various deviations therefrom are but expressions of varying quantitative aspects of this relationship which at certain crucial levels results in qualitative changes. The implications of this conception are discussed, and some of the preliminary conclusions reached are: (1) No person is absolutely immune to mental illness, either neurotic or psychotic. (2) Neurosis, psychopathy, and psychosis are potentially reversible, except where actual destruction of brain tissue has occurred. (3) The various forms of mental disorder merge into one another by an infinite series of gradations, and there are no sharp lines of demarcation between them. (4) The same individual can, under different conditions, suffer from a neurotic illness or psychopathy at one time, and a psychosis at another. (5) Psychosomatic disorders may occur at varying levels of mental disorder. (6) There are no truly "endogenous" mental illnesses. (7) Constitutional or hereditary factors condition the development of so-called functional mental disorders, but never in themselves determine it.

References

1. Aichhorn, A. *Wayward Youth*. New York: Viking, 1945.
2. Alexander, F. Fundamental concepts of psychosomatic research. *Psychosom. Med.* 5:205, 1943.
3. Bender, L. Quoted in D. Levy, *Maternal Overprotection*. New York: Columbia University Press, 1943. P. 224.
4. Benedict, R. *The Chrysanthemum and the Sword*. Boston: Houghton Mifflin, 1946. P. 187.
5. Caldwell, J. M. Schizophrenic psychoses. *Amer. J. Psychiat.* 97:1061, 1941.
6. Campbell, C. M. Two cases illustrating the combination of affective and schizophrenic symptoms. *Amer. J. Psychiat.* 6:243, 1926.
7. Fromm-Reichmann, F. Transference problems in schizophrenics. *Psychoanal. Quart.* 8:412, 1939.

8. Grinker, R. R., and Spiegel, J. P. *Men Under Stress*. Philadelphia: Blakiston, 1945. P. 327.
9. Halliday, J. L. Principles of aetiology. *Brit. J. Med. Psychol.* 19:367, 1943.
10. Hoch, P., and Rachlin, H. L. An evaluation of manic-depressive psychosis in the light of follow-up studies. *Amer. J. Psychiat.* 97: 831, 1941.
11. Jung, C. G. Psychogenesis of schizophrenia. *J. Ment. Sci.* 85: 999, 1939.
12. Kardiner, A. *The Individual and His Society*. New York: Columbia University Press, 1939. Pp. 80, 168, 266.
13. Kasanin, J. The acute schizo-affective psychosis. *Amer. J. Psychiat.* 91:97, 1933.
14. Kinsey, A. C., Pomeroy, W. B., and Martin, C. E. *Sexual Behavior in the Human Male*. Philadelphia: Saunders, 1948.
15. Levy, D. *Maternal Overprotection*. New York: Columbia University Press, 1943.
16. Miller, W. R. Relationship between early schizophrenias and the neuroses. *Amer. J. Psychiat.* 96:889, 1940.
17. Myerson, A. Neuroses and neuropsychoses. *Amer. J. Psychiat.* 93:263, 1936.
18. Novikoff, A. B. Concept of integrative levels and biology. *Science* 101:209, 1945.
19. Rachlin, H. L. Follow-up study of Hoch's benign stupor cases. *Amer. J. Psychiat.* 92:531, 1935.
20. Rachlin, H. L. Statistical study of benign stupor in five New York State hospitals. *Psychiat. Quart.* 11:436, 1937.
21. Schilder, P. Psychology of schizophrenia. *Psychoanal. Rev.* 26:380, 1939.
22. Smith, E. R. Neuroses resulting from combat. *Amer. J. Psychiat.* 100:94, 1943.
23. Spitz, R. A. Hospitalism. *Psychoanal. Stud. Child* 1:53, 1945; 2:113, 1946.
24. van Bark, B., and Baron, S. Neurotic elements in the Rorschach record of psychotics. *Rorschach Res. Exchange* 7:166, 1943.
25. Wolberg, L. R. *Hypnoanalysis*. New York: Grune & Stratton, 1945.
26. Zilboorg, G. The deeper layers of schizophrenic psychosis. *Amer. J. Psychiat.* 88:493, 1931.

PART III

GENERAL SYSTEMS THEORY IN ACTION

IN THE FIRST two parts of this book we have presented papers that deal with the conceptual models that delineate or can be subsumed under general systems theory. The specific theories to be found in the analysis of transactions, or of a corporation or a community, have clear relevance to the expanding tasks of psychiatry. In this section, however, the papers are even more differentiated and specific in that they deal clearly with clinical and administrative tasks. The authors, defined by their vantage point, have been actors at national, state, local, and familial levels of systems and write from the context of general systems theory of the experiences encountered in their work.

The first paper, originally the 1966 American Psychiatric Association Presidential Address by Howard P. Rome, is a vigorous call to action as well as an overview of the past, present, and future tasks that face psychiatry, stemming from Dr. Rome's understanding of general systems theory. We consider this paper required reading for any psychiatrist concerned about his identity and role in the next ten years. It is the product of Dr. Rome's wide experience and thought as Chief of Psychiatry at the Mayo Clinic and as President of the American Psychiatric Association.

Similarly, Leonard J. Duhl is one of the few psychiatrists who have had the opportunity to view the developments in the field of mental health from a national vantage point. In his earlier positions as Chief of Planning of the National Institute of Mental Health and Special Assistant to the Secretary of Housing and Urban Development, he has consistently predicted trends in the field of mental health and has developed his interests and thoughts accordingly. Although not formally a general systems theorist, his actions and writing represent an intuitive and experiential understanding of the concepts. His paper on planning clearly presents an extension of the psychiatrist's skills into social planning, a development he obviously feels to be in the making. For the curious, it should be acknowledged with pride and pleasure that he is a brother of one of the editors.

In the next papers, which move from the national level down through the community level to the family, it is of interest, and not very surprising, to note that most of the authors have been identified as family-oriented psychiatrists, for it is at the family interface that so many of the systems of the community meet ecologically. The family-individual systems interface is also a critical one for psychiatric awareness. Thus, clinicians who have worked with families have extended their conceptual thinking more and more into general systems theory.

John P. Spiegel, one of the original group in Chicago with Roy Grinker, is represented in Grinker's book *Toward a Unified Theory of Human Behavior* [2]. Spiegel's family studies relating culture to family style are classics. His interest in family studies has been enlarged, but not supplanted, by his interest in the study of violence. His role as a social critic is revealed in his paper on social planning, an unusual role for a clinician and researcher, but unmistakably following from his interest in the theoretical concepts of systems.

Less the overt critic and more the example of a skilled administrator, Norris Hansell is a psychiatrist whose programs are based on a clearly presented conceptual model of general systems theory. The H. Douglas Singer Zone Center of Rockford, Illinois, developed its unique system of mental health care delivery through his leadership, and this paper is one of a number he has written outlining his work. To the best of our knowledge, only Hansell and Auerswald, who is also represented in this section, have developed clinical systems of mental health care based on general systems theory. Interestingly enough, both their delivery systems are strikingly similar despite the fact that they were never in communication with each other until after their programs were well under way.

Although their programs of delivery and care are similar, Drs. Hansell and Auerswald come from different backgrounds. Edgar "Dick" Auerswald is another of the trained psychoanalysts and family therapists whose world view has extended as their responsibilities have enlarged. He is known for his theoretical work in cognitive development as well as for his research with the unusual group, which included Salvador Minuchin, who worked together at the Wyltwyck School for Boys. His paper, which touches upon only one aspect of his complex and successful program, is one of the best humanistic statements of the general systems clinician.

Like Auerswald, the late Don D. Jackson was a humanist and a family psychiatrist. He changed psychiatry in the late 1950's and early 1960's as much as any one man has ever done. His leadership not only brought forth one of the classic works in schizophrenia and family communications [1], but also focused the field of psychiatry as it never had been before upon family processes and communication. Where Ackerman explored family dynamics in isolation for years, Jackson institutionalized its study and the dissemination of information about it. Where psychiatry had concentrated on the patient and his illness, Jackson and his group attended to the transactions that involved the person in the role of patient and those which evolved him out of it. The paper that appears here, one of his last, is filled with his vigorous concern for the interactional vantage point he has championed. Since his untimely death he has been sorely missed by those who are stirred by his intense commitment. Those of us concerned with general systems theory can only fantasy at what he might have continued to add to the field.

The following paper, by one of the editors, Frederick J. Duhl, is obviously the work of a clinician and teacher who was influenced by Erich Lindemann and his concept of crisis theory, but concerned with its extension into all systems touched by the psychiatrist. This paper tries to make clear the vantage points from which the psychiatrist must work, the need for the conceptual models from which he takes action, and the awareness that he is basically an active intervener in systems regardless of the level of systems at which he chooses to work. The training program at the Boston State Hospital, where Dr. Duhl is Director of Education, is one of the few to be organized with general systems theory concepts in mind.

The heightened awareness that is necessary for the clinician if he is to administer to a complex system, as noted in Frederick Duhl's paper, is achieved in depth by H. Peter Laqueur in the area of family therapy. As a clinician he has had wide experience in insulin therapy, family therapy, psychotherapy of schizophrenia, and psychiatric training. Using general systems theory concepts, Laqueur clarifies the therapeutic process of multiple family therapy. His obvious capacities for both theoretical constructs and clinical practice reveal themselves in this important paper.

Norman Paul's extensive writings about family therapy provide only one glimpse into the breadth of his interests. He has also been concerned with the development of educational programs about family process through the use of mass media such as radio and television. Beginning with work in child psychiatry, he extended his interests to psychotherapy of schizophrenia and then into family therapy of families with schizophrenic members.

Our last author, Edward J. Carroll, is yet another psychoanalyst who has pioneered in family therapy. It should be acknowledged at this point how much Henry Brosin has brought to the field of general systems psychiatry while Director of the Western Psychiatric Institute in Pittsburgh. Not only Edward Carroll but Joseph Charny and Albert Scheflen, who are also represented in this volume, have worked under Dr. Brosin's leadership. Dr. Carroll's paper is one of the few concerned with psychiatric research that has been based upon general systems theory. We trust this is only the first of many in the field.

F. J. D.

References

1. Bateson, G., Jackson, D. D., Haley, J., and Weakland, J. Toward a theory of schizophrenia. *Behav. Sci.* 1:251, 1956.
2. Grinker, R. R. (Ed.) *Toward a Unified Theory of Human Behavior* (2d ed.). New York: Basic Books, 1967.

16

Psychiatry and Social Change *

HOWARD P. ROME, M.D.

I AM INDEBTED to the prescience of two men for a prologue to this account of the nature and the effects of social change on present-day psychiatry. The first is epitomized in the observations of Adolf Meyer —psychiatric theoretician, teacher, clinician. We celebrate his centenary at this meeting. Dr. Meyer gave his presidential address to this Association at the annual meeting held in Minnesota 38 years ago [21]. At that time he chose to review the tide of those events which were, in his judgment, the mainstream of modern American psychiatry. He spoke of the men and the trends that made those generations seminal for us.

The relevance today of other observations he made in 1907 is found in a still-valid prospectus [22]:

> One of the most important lessons of modern psychiatry is the absolute necessity of going beyond the asylum walls and in working where things have their beginnings; and experience shows that there only organized cooperation will achieve success. It might be thought that psychiatry is as yet in rather too primitive and turbulent a condition to rival in interest those domains of hygiene which have become so lucid and inviting through the wonderful discoveries of the last 25 years. We have now definitely settled more facts and methods than our community is willing to deal with unless it is properly guided. Our first aim is to make accessible the main facts with which we work.

* Presented as the presidential address at the 122nd annual meeting of the American Psychiatric Association, Atlantic City, N.J., May 9–13, 1966. Reprinted with permission from *Amer. J. Psychiat.*, 123:1–11, 1966.

My second debt is to another great and kind man. Alan Gregg was the skillful *accoucheur* to the fertile ideas of our parent generation. At another centenary—the 1944 meeting of this Association—he too displayed a remarkable talent for clairvoyance. Like our other gifts from benevolent mentors, it too has set us upon our present path. Dr. Gregg looked into the future which has become our present when he said [12]:

> Before our 200th anniversary of this Association, psychiatry will find great extensions of its content and of its obligations. There will be applications far beyond your offices and your hospitals of the further knowledge you will gain, applications not only to patients with functional and organic disease, but to the human relations of normal people—in politics, national and international, between races, between capital and labor, in government, in family life, in education, in every form of human relationship, whether between individuals or between groups. You will be concerned with optimum performance of human beings as civilized creatures. Soon after you are qualified to observe and describe the forces really determining human behavior, you will be able to explain it. And then in some measure, you will be able to predict it. And since to foresee is to be governed, when prediction is within the competence of any considerable number of psychiatrists, there could come a demand for guidance and advice in man's ways with man.*

It has been said that description and prescription are hard to separate inasmuch as our values shape our knowledge. In this connection, I propose to describe briefly some of the historical and modern influences that are the strands in the Gordian knot which ties psychiatric values, descriptions, and prescriptions to those of the community.

As you will see, one cannot follow a given strand without being stopped repeatedly by what seems to be the continual interference of entangled relations with other strands. Obviously, a plan of orderly procedure depends upon an extensive dissection of the entire network of mixed relations—sufficient at least to see its crisscrossing design as a procedural guide.

An attempt to unravel the Gordian complex of the human system requires that certain assumptions be made. One has to assume that here too there is an evolutionary process which will determine the emerging forms and order. There is no certainty that any of the next stages in sequence will be a single self-contained order; they might well be a composite of a number of integrated ordering subsystems.

One can be reasonably sure that as the details of successive stages emerge, they are likely to be at variance with one's previous assumptions. One must start with the fixed position of his anchors but if *he* is to move, *they* will have to be moved. He must assume the risk of change, safeguarded by the belief that with sufficient data and the wherewithal with which to correlate them, he will map the course of the next steps. He must recognize also as he progresses that unless he is guided by cues from external guidepoints, his personal values and judgments will become progressively more persuasive.

In this regard Dr. Meyer and Dr. Gregg, for all their practicality, have recommended to us the same quixotic feeling that moved Justice Holmes when he rued the possibility of a world cut up into little lots, each with a man on it, well fed and housed but no one feeling the folly of honor or the senseless passion for knowledge outreaching the flaming bounds of the possible [27].

The magnitude of my present task is similar to that Dr. Meyer acknowledged when he too—bound by the limits of space and the strictures of time—was prevented, as he put it, from doing "justice to what wells up in reminiscence and visions of possibilities." Dr. Meyer and Dr. Gregg termed their visions of possibilities speculative and limited. Nonetheless, they have become our realities.

Psychiatry is now in the community. It is involved in the community's territorial disputes. It is engaged in the community's jurisdictional politics. It is preoccupied with the creation of more congenial interfaces, for its intentions are melioristic. To this end, it is engrossed in learning the rites of passage between the domains of vested interests. It debates the propriety of social action by proper professionals.

No Escape but Confrontation

However, as an enfranchised member of the community, psychiatry may not discharge itself from such involvement. Hannah Arendt [1] has termed it: ". . . the *vita activa*—the essence of the human condition." It is a commitment to engage in the *Sturm und Drang* against which earlier generations mistakenly built asylums, havens, and retreats for those alienated by struggles with it. Many have learned painfully that there is no escape but confrontation; that asylums, havens, and retreats are illusory; they afford neither shield nor exit.

Psychiatry was foreshadowed in the community before the return of its institutions. It was heralded by the persuasive conviction that every man has an equal right to dignity. These tenets were established first in the ideological space of the community. They were the products of a series of revolutions—radical social changes in ideas and action. Paramount among them was the asserted right of the governed to have a say in their government—an idea first expressed in the deceptively quiet voices of philosophers. It reverberated in the emancipation of art; it was spoken in the lines of social drama and written in the exhortations of the literature of the times. The seeds of present social change were sown in the eighteenth century. They flowered in the nineteenth century's age of psychological man.

The ideas of those times begot psychiatry as a social action movement. In this sense psychiatry was a spin-off from political philosophy. For better or for worse psychiatry chose to espouse the rights of the least of men. In our time its canons have become the metaphysic of the age of secular science. They are the raison d'être of modern humanism.

Cox [8] has pointed out that "the rise of urban civilization and the collapse of traditional religion are the two main hallmarks of our era and are closely related movements. . . . Secularization . . . marks a change in the way men grasp and understand their life together." In contrast to the social perceptions and values of times past, illuminated by a new awareness of the importance of space, Cox says now: "We experience the universe as the city of man."

This appreciation of environment as a system for the expression of man's needs and the development of his capacities has rearranged social space as it bridges and laces and spans physical space. We sense the tremors as it shakes the ancient values of the world's marketplaces, rocks the decorum of behavior, threatens to topple the guideposts of monolithic ethics, and dares to overturn the moral criteria of absolutism. These were the pillars, the bastioned defenses of yesterday's society, the foundations of its system of values.

It was the radical positivism of the 19th century that opted for man's complete responsibility for his own brand of perfection. Its iconoclasms are now commonplace echoes in today's public media, although in former times they were the subjects of muted academic discussions. One rarely heard public discussions of political issues such as those raised by John Stuart Mills's essay "On Liberty." They were the intellectual fare of a small elite who questioned the propriety of criticizing established social institutions and traditions [15].

Sir Julian Huxley [18] has called the process by which man synthesizes social psychological space "psychometabolism." It is a process for the systemization of "the raw materials of subjective experience. It elaborates them into psychosocially operative organizations of thought and feeling—including principles like causation, categories like space, abstractions like truth; precepts and concepts, poems and gods, myths and scientific theories, moral commandments and legal codes."

Others [23, 31] have visualized the proper study of man in different ways. Some [25] conceive it to be the anthropology of general systems theory. They see human behavior as a network of human ecological relationships. This is a facile way to look at the whole of man and his institutions. It provides a comprehensive frame of reference—a common denominator, as it were—for an examination of what otherwise would be an incomprehensible and hence disorderly mixture of people, situations, objects, and events. The system contributes an ordered way of studying the meaning and function of otherwise separate parts in the context of their total relations.

A Focus for Psychiatry

The relevance of this perspective for psychiatry is its provision of a focus, widened beyond the usual clinical view of intrapsychic pathology. Within its sights is that panoramic array of ecosystem influences of which Dr. Gregg and Dr. Meyer spoke. As some see it, this is a diffusion of interest, having the prospect of obscuring the psychiatrist's job as a psychiatrist; they contend it assigns tasks beyond the psychiatrist's purview and demands skills beyond his competence.

Admittedly it does include within its focus the distant objective of primary prevention. However, notwithstanding the prospect of reaching that goal, as a profession dedicated to the understanding of the vagaries of man, how can we say that any part of the world is too much with us, that the psychodynamic consequences of societal forces or neurochemical discoveries are outside our pale? Erikson [9] has observed that "a clinical science of the human mind will eventually demand a special historical self-awareness on the part of the clinical worker and scholar."

Latterly our graduate curricula have extended their boundaries beyond the physical sciences to include data that deal with man as a social creature. The next move is one of synthesizing the hybrid concepts of these subsystems concerned with human behavior. The data on which new hypotheses can be tested are to be found in the roiling currents of social change. The imagination required to grasp the protean character of the human condition demands this empiricism and not a dogmatic adherence to yesterday's convictions.

Surely, the job of psychiatry is more than a blind grappling in a social vacuum with the intrapsychic precipitates of its conflicts with it. A multilevel approach can accrue from encouraging an acculturation among the different ideas and scientific technologies. Acculturation is assimilative; it is open-ended and conduces to experimentation, innovation, and institutional mutation. Such are the optimal characterestics of milieus with maximum potentialities for change.

Norman Cousins [7] in a recent editorial on "The Third Culture" wrote of the "total connection between total cause and total effect," saying, "the common and tragic failure of both the arts and the sciences is that they have given most of their energy and focus to the immediates and the intermediates and very little to the ultimates."

Psychiatry has become more generic as investigations of behavior reveal factors common to the dynamics of the entire human condition. Its various clinical combinations are formulations and manipulations of selected data akin to isotopes. Each isotopic form is recognized by its different combination of elements, a unique cluster that gives it a distinctive categorical appearance and orientation.

A Generic System

A generic system of psychiatry is in fact a periodic table of psychiatries, each with a different focus, emphasis, technique, and region of competence. The utility of this concept depends largely on its freedom from doctrinaire commitment. It is invidious to contend that there is one and only one epistemology of human behavior, one and only one subsystem both necessary and sufficient to account for the vagaries of the human condition.

Thus the aims of individual psychotherapy are not those of psychopharmacology, although both are directed toward the same target. Similarly, the study of social man by an assay of his environment seeks, by a different route, the agency of change intended by a physical treatment of him as the target element in that environment. By this token, a mental hospital has the ultimate objectives of a community mental health center. It is obvious also that the magnitude of the correlation matrix and the difficulties in definition and assignment of weights to the elements require a better technology than has been available up to now. Common sense suggests that the stores of data on human behavior are beyond the reach of a single channel.

Psychological, sociological, and physiological insights individually and collectively have the potential of catalysts and synthesizing forces. They can foment the change of adaptation and bring about the formation of new amalgams among all the elements that comprise collectively the indispensable parts of man-in-society. Generic psychiatry must accept the messenger role of a leaven to accomplish these objectives on a scale adequate to the need and the demand. Then it can continue to contribute to the clinical practice of medicine and its basic biosciences on the one hand and the burgeoning social sciences and the humanities on the other. Only in the orchestration of intra- and interdisciplinary cooperation can these parochial views lead to an understanding of the complex phenomenon of man [29].

On the microcosmic scale of individual patients it follows that the alternate strategies suggested by such an assignment demand a repertoire of techniques. Their dynamics in different combinations and with different emphasis are possessed of the same elemental character required by society if it is to solve community problems on their macrocosmic scales. But as Dr. Meyer said: "Our first aim is to make accessible the main facts with which we work."

Teilhard de Chardin [4] wrote of this necessity to conceptualize the universe as a system of coordinated parts:

> Considered in its physical, concrete reality, the stuff of the universe cannot divide itself but, as a kind of gigantic "atom" it forms in its totality . . . the only real indivisible. . . . The further and more

deeply we penetrate into matter . . . the more we are confounded by the interdependence of its parts. . . . It is impossible to cut into this network, to isolate a portion without it becoming frayed and unraveled at all its edges.*

There is a related example seen in the developing coalition of chemistry and physics. When their elemental forces are examined at the molecular level, one must be impressed by the unreality of conceptual boundaries that arbitrarily separate a single body of information. The distinction of ostensible difference that typifies appearance at molar levels disappears with a shift to another level of the system.

The presence of psychiatric patients in the community has many connotations. Like the hand of Ishmael, their hand had been against every man for they have suffered at his hands. Hopefully, "dwelling in the face of all his brethren" will change this estrangement, for times have changed. The new encounter demands a new look consonant with new expectations; a new look freshened by a growing appreciation of the common bond our Ishmael shares with his brothers. Even the casual observer is familiar with the indisputable evidence that the appearance of people changes when they are seen in different lights. At one time they may appear formidably hostile; at another, this appearance is recognized as defensive protectiveness.

Then, too, there are practical demonstrations of the potent influence of antecedent events. Moreover, many can be measured by such markers as ratios, rates, and indices. For example, one can follow a decrease in a hospital's inpatient population encouraged by a change in its administrative policies. One can relate an increase in the rate of patients who are discharged as improved to an increase in the ratio of a hospital's staff to its patients. The formation of patient government can be an index of the determined effort by a staff to offset the nondemocracy of institutionalism.

Similarly, the fact of regular community meetings, the availability of crisis consultation, and related changes in institutional practice are the palpable evidence of a link between expressions of care for the rights of others and an increasing rate of disappearance of charted symptoms.

Changes in Political System

In point of fact, in many subtle as well as gross ways the administrative philosophy and the operational practices of psychiatric institutions are being reshaped. In large measure these reflect the social changes in the political system. Colonialism has disappeared. There have been demonstrations for and redress of civil rights. Concerted efforts are being made at many levels to enfranchise every citizen. Orthodoxy and autocracy are being challenged. Through legal and administrative actions, monumental social changes are underway to secure equal opportunity for the underprivileged. One's associations to such names and titles as Monte Durham, Selma, Title VI of the Civil Rights Act, *Brown vs. the Board of Education, Gideon vs. Wainwright,* Operation Head Start, and Medicare are alone sufficient to trigger a recall of historically significant changes within the community.

The more obvious moves in the direction of the Great Society made by the Congress in the form of health, education, and welfare legislation are but the most recent social changes in a long series, begun in areas that at first glance appear to have no immediate connection with medicine or psychiatry or the operations of their establishments.

The dynamics of social change are illustrated by the diffusion of an innovation. The process is a common one quite independent of content. It begins when a critical mass of raw experience is articulated as a need. Eventually, sharply defined expressions of that need crystallize as a consensus among persons of like mind. They operate as a self-organizing system in which the heretofore separate values link in a purposefully coordinated fashion to optimize what they now recognize as a mutual interest.

The interaction initiates a particularly human kind of metamorphosis. It transforms percepts to concepts and affects to values, and finally it stimulates a response in the form of individual and group action. The system created by this interaction behaves as a series of relays in a communication net, by which its information is processed, converted, and transmitted. Information in this sense may be ideas or feelings or action. The system also has the capability of amplification; that is, it can enhance, expand, and distort any kind of information.

Hence, it is a social device for communication—one that can generate the motivating and organizing force of a riot on the one hand or a peace movement on the other. It is the responsible mechanism for such behavior as the hyperbole of rumor or the spread of style or the endowment of charisma upon a leader. There is much evidence to show that the modes selected for amplification may be either dissonant or consonant [10].

Economic Imbalance

Certain facets of the national economy have grown to opulence by the employment of these amplification methods in the business and technological sectors of the industrial system. Some critics [16], concerned with the effects of this upon the community as a whole, have levied a charge of disproportion. They inveigh against the allocations of materialistic priorities and emphasis. Other critics [11, 19] have deplored the deficiencies in the control and direction of the production phase of our industrial revolution, contending that it has been overstimulated and therefore generalized to the disadvantage of all else. Alfred North Whitehead told a Harvard audience in 1928 that: "It is a libel upon human nature to conceive that zest for life is the product of pedestrian purposes directed toward the narrow routine of material comforts" [5].

The production system makes and creates a demand for more and more things. The gross national product now stands at an all-time high. Yet surely the same sophisticated methods of production and management can be applied, with an equal probability of success, to the distribution of needed goods and services and resources. The absence of this countervailing balance has produced the rudderless environment Durkheim called anomie [20].

Our phenomenal success in the manipulation of some of the cybernetic controls of biological systems is responsible for shifts of an expanding population from country to city and from city to megalopolis at an accelerating tempo. In the agricultural subsystem we produce more than we can eat and do it with less collective effort. We thereby embarrass the production sector, for we have not yet the solution to fill the starved channels of distribution with this cornucopian surplus.

And there is evidence of similar imbalance in the other subsystems of the community. Rapid social change in numbers, rates, and locations precludes an orderly assimilation at every level of the system. It adds still more disorder to their cognate forms of anomie. Thus the operational, ethical, and sociological implications of Norbert Wiener's [30] formulations of cybernetic controls gone awry have introduced still other dimensions of value in the picture of social change.

As the flow of events changes from laminar to turbulent, the stratified norms of traditional order and control are disrupted. Old ties no longer provide a secure anchorage. A reassessment of the shifting tides is required to stabilize a system of new relations. It calls for more knowledge of all the critical mechanisms on which social survival depends. Such principles of control are most effective if they are applied at the hub of the system. The struggle for control within all human systems inevitably generates friction even when the force that is used is judiciously applied. This, of course, is the anxiety one has about the progressive centralization of power [17].

We live with a special version of this anxiety in our political subsystem. In the past quarter century it has been compounded inasmuch as the potential of the destructive power of the atom is now in the possession of rival political systems. As it is at all levels of human relations, the choice is between attraction or repulsion, a choice between us and them; for neutrality is interpreted as a state of unresolved ambivalence. On the international level, the dilemma is an awesome one and threatens to become more so as the forces of competitive systems move in ever closer orbits. All subordinate systems shift to their respective modes of defense in anticipation of an impending collision. Nationalism is the response in the political socialization system.

The economics of the production and distribution operations are the next to prepare for the impact. In turn, they signal the reallocations of manpower. These are the bellwethers of a shift in values. In the ploys of power, catch-penny slogans like "bullets or butter" obscure rather than clarify the true range of choice. The fact is that these scare lines in the communication nets are used to mobilize opposition to a consideration of the very programs which have the possibility of decelerating the mounting tensions. The same dynamics may be traced in the operations of the subsystems with which we are more intimate and for which we have a greater measure of responsibility.

"Ideology of Defense"

Coeval with the growing capacity of the economy to afford a choice, we have lived through half a hundred years' war. During this time the anxieties of survival have mobilized and maintained a costly state of preparedness. An ideology of defense is consonant with such behavior, and the needs of other sectors of the system accordingly are considered in dissonance. Thus the major mode in total allocations for the past quarter century has been the one of defense. The second-order effects of this decision determine the direction of the next branch in the decision tree. Its content has furnished the ideological supplies that have nourished the expanding role of the federal government as it has moved beyond the conduct of foreign affairs and national defense to become more involved in the dependent sphere of domestic affairs.

There are still other derivatives that inform the shape of public policies. An excellent example is seen in the relationship between the posture of national defense and the missile program which supports this posture, and the extraordinary allocations of money and scientific resources for the exploration of outer space. This cluster of associations derives from their joint occupancy of the same ideological niche in psychosociopolitical space.

Social historians have suggested that although the influence of power determines the route and the associations of the elements in a sequence of system decisions, there are important collateral effects too. Another instance is evident in what Daniel Bell [2] has called "eudaemonism—the proposition that each person is entitled to happiness and that it is one of the functions of government to try and assure him at least of the preconditions of happiness."

The affective and ideational transformation of defense to protection and protection to security and finally security to happiness seems to make this eudaemonic expectation of the role of government appear to be the logical next step in the progression. Bell explains this by ascribing it to:

> The simple sociological fact . . . that in a complex, interconnected society, the conditions of happiness for the mass of people—social and economic security, education, equal opportunity, decent housing, medical care and the like—involve a high degree of collective action through the agency of government, national or local.*

This and similar hybrid correlations create a climate of multiple expectations. The federal government's response in these post-Keynesian times has resulted in an unprecedented rise in the standard of living and the beginnings of programs for the fulfillment of the material preconditions of this expectation. Among the particular preconditions, there is the one of well-being. Over 100 years ago de Tocqueville [28] recognized this in his famous treatise on *Democracy in America* when he wrote: "In America the passion for physical well-being . . . is general." He saw this passion for bodily comfort, for a sense of individual ease, as an inseparable part of a desire for material well-being.

While health expenditures are not large in comparison with the size of the total economy, they do amount to about one-seventeenth of the gross national product and one-eleventh of consumption expenditures. The rise since 1950 of some $19 billion is an increase of more than 1½ times—an increase greatly exceeding that of the gross national product. This reflects both the magnitude of the demand and the market response in cost, incidentally a well-established relationship in the economics of reciprocity.

The nature of the present climate of expectation can be gauged also by the large volume of outlays for welfare activities. They are approximately three times the percentage of expenditures for health. Similar outlays for education further confirm the direction of future changes. President Johnson's decision to honor Mr. Truman by signing into law the Medicare Amendments to the Social Security Act in Independence, Missouri, underscored the fact that it took 25 years for the climate of that expectation to produce demonstrated social action. Further, its thrust at the provision of health care for the aged and the indigent makes it clear that a balance of goods and services is the direction of its intent, for these are among the least advantaged sectors of the community.

* From D. Bell, The study of the future. *Public Interest* 1:119, 1965.

Medicine in Throes of Change

As a consequence of these trends, the institution of medicine itself is in the throes of significant change [14]. There is the implication that its nonpareil status as the *eminence blanc*, the sole arbiter of what is best for the health care of the nation, is being challenged by another aspect of the secularization process. Welfare legislation, drug and other regulatory commissions, and licensure and accreditation standards have established the fact that the consumer of professional goods and services wants a voice in the quality of what he purchases. Up to a few years ago, this was a reserved territory.

One sees reflections of the same demands for a voice in the conduct of the affairs of the educational system, in the process for the resolution of labor-management conflicts, in the right to dissent from positions taken by those who have the responsibility for decisions that affect the public weal. It raises serious questions about the democratic process in a vast and technically complex society. The nature and quality of response to this kind of social change depends on the extent to which the climate of expectation makes it clear how much heretofore private territory must be in the public domain if our community is to maintain its traditional freedoms.

The same political struggles of governance are to be found among professional groups as well as between the laity and the professionals. Each contestant asserts his right to a share in the allocation of responsibility and resources. Recent legislation has underscored the repeated statement of official spokesmen that the health of the citizenry is our paramount natural resource, and measures to conserve and improve it are also an entitlement of citizenship.

These acts legitimized the government's stake in this larger, heretofore private territory. The next step to follow in sequence was the regulatory one. Its definition of standards, permissible procedures, and obligatory guides extended the officialdom of public health far beyond its original quarantine functions. Third parties, neither government nor professional providers, also claim a share of this territory by virtue of their collateral role in the $40 billion annual business of health care.

The interdigitation of these activities and the changes they provoke create problems analogous in principle to those previously mentioned.

The complexities of the system obviously require direction. We approach a population of two hundred million at the rate of more than three million a year. A balance must be struck between the political, the economic, and the social considerations. In the frontier society of the past, the boundaries of individual wants were set by what the limitations of personal and local resources could afford. These restrictions no longer obtain. The climate of expectation, the eudaemonic philosophy, has opened new channels. Social change involves the affluent in the affairs of the disadvantaged, and their commitment is altruistic only in part.

The case of psychiatry is in point. The historical assignment of the psychiatrist was the task of attending alienist to outcasts of the system. This was a relatively late development, reflecting a special concern by physicians. The alienated had made the mincing progress permitted by charity over long generations. Action in their behalf fulminated spasmodically an early form of protest. They were the Chartist equivalents of our recent demonstrations in Watts and elsewhere. These are the frustrated protests of the otherwise impotent.

Pinel's [24] "Treatise on Insanity" 166 years ago criticized what he called "an imperfection in the treatment of insanity." In scathing terms he said:

> . . . it would appear that every lunatic is under the control of a keeper, whose authority over him is unlimited, and whose treatment of him must be supposed, in many instances, to amount to unbridled and dangerous barbarity: a dedicated latitude of power totally inconsistent with the principles of a pure and rigid philanthropy.

Psychiatric patients were indistinguishable from what Michael Harrington [13] calls the invisible poor. Even in the day of a more affluent society than that of the eighteenth and nineteenth centuries, there has been proportionately little action in their behalf. The hospitalized mentally ill especially, and the poor, have been seen by the community as the brown-gray blur on its outskirts. Their invisibility has kept them the unincorporated sections of the community.

New Role for Psychiatry

The years of World War II launched this Association into the arena of social protest. Ours was an earlier and quieter version of the present day teach-in, preach-in, social-action movement. We have succeeded in the attainment of some of our objectives. Our patients and our services at long last are no longer in the isolated ghettos of a segregated community. Our practices are becoming a corporate part of the medical community. Our institutions have been the beneficiaries of political actions that begin to accord them equal rights to health care. We have won some battles but not the war; we have in part accomplished Dr. Meyer's first aim. Future strategy and tactics require a wholly different orientation. We are now required to deliver what has been promised.

Psychiatry must be as modest in its claims of ignorance as it is in its other claims. For too long alleged ignorance has been its immodest plea for exemption from its responsibility to systematize what it collectively knows about human behavior. The technical means are at hand to assemble and collate any amount of information required to predicate and test hypotheses, to build models and simulate, to use everything we do as a data base for critical review [26].

The same computer-based techniques are available to maximize the explicitness with which we state assumptions; to record, store, retrieve, manipulate, and display any amount of empirical data necessary to validate the logic of these assumptions. The time is come when a psychiatric information system can extend its scope beyond the head-counts of admission and discharge and readmission statistics of single hospitals and systems of hospitals. There are tactical and strategic simulation procedures for the solution of traffic problems and the mapping of the effects of decisions [3].

The plot of the demographic characteristics of populations at different risk can help select target areas in the community in need of special services. These are the means for getting at the core problem of social morbidity, for matching its gross and subtle relations with the classical and neoclassical models of psychopathology. Electronic data processing methods have joined forces with systems theory, and this new technology makes it possible to rehearse the entire range of imaginable options.

We Now Have the Tools

Up to now we have lacked the navigational tools to explore and map the *terra incognita* of social space. We have been forced to extrapolate from the data we have called pathological to the world at large. While some of our guesses have been shrewd, we now have the means to confirm them in replicated samples on a large scale. We do not know whether assumptions made in one culture are reliable when transferred to another culture. Up to now we have been forced to depend on an incomplete taxonomy for decision clues to treatment and prognosis—a taxonomy of functionless descriptors. As a consequence, we have been unable to plan with a predictable accuracy the appropriate path through a maze of alternative procedures. As a result we have been committed to the costly empiricism of trial and error.

Up to now there has been no really substantial way to capture the essence of these data so as to profit from past errors and past success. Up to now we have lacked the tools to determine the significance of hybrid events on a scale sufficient to be meaningful. The sets of data on which successful treatment and accurate prognosis depend require the grammatic rigor of a truly communicative language [6].

We have the intuition that significant additional cues and clues are hidden in information stores to be found in domains other than our own. And to attract the support we need for this amplification of data on the human condition, the verification of the predicates of sociology must be found in the subjects postulated in psychopathology. The adjectives of economics must have more than a metaphorical relation to the nouns of personal behavior. We need new lexicons and generative grammars to provide the meaningful words and the logic required to write a library of computer programs for a network of computer centers which will tie us in a common information system.

I can do no better in summary than to paraphrase Emily Dickinson's quatrain:

> Psychiatry will be
> Velocity or pause,
> Precisely as the candidate
> Preliminary was.

References

1. Arendt, H. *The Human Condition.* Garden City, N. Y.: Doubleday Anchor Books, 1959. P. 11.
2. Bell, D. The study of the future. 1. Can one predict? 2. The oracles at Delphi. *Public Interest* 1:119, 1965.

3. Bernstein, A., and Hanon, R. Artificial evolution of problem solvers. *Amer. Behav. Sci.* 8:19, 1965.
4. de Chardin, P. T. *The Phenomenon of Man*. New York: Harper & Row, 1959. Pp. 43–44.
5. Cheit, E. F. (Ed.) *The Business Establishment*. New York: Wiley, 1964. P. vii.
6. Chomsky, N. *Aspects of Theory of Syntax*. Cambridge, Mass.: Massachusetts Institute of Technology Press, 1965.
7. Cousins, N. The third culture. *Saturday Rev*. May 7, 1966.
8. Cox, H. *The Secular City: Secularization and Urbanization in Theological Perspective*. New York: Macmillan, 1961. P. 1.
9. Erikson, E. H. *Young Man Luther: A Study in Psychoanalysis and History*. New York: Norton, 1958. P. 18.
10. Festinger, L. *A Theory of Cognitive Dissonance*. Palo Alto, Calif.: Stanford University Press, 1957.
11. Friedman, M. *Capitalism and Freedom*. Chicago: University of Chicago Press, 1963.
12. Gregg, A. Critique of psychiatry. *Amer. J. Psychiat.* 101:285, 1944.
13. Harrington, M. *The Other America: Poverty in the United States*. Baltimore: Penguin, 1963.
14. Harris, S. E. *The Economics of American Medicine*. New York: Macmillan, 1964.
15. Hart, H. L. A. *Law, Liberty and Morality*. New York: Vintage, 1960.
16. Heilbroner, R. L. The future of capitalism. *Commentary* 41:23, 1966.
17. Hofstadter, R. What Happened to the Anti-Trust Movement? Notes on the Evolution of an American Creed. In E. F. Cheit (Ed.), *The Business Establishment*. New York: Wiley, 1964. Pp. 113–192.
18. Huxley, J. The Future of Man—Evolutionary Aspects. In G. Wolstenholme (Ed.), *Man and His Future*, A Ciba Foundation Symposium. Boston: Little, Brown, 1963. P. 3.
19. Lerner, M. Six Revolutions in American Life. In T. R. Ford (Ed.), *The Revolutionary Theme in Contemporary America*. Lexington, Ky.: University of Kentucky Press, 1965. Pp. 1–20.
20. Merton, R. K. *Social Theory and Social Structure. Toward the Codification of Theory and Research*. Glencoe, Ill.: Free Press, 1949. P. 128.
21. Meyer, A. 35 years of psychiatry in the United States and our present outlook. *Amer. J. Psychiat.* 85:1, 1928.
22. Meyer, A. Modern Psychiatry: Its Possibilities and Responsibilities. In A. Lief (Ed.), *The Commonsense Psychiatry of Dr. Adolf Meyer*. New York: McGraw-Hill, 1948. P. 293.

23. Miller, J. G. Living systems: Basic concepts. *Behav. Sci.* 10:193, 1965.
24. Pinel, P. A *Treatise on Insanity*. Sheffield, Eng.: Todd, 1806. P. 66.
25. Sorokin, P. A. *Social and Cultural Mobility*. Glencoe, Ill.: Free Press, 1959.
26. Steffire, V. Simulation of people's behavior toward new objects and events. *Amer. Behav. Sci.* 8:12, 1965.
27. Sutherland, A. E. *Apology for Uncomfortable Change: 1865–1965*. New York: Macmillan, 1965. P. 106.
28. de Tocqueville, A. *Democracy in America*. New York: Oxford University Press, 1946. P. 398.
29. Weller, T. H. Questions of priority. *New Eng. J. Med.* 269:673, 1963.
30. Wiener, N. *The Human Use of Human Beginnings: Cybernetics and Society*. Boston: Houghton Mifflin, 1950.
31. Wittgenstein, L. *Tractatus Logico-Philosophicus*. London: Routledge and Kegan Paul, 1955.

17

Planning and Predicting: Or What to Do When You Don't Know the Names of the Variables*

LEONARD J. DUHL, M.D.

FORECASTING was once an honorable occupation for seers and magicians. In more recent times we have tried to take it out of the area of magic; it may now range from an educated guess to a genuine attempt to find methods that make accurate prediction possible. Essentially, forecasting is a kind of game, even though it has become a very serious game with business and government alike pouring enormous amounts of money into attempts to predict the consequences of given sets of policies, actions, decisions. It would be advantageous if we could truthfully say, "Business and government alike are pouring enormous amounts of money into attempts *to learn how to predict* the consequences. . . ." In some instances, the second version would be correct, but not in all, and most certainly not in enough instances.

Some forms of prediction are feasible now. We can predict with reasonable accuracy what kinds of people will succeed in certain types of jobs. Even the number of tests that must be used for this kind of prediction has been reduced to a manageable lower limit. It is possible to predict within varying margins the probability of sales going up or down in certain consumer areas at certain times of the year. The variables for that sort of analysis have become available to us out of experience, hunch, and some research. Given reasonably good conditions for weather forecasting and a knowledge of the current

* From Toward the Year 2000: Work in Progress, Proceedings of the American Academy of Arts and Sciences, *Daedalus* 96:779–788, 1967. Reprinted with permission from *Daedalus*.

economic state of the nation, we can predict approximately how many people will die in automobile accidents over a Labor Day weekend. We cannot predict with any reasonable accuracy, however, what the long-range effects of large-scale automation are going to be on the economic and cultural patterns of the nation, although it is now possible to begin observing and even measuring certain kinds of indicators. In an area as broad as this we are still *speculating*.

We can predict well in situations that involve few—and identifiable—variables and with which we have had many experiences. But in most areas of social planning, we cannot even extrapolate with reasonable reliability how a single policy statement on housing will affect educational and segregation patterns, mortgage interest rates, or the building trades twenty years from now. But it is this kind of prediction—or extrapolation, to use a term I prefer—that the social planner is attempting.

Role of the Planner

In my files I have a compilation of students' views on planning. The students ask, with beautiful candor: What is a planner? What is the process of planning? What are the limits of planning, of the planner's potential effectiveness? What are the ties between social concerns and housing, between social concerns and transportation? *What is the nature of our society, and where and how should it be altered?*

This final question would be easy to confront if one were writing a utopian novel. Like Plato or B. F. Skinner, one would decide on the ideal citizen in an ideal life situation, and then construct the social system that would produce such individuals and citizens.

Unfortunately, the planner is not in the position of constructing a new and ideal social system. He is involved in a very complex task: He must decide what in the system should be changed; persuade a majority that his decision is valid on empirical, moral, and legal grounds; and, finally, find ways to involve large segments of the affected population in implementing the actions he is advocating.

The planner is an agent of change, and any agent of change is a planner. It does not matter whether he is a politician, a producer, a businessman, an administrator, or an educator. There are many kinds of planners, each performing different functions and fulfilling different roles, depending upon the specific problem or situation with which he must deal. What is essential to his definition as a planner is that he be concerned with instituting change in an orderly fashion, so that tomorrow something will be different from what it is today.

The short-range planner has immediate objectives, and his function is to delineate a step or a series of steps that will achieve these immediate and pressing objectives. The long-range planner has more scope and more time. His goals may be less specific and call for an evolving policy that permits and encourages certain actions. He also runs much greater risks, because the distance in both time and space from his objectives increases the possibility that some of his planning may have unexpected consequences—consequences that are at serious variance with his original intentions. The short-range planner runs the risk of fighting brush fires without reference to what his immediate actions may mean for the long run; the long-range planner can be tempted to treat planning as the utopian novelist treats his material—to assume that he can structure a series of steps that will *surely* result in a particular long-range consequence.

But in truth the planner's information sources are the critical element in his functioning. The more complex the problem he is attempting to analyze—in terms of present realities, discernible trends, and so forth—the more tempting it becomes to assume that the future can be structured just as we want it if we simply develop the proper scientific techniques. Thus, the planners turn to systems analysis and ecological models in their attempt to identify and manipulate the complex of factors underlying causality. Unfortunately, the information available is that all too familiar top of the iceberg, and reliance upon it alone can be dangerous.

No matter how much information can be collected by the techniques of the hard sciences and mathematics, we are a long way from being able to computerize the interplay of forces that operate in social conditions, whether the social laboratory be a settlement house, a neighborhood, a nation, or the United Nations. Alter a stress here—merely a minor stress—and you may find that the alteration has produced nine other stresses, different and perhaps not so minor, at nine other points in the social structure.

Perhaps the most obvious example of the American tendency to let enthusiasm guide progress toward disaster has been our superhighway system. In general, superhighways were calculated on the basis of what transport engineers determined to be the current demand. No one stopped to wonder whether the existence of the superhighway might change the pattern of the demand—and the patterns of emigration and industrial sprawl—so drastically that within two years of completion a superhighway was a very wide expanse of road on which fifteen thousand automobiles were stalled for two hours until the car that had broken down in a middle lane was towed away.

By now we are less naive in many areas, and planners attempt to find ways to anticipate the social consequences of planning and policy decisions. But as long as systems continue to open up, as long as information is channeled to all portions of the globe and creates increased aspiration without an accompanying increase of training for special skills, as long, in fact, as messages conflict, the actual predictability of our future will be faulty. The input of information constantly redefines the situation, the problem, and the possible range of solutions.

This essay is not an attack on the scientific method. We must continue to use technology and science to try to evolve a theory of prediction that incorporates *process* as a constantly changing—but definable—variable. In the meantime, however, we must remember that planning for the future will be inaccurate to varying degrees so long as the planning process deals with human beings—with their idiosyncratic, nonrational behavior. For it is the nonrational and the idiosyncratic that turns the planning process into a technique with results somewhat less predictable than those of roulette.

The planner or forecaster must, therefore, be not only an agent of change. He must know how to guide change as the therapist guides his patient. He must be concerned with processes, with assuring that a dialogue is initiated and continued. A never-ending process of interaction—confrontation and counterconfrontation—must take place so that both specific and general goals, as well as trends, can be identified.

In one sense this makes it necessary for the planner to be an astute politician. In our system, politics is a process of conciliation and reconciliation, of accommodation between competing interests. The political system limits the real power of one man to put a plan into practice. The system is both more and less controlled than most

Americans like to admit. Politicians do in reality have the power to put bad plans into effect; but planners, if they are bad politicians, may not have the real power to put good plans into operation.

The Community Therapist

Few would argue that change is not critically necessary. But it is equally critical that change take place in a democratic fashion; that it be the product of a wide consensus, not the fiat of an expert. To function as an agent of democratic change, the planner must conceive of himself as a kind of "therapist" to the community. The therapist is presented with problems of the present, born of the past, with the request that he help the patient alter his system of response so that the future will be different and bring different results. If the therapist cannot or does not help the patient, the patient will search elsewhere for answers.

If the therapist concentrates only on the present, he will fail to anticipate the future. The problem must be understood in terms of how circumstances and processes in the past have led to circumstances and attitudes in the present, and which aspects of the process must be altered in order to alter the outcomes in the future. For this, information and data are vital. If, however, the information is confined to what is readily available, the therapist will be no better able to propose solutions than the patient. The therapist must gather data on the interactions between the patient and his environment—using environment in the sense of situations and persons; the more adequate the data gathered by the therapist, the deeper his understanding of the problem and the wider the range of alternative solutions he can find. But equally important is the therapist's ability to interpret his data to the patient and to help the patient redefine the problem as new data are gathered. As the problem is redefined, the actions necessary to cope with the problem may also need to be altered; both patient and therapist must always be aware of this.

This is essentially the process the planner faces. Let us take, as a case in point, the war on poverty. When the President focused the attention of the American public on the problems of poverty—vast issues of economics, housing, voting rights, transportation, education, and health—he began certain irreversible processes that were to have a profound impact on every town and citizen of this nation. The responses to the challenge posed by this *definition of a social problem* varied dramatically, from absolute apathy or hostility to some genuine action programs. But the problems were posed in such general terms and the responses made so hastily that planning in a scientific sense did not take place. The implications of the public statements about the problems and the steps that would be taken to cope with them did, however, have measurable results. They vastly increased expectations; they momentarily dissipated apathy and lethargy among the dispossessed; they stimulated hope.

With the best intentions in the world and an inadequate budget, various agencies on the federal, state, and local levels began to do planning. But for the most part it was planning without research. Inevitably, the frustrated expectations found expression in outbursts of rage; Watts is, perhaps, the best example of this process. The Watts outburst was a crisis—and change often comes only after a crisis. In Watts the crisis brought forth declamation, acclamation, and disconnected federal funds, but not much in the way of concrete research for effective planning action. Where was the forecaster, the planner, the agent of change, the therapist who predicted what would happen in Watts before it happened? Studies had been contracted and effected, and people had been concerned, but planning had not produced action leading to change; and without such change, there was anger. There were many reasons for this, a primary one being that the planning did not involve the "patient"—the residents of Watts and the power structure of the city and state. The therapist cannot produce change in the patient by an act of will. The process requires the patient's participation not only in the actions that lead to change but in the evaluation of what in the current situation needs to be changed. Even with such joint effort, planning actions may fail, but at least their initial chances for success will be improved.

Tools and Process

The planning problem is, therefore, twofold. First, we do not yet have the tools to plan with scientific accuracy, nor are we finding systematic ways to develop them. Second, accepting this limitation, our planners have not understood that their primary concern must be to develop a sensitivity to *process* among all the people who must be involved in planning activities. This sensitivity implies an ability to see the society or aspects of it as an *ecological system* and to recognize that balances shift with every change in the system. The planner must be able to analyze these shifts quickly and to readjust the plans if the shifts are in the wrong direction.

This implies that the planner must be as much involved in educating as he is in planning. He must attempt to get the fullest participation in the planning process from the widest representation of the society—in other words, to keep in mind that the planning process must go hand in hand with the democratic process. The democratic process is effective and fair to the degree that it is operating among an enlightened electorate.

The recent confrontations of the society by civil rights advocates and members of other heretofore "slumbering" depressed groups are both exciting and disconcerting. In one sense, they demonstrate how people respond to an opportunity for participation. They have stirred up the system. They have also made many Americans (themselves included) aware that true democratic participation is uncomfortable and threatening, slow and arduous.

The planner must bring to the forefront ideas, thoughts, and concepts heretofore not part of the public consciousness. He then must provide the guidance and the education that will help the newer participants learn to use the political process effectively.

True democratic participation means that society must seek as a legitimate activity the education of the poor so that they can confront the system through political organization and command a response. Moreover, the system must also confront the professionals, and the professionals must counter-confront the system. One group or faction has to accept the danger implicit in the growing power of another group or faction.

Democratic participation does not mean that there must be a poor man—black, brown, or white—in every office or on every organization board. It does mean, however, that representation on all levels must be assured. It follows that lawyers, planners, architects, educators, and doctors will, at certain times, assume advocacy for special-interest groups—be they the poor, the Negroes, the Indians, or the Mexican-Americans.

The crucial concerns relate to ethics and values—the ability of the therapist or planner to be nonmanipulative, yet to set the stage for people to participate in the solving of their own problems. Any professional in any field is as subject as the slum dweller to preoccupation with internal needs, his own perceptions, views, and conceptualizations of the issues at hand. Confronting our problems and then expanding our understanding to see how other people's perceptions of these problems impinge on ours are frightening experiences at first. The planner must, in a practical sense, demonstrate that just as there is no single therapy for heroin addiction, there can be no single solution for any of society's ills—he must do this even while he is diagnosing the ill and insisting that some therapy be instituted.

Current policies, programs, proceedings, even agency personnel become subject for question, for examination, for evaluation and analysis. At all levels of society we find unwillingness to relinquish or even alter familiar and functional patterns of behavior. The paramount objective of the planner is to help the members of any group reconceptualize current problems in terms of long-range goals, to show them how, in terms of known and familiar situations, the planning goals are both necessary and practical, and of value to everyone in the system, not just to the immediate objects of the planning. In short, he must minimize fear and anxiety while urging the "patient" (or institution) to look ahead even while he (or it) is performing his daily functions.

The political nature of the planner's efforts becomes clearer as he collects, controls, and feeds information into the system. The process can seem overwhelmingly threatening. Innovative collection and correlation of information imply that previously independent organizations within a system have become integrally involved with one another. Wasteful duplication and repetition could be prevented if lines of communication founded upon similar directions, values, and goals could be established and made responsive to feedback. It is obvious that the planning body must be centrally involved in the running of an organization. But, even though the project programmers and effecters should be closely involved in the determination of goals and in the reorientation of the organization, the planning sector needs to remain distinct from the sectors of the organization that expedite the programs.

Secretary of Health, Education, and Welfare John W. Gardner terms a "self-renewing agency" one in which the total organization (and the consumer of the services) participates in redefining its own problems. Successful self-renewal depends upon employing the model of complex causality and remaining alert to ever-changing goals. Maximum involvement by all who have mutual or complementary interests must be sought. Often a disinterested outside consultant—an experienced, knowledgeable, unprejudiced, and respected person—can best initiate and aid in the development of this process.

"Invisible Colleges"

But beyond formal organizational structures there are "invisible colleges"—the loose aggregates of individuals scattered throughout the nation and the world who periodically communicate with one another. They are sociologists, architects, lawyers, doctors, teachers, and others whose avocation is "change" and how it might be effected. All are intimately involved in reality—some participate quite actively in the affairs of an organization; others have removed themselves from decision-making by becoming advisers, consultants, or assistants.

Their communications are via the telephone, the Xerox machine, and the jet. They meet, exchange information, ideas, theories, and concepts. Tied neither to time, place, nor position, they operate on many different levels at the same time. They are a link between industry and government, between the public and private sectors, between the federal, state, and city governments, between the governments and neighborhoods, between the money givers and money receivers, between the theorists and activists. Their value lies both in their access to information from many sources and their rapid dissemination and utilization of that data. Differing combinations of these agents of change may assemble for many purposes: to explore the possibilities of and to launch a New Town, to discuss a Watts and its implications for planning, or even to weigh the impact of systems technology upon forecasting. The long-range planner must connect informally with one or another level of these "invisible colleges," for the information developed and passed on in them is not of the typical census type, but part and parcel of the day-by-day reality of social systems and the people functioning within them.

These planners are not dreamers. They have cultivated what Sir Geoffrey Vickers has called "the art of judgment"—the process of making decisions in the present that dramatically affect the future.

They are experts in combining and reformulating data and information, in redefining the problem, and, most importantly, in causing others to feel they must do likewise. They achieve this by presenting additional information relative to the issues at hand in a way that convinces others. They are experienced in working imaginatively with performance criteria or specifications and rebel against performance standards that are not potentially multi-applicable. They have the ability to "feel" data. They have an appreciation of the implications of decisions and how they might affect a staff as well as tangential activities.

No mechanism can keep track of all of the irrationalities and the idiopathic responses of the many complex systems that affect human behavior. These unforeseen irrationalities may be a greater determining factor for future events than any of the more rational activities. The "invisible colleges" may, therefore, prove more effective in the long run than more formalized mechanisms. Marshall McLuhan has asserted that sense will be made out of the irrationality of input not by making the input more rational, but by creating mechanisms that are competent to deal with the irrationalities.

The planner-forecaster who has a goodly share of philosopher-educator-therapist in his personality will be the most understanding and the most effective. If he also has political skill, fortitude, and persistence, he will be the most adept. His talent for linking, connecting, and communicating makes it possible to create mechanisms for planning that respond to a society of idiosyncratic human beings. In essence, the planner-forecaster's role is that of orchestrating people and institutions and organizations. In performing this role, he operates to restore to a complex society something of the "democratic" and "representative" flavor it possessed in simpler times. He operates, in other words, to return the franchise to society's membership to some extent.

We too often find ourselves preoccupied with attempting to divine what will be our state 10, 25, even 50 years hence. But what we often fail to realize is that every decision made, every voice heard and not heard, and every success and failure drastically affect that future. Instead of speculating on what the world might be like in the year 2000, we would do well to consider what mechanisms, what people, and what decisions must be attended to today in order to shape all the years to come.

18

Environmental Corrections as a Systems Process*

JOHN P. SPIEGEL, M.D.

GENERAL SYSTEMS THEORY is often presented as a novel approach involving a kind of high-level conceptualization—as a new science, so to speak. It is certainly true that von Bertalanffy introduced an original idea by generalizing the biological principle of "open, self-regulating systems" to explain, by analogy, the behavior of personality, group, and social systems. But the chief utility of the "systems approach" does not derive, it seems, from any revolutionary conceptual properties it may contain; rather, its advantages flow from a very simple and practical effect of the emphasis on systems qua systems. As a result of this emphasis, it is impossible to ignore the complexity and interrelatedness of the phenomena of behavior, from the biological through the psychological and social to the cultural dimensions. There is an obvious corollary to this result: If all behavior systems are interrelated, no one discipline or profession can play a significant role in effecting a *change* of behavior without taking into account other disciplines and professions concerned with related systems. For, in making a change in any one system, the change-producer has to take into account the effect of that change, and the resistance to it, in a multitude of other dependent systems. In other words, each system coming under the influence of change must be specified as precisely as possible and the predicted effects of change must be spelled out in advance.

I would like to apply this simple, almost common-sense idea to

* Presented at the 123rd annual meeting of the American Psychiatric Association, Detroit, Mich., May 8–12, 1967.

two sweeping national efforts to effect change. The first is the current planning for community mental health programs which, in one way or another, has involved many of us in the psychiatric profession. The second is the planning for new approaches in law enforcement and the administration of justice, initiated recently by the President's Crime Commission. The latter is a newborn infant, while the former is at the toddling stage, so that both may be considered together with regard to their weaknesses, strengths, and chances for healthy growth. Both are of so broad a character that they touch upon—almost by definition—innumerable systems of organization and behavior. Both appear to be characterized in large part by the absence of a systems approach on the part of the initiators of these programs. And both seem to us to be destined for unnecessary difficulties because of this deficiency.

The proposals for community psychiatry and for new procedures in the administration of justice share another group of features which require discussion. Both are based upon the notion that the traditional institutions and agencies established by society for the purpose of coping with the behavior of mentally ill or criminally inclined persons are overcrowded, undermanned, underfinanced, and in need of fresh, innovative, experimental programs. They both endorse the principle that these new programs originate within the community and that primary emphasis should be placed on prevention. Moreover, both recommend that the Federal Government use its enormous powers of persuasion, through financing and legislation, to promote planning at the state and local levels by professionals working with ordinary citizens, the representatives of the community. Finally, the constant stress on the importance of the community to the patient and to the criminal means that the social environment is seen as the primary agent of change. The question we would now like to examine is this: To what degree is environmental correction currently being understood as a systems process?

At first glance there seems to be considerable evidence that the planning efforts do actually take into account the mutual influence between various environmental systems and organizations. An example of this apparent insight is the importance given to placing ordinary citizens along with a variety of professionals on the planning boards. If community representatives have a part in formulating the plans, then it is logical to expect that they will be able to interpret the plans to the community and gain a wide acceptance of proposed changes within the community. Further, if an assortment of professions and disciplines which have in the past tended to be antagonistic to each other find themselves working shoulder to shoulder on problems of innovation, then, presumably, they will understand each other better and will learn how to cooperate.

Another example of attention paid to systems effects is the emphasis given to the continuity of services to patients and criminals. The programs being planned insist on a comprehensive schedule which entails a coordination of agencies and a network of consultations binding together all the influences within the family, within such institutions as hospitals, courts, jails, and places of employment and within other public and private welfare facilities. The expectation that a fruitful increase in contacts between organizations will take place requires, it seems, a considerable faith in the ability of systems to coalesce or gear into each other's operations once they have been obliged to interact.

The position held here is that these aspects of program planning are not really signs of a systems approach to problems of change. On the contrary, they tend to mask a pervasive and unacknowledged lack of understanding of how systems actually operate within personality and social structures. There is something deceptively rational and naively oversimplified about the way planning is currently taking place. The very concept of large-scale, expensively financed planning is itself a peculiar denial of some of the fundamental properties of systems, especially of their conservative or change-resisting features.

Nothing can be gained by ignoring the *legitimately* rational aspects of current planning efforts. In a country such as ours which undergoes extremely rapid, involuntary, and unplanned social change, it certainly makes sense to try to control the direction of change, to make change happen rationally rather than to suffer it passively. In a country which so highly values the future, one must indeed look ahead to pressing national problems. But it must also be said that nothing can be gained by using planning as a slogan, as a cover to hide the deep-seated conflicts of professional and lay opinion and the firmly entrenched disagreements in such areas of policy as the treatment of criminals and mentally ill patients. Differences of this sort are too persistent and nonrational to be washed away by the proclamation of "bold new approaches."

Planning Versus Social Reform

Before the development of the more sophisticated languages of the social sciences, what is now called "planning" was known as "social reform." The advantage of social reform as a concept was that it implicitly connoted political struggle. Those who involved themselves in pushing the reform movement would normally expect intense opposition. One could declare oneself openly for or against, and the adversaries, as well as the uncommitted public, could easily keep track of the victories and defeats on both sides. Women's right to vote, prohibition, antitrust legislation, the organized labor movement, the civil rights movement, birth control, abortion, and capital punishment have been handled in this "old fashioned" way. Another advantage of the reform label was that its ideological base and its social targets were clearly in evidence. The struggle in the labor movement for the closed shop, workmen's compensation, and the like was clearly in the interests of the working class and against the middle and upper classes. Similarly, antitrust legislation was initiated in the interests of the independent or small businessman against big business, or the monopolies.

There is some evidence that reform movements in the health and welfare fields have always been characterized by a lack of clear-cut delineation of social class interests. The social security program, for example, cuts across social class lines. Similarly, public health programs have been pushed on the egalitarian ground that all people should have equal access to quality health services regardless of ability to pay. Nevertheless, the main thrust of these programs has been for the benefit of the lower-middle and indigent classes. Another chronic characteristic of these reform movements, particularly where public health is concerned, is that the research cited in support of the programs is largely initiated *after* the policy decision about the main components of the program have already been made. The decision itself is most often made on the ground of a set of values about which there is bound to be some controversy. The subsequent research and fact-finding is then used not only to support the program but also to convince those who resist the proposed changes that the program itself grew out of a firm knowledge base accumulated by prior research.

This last point is so important to our present discussion that I would like to quote from a recent review by Odin Anderson [1] entitled "Influence of Social and Economic Research on Public Policy in the Health Field." Speaking of research on general health policy rather than the special field of mental health, he states:

> Research reports have been pervaded with a certain naiveté that an optimum health service may be attained if doctors become more altruistic, patients wise, and health insurance would reduce cost of administration, and so on. No sophistication is discernible in assessment of possibilities, of tolerance of providers of services, the recipients of services and the health insurance agencies as to what is a workable equilibrium between them . . . The field is positively rife with incompatible objectives which research can elucidate and expose. But the results of research are not applied rationally in public policy because of overlapping needs and inherent frictions between the parties mentioned. Sometimes research is expected to answer questions which, at the moment at least, are unanswerable.

If these statements are true for the public health field in general, they apply with particular cogency to planning for community mental health and the administration of justice. Looking at the community mental health movement for the moment, we can see that it was formally initiated as a public policy decision by Congress in the 1955 Mental Health Study Act, which set up the Joint Commission on Mental Illness and Health. The act stated that Congress was responding to a sense of "critical need" for an objective, thorough, and nationwide analysis and reevaluation of the human and economic problems of mental illness and of the resources, methods, and practices currently utilized in diagnosing, treating, caring for, and rehabilitating the mentally ill.* The Joint Commission then carried on the research and fact gathering, over a five-year period, which formulated the principal items of mental health center programs and which was followed by the federal funding for the current planning phase of the movement.

Direction of Change

Just as with public health in general, the character of research on community mental health has obscured the "incompatible objectives"

* For a review of the antecedent and subsequent developments leading up to the current planning and beginning implementation phases of the community mental health programs, see John J. Stretch [5].

and "inherent frictions" which are associated with the direction of change growing out of the policy decision. Although the subject deserves a more penetrating analysis, we can here briefly outline some of the hidden conflicts. Like previous public health movements, community mental health is aimed at the poor and the indigent who have not been able to receive adequate psychiatric care under previous arrangements for private practice and public mental hospitals. This brings the program into conflict with the established psychiatric institutions, both public and private, which had previously settled largely for quality care for the reasonably well-to-do and inadequate or "custodial" care for the poor. If the program had been clearly labeled for what it is—a reform movement—one could easily have anticipated a strong resistance from the traditional psychiatric establishment which is oriented, on a fixed-fee basis, toward the treatment of middle and upper class strata, the poor having been seen by psychiatrists in private practice mainly for research and training purposes.

This shift in the *right* to claim and obtain adequate psychiatric care from the middle class to the economically deprived segments of the community also implies a strain in the old, established ways of financing therapeutic help. Outpatient and child guidance clinics, family service agencies, and vocational guidance and half-way house facilities, along with others, have traditionally been financed by the "Red Feather" combination of public and private funds; and their policies have been decided by lay boards composed of middle class nonprofessionals. Community mental health policies, especially the emphasis on comprehensiveness and continuity of care, tend to reorganize such agencies, placing more power of decision in the hands of lower class nonprofessional persons who can now lay claim to being placed on their boards. This rearrangement produces a conflict not only between the class interests controlling decision-making, but also between the old and the new definitions of the roles of staff within the agencies. Community mental health programs require agency staff personnel to stop thinking of the agency as providing only one type of service to clients. Agencies will have to begin—in some cases have already begun—to realize that they must take the total needs of clients into consideration. I will discuss later this shift in role definitions from the point of view of systems theory.

A third source of incompatibility derives from the apparent conflict between two theories of the origin and treatment of mental illness. Psychoanalytic theory, on the basis of which therapeutic practice in medical schools and hospitals has been taught for the past two decades, assumes that mental illness is caused by intrapsychic and mainly unconscious conflicts. Accordingly, treatment takes place in one-to-one, therapist-patient arrangements, or at best between one therapist and a small group of patients. The treatment focuses on relieving the emotional conflicts within the patient.

One-to-One Psychotherapy

But, as has already been pointed out, community mental health programs are based on environmental corrections, a procedure that is often difficult to fit into psychoanalytic theory. In fact, the principal theoretical impetus to the environmentalist approach derives from the social and behavioral sciences. Social psychologists, sociologists, cultural anthropologists, and behavioral psychologists have for many years emphasized the importance of community strains and stressful family interactions to the processes of mental illness and recovery. Although it has been claimed by some, including the author,* that there is no inherent or necessary incompatibility between these two theoretical approaches, training facilities for psychiatrists, social workers, and clinical psychologists are nevertheless currently undergoing an intense struggle between the traditional psychodynamic and the newer community organization approaches to the prevention and treatment of mental illness. Faculties of schools of social work are especially divided on this issue at the present time, while graduate departments in clinical psychology are characterized by an almost paralyzing indecision about which direction university policy should take. Meanwhile, the American Psychoanalytic Association, in a recent "Statement on Community Psychiatry" circulated for discussion to local societies, complains that, "Traditional one-to-one psychotherapy is sometimes disparaged as contradictory to the needs of community psychiatry. Little respect is shown for the particular skills of the psychoanalyst who administers to the needs of the individual without more than the usual regard for social or community considerations. This is part of the uniform ambiguity and ambivalence

* See, for example, John P. Spiegel [3, 4].

displayed toward psychoanalysis despite the fact that psychoanalysts are among the most active and prominent workers in the field of community psychiatry."

Complaints about the ambiguity and uncertainty of the community mental health programs are by no means confined to the American Psychoanalytic Association. There are probably many who, like the author, have had to deal with the plaintive question, "What do they mean by community psychiatry? What is it supposed to be?" Psychiatrists themselves are becoming too embarrassed to articulate such questions, since by this time they are supposed to know the answers. But ancillary professionals and laymen still register confusion as an alternative to the discomfort of pretending they do know the answers.

In part, this "Emperor's clothes" situation is due to the vigorous endorsement of community mental health planning at state and federal levels. It is difficult to oppose—or even to fail to understand—a program in which so much federal money is involved.

In part, also, the problem arises from the aforementioned failure to build a systems approach into the planning process. The strains outlined above could have been predicted from a systems analysis of the direction of change proposed by the new public health policies. Enough is known about the nature of decision-making in large-scale organizations and about the effects of organizational change to have introduced a note of caution with respect to the optimum speed of change expected in hospitals, schools, and agencies. It is known, for example, that the behavior of people within bureaucratic structures has less to do with their personal characteristics than with their roles as members of the bureaucracy. Such roles are learned slowly and sometimes painfully. Accordingly, they cannot be unlearned without a decided wrench, which usually means intense resistance to change. Moreover, members of organizations cannot be expected to adopt a change unless they can see that the change is to their own advantage and to the advantage of the agency as well. Loyalty to organizational tradition—or, to put it differently, identification with an institution—functions as a source of "irrational" resistance to the assumedly "rational" plans being proposed for adoption. Finally, a detailed systems analysis would show that if changes are actually adopted by one institution, say, a hospital, then all the other agencies with which it normally does business would have to make one of three decisions:

(1) to resist the change by vigorous protest; (2) to withdraw its services or referral procedures; or (3) to adopt a change which adequately meets the new conditions. Accordingly, the change in one part of the medical care system may set up reverberating cycles of strain and resistance in other parts of the system.

The danger of unattended forms of resistance should not be exaggerated. Given enough time—and under the lure of the federal "carrot"—mental health and welfare facilities will probably stumble in the direction of community psychiatry. The danger will likely take the form of lip service, of empty gestures, of assimilation of the language and formulas of community mental health to the traditional procedures in the style of *plus ça change, plus c'est la même chose.* Practices set up within the community for the purpose of effecting environmental corrections will be reinterpreted as offering insights into the intrapsychic problems of the culturally and economically deprived. At the present time no one can say to what degree such concealed forms of resistance will inhibit the growth of real innovation and change—plus local variations—in the healthy development of community mental health programs.

Insufficient space remains to analyze the recent recommendations of the President's Commission on Law Enforcement and the Administration of Justice from the systems point of view. The commission made a great number of suggestions for change in the direction of environmental corrections. It proposed following the planning model used in the development of the community mental health programs in order to get such suggestions discussed and adopted. To avoid, as much as possible, the kinds of errors inherent in that model, it would be advisable that the planning be conducted, on a considerably reduced scale, in a step-by-step process. It is not that the stimulus of the federal banner with its financial powers is wrong in itself. Rather, I would describe the error as consisting of attempting too much too fast. The proposed changes in law enforcement will, as in community mental health, encounter strong resistance. They will need outspoken "champions of change" at every level, federal, state, and local. In order to conduct an optimal systems analysis of the process of change, planning and implementation must occur together and must be given enough time that the sources of resistance and the required innovations can be carefully studied.

Four-Stage Program

Accordingly, a long-term program of change should be undertaken through simultaneous planning and implementation, conveniently divided into four stages:*

First, a "finding and seeking" stage designed to select individuals capable of carrying out a group of systematic pilot projects incorporating some of the changes suggested by the President's Commission.

Second, a "nurturing" stage in which the first pilot projects are established, supported, sustained, and tested over a sufficient length of time to determine their merits.

Third, a "growing and showing" stage in which the pilot projects are expanded and some of them begin to be used in other communities as demonstrations of what can be accomplished with the new methods.

Finally, a "reproduction" stage in which new projects are initiated in other communities and states, based upon and employing individuals who have been trained in the first pilot projects. Such a process of reproduction or transmittal will be less at the instigation of federal agencies and more the product of genuine interest within the communities in the productiveness and utility of the new programs of law enforcement.

The report of the President's Commission is of the greatest importance to all who are interested in improving the administration of justice. It should be carefully studied by psychiatrists and mental health professionals in general. It is to be hoped that the planning for its implementation will take place in a more cautious manner than has been the case for community mental health.

References

1. Anderson, O. W. Influence of social and economic research on public policy in the health field: A review. *Milbank Mem. Fund Quart.* 44: No. 3, Part 2, 1966.
2. Report to the President's Commission on Law Enforcement and

* These suggestions have been adopted from a report prepared by the Organization for Social and Technological Innovation [2].

the Administration of Justice. Cambridge, Mass.: Organization for Social and Technological Innovation, 1966.

3. Spiegel, J. P. Some Cultural Aspects of Transference and Countertransference. In J. H. Masserman (Ed.), *Science and Psychoanalysis*. Vol. 2, *Individual and Family Dynamics*. New York: Grune & Stratton, 1959. Pp. 160–182.
4. Spiegel, J. P. Formal and Informal Roles in Newly Acculturated Families. In D. M. Rioch (Ed.), *Disorders of Communication*, Vol. XLII. Research Publications, A.R.N.M.D. New York: Association for Research in Nervous and Mental Disease, 1964.
5. Stretch, J. J. Community mental health: The evolution of a concept in social policy. *Community Ment. Health J.* 3:1, 1967.

19

Patient Predicament and Clinical Service: A System*

NORRIS HANSELL, M.D.

THE TERM *SYSTEM* as used in this report refers to (1) a method for analyzing a problem: cataloging its parts, the setting, and their interrelationships; (2) a method for designing a response to that problem, taking into account the several parts to the whole; and (3) a method for monitoring, controlling, and changing the response so it may continuously meet changes in the reality of the problem. System concepts are a way of describing reality drawn so as to conceive a certain "whole" divided into "parts" in specified "relationships." The description is designed so the behavior of the parts is generally and substantially explained by a small number of specified relationships of the parts, or "states," of the whole. A system has a "boundary," includes certain elements which are "in" it, and excludes a much larger number of elements defined as "outside." The definition of system in this general way is similar to the following statement in other language:

> The problem we are facing is composed of the following aspects . . . and any management of this problem will need to understand the following relationships between aspects of this problem . . . understood within the following environment. . . .

There is an important and growing body of material relating to the consideration of living systems [4, 8, 9], social systems [1, 10], health support systems [3, 7], and mental health operations [2, 5] under the rubrics of general systems theory.

* Reprinted with permission from *Arch. Gen. Psychiat.* 17:204–210, 1967.

It is not the purpose of this report to review that literature, but to give an example of a specific agency, the H. Douglas Singer Zone Center of the Illinois Department of Mental Health, and a consideration of a particular clinical design growing out of a general systems approach.

The System in Overview

The Winnebago-Boone Catchment Area Service Unit of the H. Douglas Singer Zone Center is an incare and transitional outcare services unit relating to two counties in northern Illinois. It is part of a larger developing community mental health and mental retardation services agency, the H. Douglas Singer Zone Center in Rockford, Illinois, and is not far from the population center of the catchment area it serves. Winnebago and Boone Counties have a total of 245,000 citizens, and are responsible for an average resident population of 515 persons in Department of Mental Health hospitals, with an average yearly admission of 370 persons. The same population is responsible for an average resident population of 205 citizens in state schools for the mentally retarded, and an average yearly admission to the state schools of about 15 persons. These counties are also annually responsible for 160 state and federal prisoners newly enclosed each year, 35 juvenile commitments to correctional institutions, 200 marital dissolutions, 22 suicides (estimate), and 1,300 retarded children of school age (estimate).

The typical patient who enters the auspices of the Winnebago-Boone Unit has clear and substantial clinical mental illness, is poor, is lacking a job or job skills (or both), and is a member of a family which is coming apart, or else not a member of a family. In addition, he may be old, in trouble with the law, and may have developed habits which make him difficult to relate to in conventional social roles.

The system of which the mental health casualty patient is a member includes himself, the repertory of problem-solving skills he has available, his family, the groups of which he is a member, the agencies he could or does relate to, and the particular health service professionals and social services professionals he could, or does, relate to. It also includes the opportunities for socialization, education, employment, residence, money and income, etc., as well as the relevant environment of attitudes linked with mental health casualty persons relating to their human value, attributes, potential for reentry into society, and so on. The system also includes judges, the courts, and police: the symptoms and trouble the patient is a part of bring him to the attention of these social agents. Their role in linking him to a management system or, alternatively, extruding him to a containment system is crucial. The H. Douglas Singer Zone Center, and in particular the Winnebago-Boone Catchment Area Service Unit, is designed and operates as *convener* to the agents and agencies which relate to the patient's predicament so that they may behave as a system. No single operation of the Zone Center is as important as its function of convener to the parts of an interrelated system. Convening the parts can bring to bear on the predicament of the patient latent conjoint (system) potentiality in the total field.

The problem system is a patient at high risk of institutionalization, high risk of death, or the cause of social offense or injury, who has a package of trouble including old age, poverty, deteriorated or absent social skills, substantial alienation from the mainstream of the community, as well as certain "psychiatric" symptoms which may include thought disorder, altered feeling states, and altered control of behavior.

Operations in Response to Problem System

The Winnebago-Boone Catchment Area Service Unit has a staff of 39 persons with a psychiatrist as clinical director and a social worker as administrator. The Unit is responsible for incare operations in 30 beds, as well as case-specific consultation, transitional outcare services (preceding, following, or in place of incare), and the development of system-linked relationships with a network of agencies and agents in the field.

The incare phase of treatment consists simply of a socially activating environment, with an average length of stay for all patients of 26 days. The staff work has an emphasis on testing social skills, group discussion methods, patient government, and phenothiazines and antidepressants.

All persons who are proposed to the staff as being in need of incare by any portion of this system (the patient, his family, a judge, a physician, or another social agency) are each scheduled for a screening-linking-planning conference. These conferences are held twice each morning and twice each afternoon. Their purpose is twofold: on the one hand, they identify with precision persons who are at risk of institutionalization and, therefore, proper for Unit services; on the other hand, they identify the network of persons and agencies which can be linked to the predicament of this particular person in trouble.

An example of the usefulness of the screening-linking-planning conferences can be given by the change in the system activity of the courts and judges. Until the opening of the Winnebago-Boone Unit, the principal mental health roles of the courts were: (1) to receive distress messages from families and individuals and reports of persons with grossly disturbed behavior; (2) to arrange for the apprehension, transportation, and holding of such people; (3) to arrange for the examination of such persons by two physicians; and (4) when a sufficient number of persons had been certified in need of inpatient care, a station wagon would transport a full load to the nearest state hospital which was, depending on the case, 35 or 100 miles distant. Since the opening of the Winnebago-Boone Unit, an important evolution has occurred in the system activity of the judges and courts. Judges continue to receive distress messages from family members and from patients, but instead of converting them into a request for commitment, they convert them into an appointment at the Singer Zone Center for a screening-linking-planning conference. The judge, along with the unit staff, insists that the patient, members of the family, and representatives of involved agencies participate in such a conference. The general practitioner of medicine, or the family doctor, is involved; if he cannot be there in person, he is included by a conference telephone call so that he can be a party to the planning and is able to consider being a portion of the aftercare apparatus.

Through the Singer Zone Center function of acting as convener to the several parties in planning for a patient's predicament, including the judge, the family, and the physician, a series of possibilities is raised for the management of the patient's problems which are not otherwise truly available. Equally important, *changes occur in the behavior of parts of the referral system* because they develop a more intimate knowledge of each other's behavior and of the system aspects of each other's behavior. The courts now collaborate actively in utilizing the Zone Center as a convener to the parties in planning for management of the problem, rather than as a place to deposit troubled persons.

The judges became informed of the operation of the screening-linking-planning conferences partly through our staff's describing it to them, and from hearing about it from patients and families. But a surprising and most important step of growth occurred when the judges directly observed the conferences through a one-way mirror and sound system. They developed an appreciation for new issues in the complexity of the overall predicament, and an understanding of the Unit staff's particular way of appraising the problem and what could be done about it. They also came to an appreciation of how the court could become a constructive collaborator toward a community reentry outcome rather than a participant in the process of alienation and deposition in a distant institution.

The mental health expediter is another important element in the design of the Winnebago-Boone Unit. The expediter's job is to sit in on the screening-linking-planning conferences at intake and to establish and list those elements of the patient's predicament for which complicated extrainstitutional arrangements need to be made. These arrangements might include job training, getting a room in a boarding house, getting the patient joined up with an aftercare group, helping him get a job, etc. The expediter establishes target dates for the completion of the several tasks and starts on them right away, i.e., at the time of admission to incare. The function of the expediter in this mental health setting is precisely the same as in the industrial setting: he does a variety of things in order to cause an overall package of events to happen so that the successful outcome can occur. He is the agent responsible for attaching people to families and planning for an appropriate mix of programs to support a human existence at a maximum contribution to the community. He links a system to a troubled person.

The expediters are persons with a baccalaureate degree, some of whom are returning Peace Corps volunteers and some of whom are other persons interested in innovative social service. The Illinois Department of Mental Health supports a one-year educational program at Northern Illinois University in DeKalb, ending in a masters' degree in behavioral science, social science, or psychological counseling.

During the incare phase and the transitional outcare phase, many patients are linked to several kinds of transitional groups. As is generally true for many agencies, the identified person in trouble is not the only one in his family with difficulties; headway must be considered family headway. Usually several family members are brought into the process in the same, or different, groups. All of the treatment groups are of relatively short duration, eight to ten sessions, and are specifically aimed at identifying problems, bringing them to group discussion, and asking the group's reaction, advice, and help, on particular *decisions*. Some patients have never been members of groups and have substantial difficulty in deploying group skills. We have, therefore, established transitional groups and spin-off groups. Spin-off groups have ten to twelve members, meet eight to ten times, and quickly learn such functions as "the convener," "the arbiter," "the social chairman," "the recorder," and the like. The task of these groups is to become a collectivity expected to function autonomously, without a staff member being present, and to be a place where persons without a group support system can meet and help each other in crises. Many of these groups do not actually spin off to independent, leaderless function; but in the process of attempting to form a group, to allocate the various jobs which need to be done, and to set up group sanctions and behavior influencing systems, a great deal of learning takes place which may be helpful in learning how to remain in a noninstitutional position in society.

The officer of the day is a person of the Unit staff charged for 24 hours with receiving all telephone calls and requests for service. It is his responsibility to determine (1) whether there is a serious question that the inquiry relates to a patient at high risk of institutionalization, and (2) whether the patient is in trouble and needs referral to a mental health resource, but is not in immediate danger of institutionalization, hospitalization, or event of violence. He is charged with maintaining the precision of focus on a particular risk group, those at risk of institutionalization; he also determines the first opportunity that the persons to be convened at the screening-linking-planning conference can be gotten together. He makes arrangements for a screening conference bringing together the persons he thinks should be involved, convenes the conference, and is charged with drawing from the present situation and from historical information data necessary to the assessment of risk, as well as bringing in such special examinations as may be needed, e.g., a physical or mental status examination. Always the expediter is present, and always a member of the incare staff on the unit to which the patient may be admitted is present.

Community organization, apart from case-specific convening, is also a responsibility of the Winnebago-Boone Unit. The Unit developed and regularly reconsiders its intake policy and its services design in a dialogue with a mental health planning authority for the territory, an authority which lies outside the Unit. In order to make sure that a growing group of persons is aware of the shortage of services, of the poor connections and coordination between services, and of other problems that lie in the casualty management system, the Winnebago-Boone Unit established a blue ribbon group of power-linked citizens and agency professionals with whom it meets regularly. They examine: (1) the type of calls received by the officers of the day; (2) the disposition of such calls; (3) the composition of troubles that persons admitted to the incare unit or to transitional services seem to have; (4) the problems in finding aftercare and rehabilitative services for persons with complex packages of trouble; and (5) the continuing residual alienation of community attitudes in agencies, policies, insurance, general hospital care, etc. By examining these troubles with a group of citizens and professionals outside of itself, a group charged with territorial planning for the catchment area, the Unit establishes a constituency interested in advocating improved care in management of the problems of its particular risk group of patients. One of the difficulties has been that there was no constituency for the most severely alienated mental health casualties. The most vocal mental health constituency was related to minor grades of mental health disability and to persons substantially still members of families, with jobs, and members of the middle- and upper-income groups.

The Winnebago-Boone Catchment Area Service Unit designed its services and the risk group precision of its intake policy to attain a maximum amount of linkage between community organization goals and direct patient service goals. By focusing on the most seriously ill group of people, involving the largest number of potential contributors to the management of that group of persons, operating the screening-linking-planning conferences at a high level of visibility, and regularly apprising community groups within and without the casualty management system of difficulties within that system, a maximum amount of community education is obtained. Commitment to improving the predicament of the severe casualty group is gradually established. The system aspects of an intake policy, of the screening-linking-planning conferences, and of the design of an intake policy are crucial to the continuing effectiveness and improvement of a casualty management network.

Results

At the time of this writing, the Winnebago-Boone Catchment Area Service Unit had been open less than a year; but the longer-term good results of an agency's pattern of operations are not assured by shorter-term good results. We can, however, report good short-term results, with a promissory note for later follow-up.

Keep in mind, in considering these results, that the risk group being taken care of by the Winnebago-Boone Catchment Area Service Unit is the same risk group, as close as we can know, to that previously cared for by the several state hospitals which have serviced this territory. That is, *any* person from Winnebago or Boone Counties who requires inpatient care and is admitted to Department of Mental Health inpatient care can only be admitted to the Winnebago-Boone Catchment Area Service Unit; the state hospitals formerly serving them are no longer available. Patients arriving at these state hospitals by mistake are brought back to the Winnebago-Boone Catchment Area Service Unit.

The average age of persons previously admitted to Elgin and East Moline state hospitals from Winnebago and Boone Counties was 42.2 years; the average age of persons admitted to the Winnebago-Boone Catchment Area Service Unit is 41.5 years. The average length of stay of persons who were admitted to Elgin and East Moline state hospitals was 38 weeks; the average length of stay for persons admitted to the Winnebago-Boone Service Unit is 26 days. The package of trouble that the patients arrive with is presumably the same: clinical symptoms, poverty, lack of job skills, deteriorated or absent social skills, absence of a family or a small group support network, and absence of a place of residence.

Because of the cooperation of the courts, social agencies, and families in the screening-linking-planning conferences, and because of the good results of the expediters in achieving effective aftercare programs, the treatment power of the 30 beds of the Winnebago-Boone Unit is equivalent to 420 plus beds of the Department of Mental Health formerly deployed for the care of the same population. The cost of operating the Winnebago-Boone Unit is about $60. a day, allocating a fair fraction of the administrative costs of the larger agency of which it is a part; the daily costs at the state hospitals it replaces approximated $6. a day. However, the length of stay in the Winnebago-Boone Unit is less than one-tenth as long as the previous hospitalization. The cost per episode of incare is therefore about the same. The delivery effectiveness is much better in the Winnebago-Boone Unit because the patient is more likely to be kept as a member of the family, or to be kept on the job if he has one. The process of alienation and extrusion to distant institutions is aborted at an earlier state.

Comment

The design and operations of the Winnebago-Boone Catchment Area Service Unit are presented as an example of a systems approach to the design of a casualty management apparatus. The elements of the system include the patient, his skills, his social predicament, his family, and the agencies, employment, and residence which are or can be related to this predicament. The problem system also includes his poverty, substantial alienation from the community, and lack of a constituency and community embrace. The Zone Center is conceived as an agency which inventories the problem and convenes the portions of the system which can relate to one another over the design of an individual person's or a group's care. The Zone Center is also the supplier of a particular portion of that care, a portion which is designed to gain maximum advantage in community organization and community education.

Negotiation of interagency agreements between the Zone Center and other agencies has been a source of great opportunity in bringing the casualty management system together. Our negotiations aim for a reciprocal and symmetrical *nondecline* interreferral agreement. We attempt to identify the risk group we attend to; then we give other agencies permission to identify persons who are members of that risk group for us, and refer to us.

Referral is always to the screening-linking-planning conferences, rather than directly to incare, so that no opportunity is lost for linking persons in at the moment of intake. It is our experience that the moment of intake is one of the most flexible times in the career of a mental health casualty because of the high "crisis" element in it. If this opportunity for linking the person to his family, agencies, an aftercare network, job skills, a job, and a residence is lost, it is a most important loss. The other agencies in our network have found that screening-linking-planning conferences are valuable to them and they participate actively. To accommodate the pressing time commitments of other agency staffs, we involve them mostly through conference telephones.

The symmetrical nondecline interreferral agreements tend gradually to bind the assembly of agencies into an interdependent system, and it becomes less possible for patients to fall between "gaps." Of equal power in filling gaps is the committee of citizens and agency representatives which is convened regularly to discuss the calls taken by the officer of the day, the intake conferences, and problems at the point of transfer from incare to transitional services.

The particular strategy of focusing the activity of an expensive and highly visible new mental health center exclusively on persons at high risk of institutionalization, although it has a great deal of sensibleness in it, is a particularly risky strategy. The risk involved is related to the public frustration of focusing on a group of troubled people whose problem is so serious that they are socially almost invisible. Much more visible and vocal people are not serviced by this new agency. In addition, closing off the avenues for easy deposition of persons who were formerly in jail or were found on the streets raises with new urgency options for the community to develop effective alternative patterns for management. The activity of this agency in frustrating community agencies, caregivers, and citizens in the practice of depositing troubled people then running has been most effective. It has greatly improved the care of a group of troubled people, but has also been associated with the development of a significant amount of criticism. An agency which takes on this particular systems approach has to be willing and able to deal with the push and strain of public criticism; the staff must understand the policy, endorse it, and be committed to it.

The involvement and commitment of a staff of 50 to 100 people in a strategy which is not well understood when they arrive on the scene, and not well understood by the community, is a particular administrative problem. The administration of the H. Douglas Singer Zone Center felt that the previous service pattern was partly perpetuated by terms such as *commitment, state hospital, mental illness, admission, discharge, aftercare,* and *referral.* These terms certainly have usefulness, but are associated with stereotypes linking the service network to a pattern of operations which is ineffective. We, therefore, took the step of attempting to crush or soften the current conceptual model and substitute new concepts. Often the new concepts and the new words have quite similar explicit meanings to previous terms; *but they are not* (yet) *linked to the stereotypes.* In addition, many of the persons one is talking with do not understand the new terms, and it provides an opportunity for drawing them into a meaningful encounter with "the whole problem" (the system).

In order to link the staff to a specific pattern of operations, we introduced a specific way of appraising our productivity; it is a numerical method and is related to reduction in the territorial per capita, per diem use of institutionalization employed to manage reduced social competence. *Because we are all appraising our work by the same system, we are able to say when we are successful and when we are not.* The terms that we use to describe our work are very much the ones such as are used in this article: average length of stay; cost of an incare day; number of persons who arrive without a job, but who have a job by the time they leave; employer or family's appraisal of comfort and effectiveness before and after care, and the like [6].

It is worthy of mention that when staff arrived at this new agency and were given a very general charge ("to operate in such a way as to link patients to families and to a support system and to improve the level of effectiveness of operation of persons with reduced social competence, etc.") and then bound to this general charge, a great deal of anxiety developed. Only over a period of six months did a gradually more detailed strategy unfold; as it did, the level of anxiety went down and the energy which was bound to the general task became bound to the specific strategy. The process of the staff discussion was structured as (1) an examination of the several options which are open to it at that point; (2) the building of a decision model for deciding among the options; (3) the making of a decision which all may not agree with, but all agree to abide by; and (4) the setting up of patterns of action with target dates.

Because the whole staff has been involved in an explicit understanding of the process of moving from a general charge to a specific strategy and pattern of operations, they follow their movement in this process with interest. They are able to objectify their own anxiety and see its relationships to the portion of the task in which they are engaged. Making the staff aware of the administrative system within which it works, the conceptual system within which it works, and the decision model within which it works in great detail, allows it effectively to monitor and correct its own behavior. The specific actions that a service operation carries out are often not as successful as first hoped; if the service operation is bound to either a concrete collection of actions or a general theory of explanations, it may become fixed and ineffective. If it has a continuous monitor of its behaviors against a system of goals, it is more likely to remain on target.

Summary

The Winnebago-Boone Catchment Area Service Unit of the H. Douglas Singer Zone Center in Rockford, Illinois, services a group of mental health casualties using 30 beds which had required 420 plus beds of the Illinois Department of Mental Health. The effective pattern of operations of this unit was developed as a systematic response to an inventory of the predicament of its typical patient and a systems statement of the agency and casualty management environment within which the unit operates. Administrative problems and problems in evaluating the effectiveness of the unit also were developed under the guidance of general systems theory. The formulation of a mental health service network as a system is an acknowledgment of the related facts that the human personality is a system, that society can be understood as a system, and that the casualty management network is a subsystem of that society.

Acknowledgment

Gerald Caplan, M.D., Harvard Laboratory of Community Psychiatry, aided in the formulation of systems concepts in mental health work.

References

1. Jackson, D. D. *The Role of the Individual.* Procccdings of thc Conference on Mental Health and the Idea of Mankind. Chicago, February, 1964.
2. Klein, D. C. The community and mental health: An attempt at a conceptual framework. *Community Ment. Health J.* 1:301, 1965.
3. Paul, B. D. (Ed.) *Health, Culture and Community.* New York: Russell Sage Foundation, 1955.
4. Richter, C. P. *Behavioral Regulation of Homeostasis: Symposium on Stress.* Washington, D.C.: Walter Reed Army Medical Center, 1953. Pp. 77–88.
5. Ruesch, J. Social psychiatry. *Arch. Gen. Psychiat.* (Chicago) 12:501, 1965.
6. Smith, W. G., and Hansell, N. Territorial evaluation of mental health services. *Commun. Ment. Health J.* 3:119, 1967.
7. Vickers, G. The psychology of policy making and social change. *Brit. J. Psychiat.* 110:465, 1964.
8. von Bertalanffy, L. An outline of general system theory. *Brit. J. Phil. Sci.* 1:134, 1950.
9. von Bertalanffy, L. The theory of open systems in physics and biology. *Science* 111:23, 1960.
10. Wallace, A. F. C. Mazeway disintegration: The individual's perception of socio-cultural disorganization. *Hum. Organ.* 16:23, 1957.

20

Interdisciplinary Versus Ecological Approach*

EDGAR H. AUERSWALD, M.D.

THE EXPLOSION of scientific knowledge and technology in the middle third of this century, and the effects of this explosion on the human condition, have posed a number of challenges for the behavioral sciences that most agree are yet to be met. The overriding challenge is, of course, the prevention of nuclear holocaust, but such problems as crime and delinquency, drug addiction, senseless violence, refractive learning problems, destructive prejudice, functional psychosis, and the like follow close behind.

Practically all behavioral scientists agree that none of these problems can be solved within the framework of any single discipline. Most espouse a putting together of heads in the so-called interdisciplinary approach. The notion is not new, of course. The interdisciplinary team has been around for some time. Some new notions have emanated from this head-banging, but there have been few startling revelations in the last decade or so.

However, a relatively small but growing group of behavioral scientists, most of whom have spent time in arenas in which the interdisciplinary approach is being used, have taken the seemingly radical position that the knowledge of the traditional disciplines as they now exist is relatively useless in the effort to find answers for these particular problems. Most of this group advocate a realignment of current knowledge and reexamination of human behavior within a unifying holistic model, that of ecological phenomenology. The implications of this departure are great. Once the model of ecology becomes the

* Presented at the 122nd annual meeting of the American Psychiatric Association, Atlantic City, N.J., May 9–13, 1966. Reprinted with permission from *Family Process*.

latticework upon which such a realignment of knowledge is hung, it is no longer possible to limit oneself to the behavioral sciences alone. The physical sciences, the biological sciences, in fact, all of science, must be included. Since the people who have been most concerned with constructing a model for a unified science and with the ingredients of the human ecological field have been the general systems theorists, the approach used by behavioral scientists who follow this trend is rapidly acquiring the label of the "systems approach," although a more appropriate label might be the "ecological systems approach."

These terms are currently being used metaphorically to describe a way of thinking and an operational style. They do not describe a well-formed theoretical framework as does the term *general systems theory*. It is with the former, the way of thinking and the operational style, that I am concerned in this paper.

The two approaches described above differ greatly. Let us examine why the difference is so profound. The ongoing accumulation of knowledge and its application to practice follows a well-known sequence. This might be broken down into steps as follows: the collection of information or data, the ordering of that data within a selected framework, analysis of the data, synthesis of the results of analysis into hypotheses, the formulation of strategies and techniques (methodologies) to test the hypotheses, the construction of a delivery plan for use of these strategies and techniques, the implementation of the plan, and the collection of data from the arena of implementation to test its impact, which, of course, repeats the first step, and so on.

The key step in this sequence is the second one, the ordering of data within a selected framework, because it is this step, and this step alone, that gives structure to the rest, all of which are operational. Not only do the nature and outcome of subsequent steps depend on this structuring framework, but so does the prior step, the collection of data. What data among the infinite variety of available natural data are considered important, and are, therefore, collected in any given arena, will depend on the conceptual framework used. It is here that the difference between the two approaches is to be found.

The interdisciplinary approach maintains the vantage point of each contributor within his own discipline. While it has expanded the boundaries of the theoretical framework of each discipline to include concepts borrowed from other disciplines, only those concepts which pose no serious challenge or language difficulties are welcomed. More importantly, I think, the interfaces between the conceptual frameworks of different disciplines are ignored, and, as a result, the interfaces between the various arenas of systematic life operation (e.g., biological, psychological, social or individual, family, community) represented by different disciplines are also ignored.

The structural aspects and the clarity of context of the data collected are lost as a result. The precise source, pathway, and integrating functions of messages passing between various operational life arenas in the ecological field cannot be clearly identified. Analysis of such data depends almost entirely on the *content* of these messages, and much distortion can and does take place.

The "systems" approach, on the other hand, changes the vantage point of the data collector. It focuses precisely on the interfaces and communication processes taking place there. It begins with an analysis of the *structure* of the field, using the common structural and operational properties of systems as criteria for identifying the systems and subsystems within it. And by tracing the communications within and between systems, it insists that the structure, sources, pathways, repository sites, and integrative functions of messages in addition to their content become clear. In my opinion this, plus the holistic, nonexclusive nature of the approach, minimizes the dangers of excessive selectivity in the collection of data and allows for much more clarity in the contextual contributions to its analysis. And the steps which follow, including prescription and planning of strategies and techniques, gain in clarity and are more likely to be rooted in concrete realities.

There are some very practical advantages that accrue as a result of the above. At the level of *theory*, for example, the ecological systems model, by clarifying and emphasizing the interfaces between systems, allows for the use of a variety of theoretical models which have to do with interactional processes and information exchange. These models form bridges between the conceptual systems of single disciplines. Information theory, crisis theory, game theory, and general communications theory, for example, represent some of the bodies of research and knowledge which become usable in an integrated way.

Knowledge that has been accumulating from the study of specific ecological systems, such as the family and small groups, the development of which lagged until recently because the systems did not fit neatly into the bailiwick of any one traditional discipline, can also be included without strain. And the developmental model of the life cycle of the individual man and of various larger human systems as they move through time in the ecological field of their environment assumes meaning in a larger context.

In addition, the use of this model in planning has demonstrated its many implications for the design and operational implementation of delivery systems, especially for community programs (e.g., "comprehensive community health" programs). The ecological systems approach insures that the entire process of planning for a community is rooted in the realities and needs of that community. The organized identification of the ecological systems making up a target community allows for the planned inclusion of information collection stations in each key system and at primary interfaces which provide feedback to the planning arena, thus setting up a servo-system which assures that planning will remain closely related to changing need. Over a period of time, as a picture of a target community emerges from such data, it will emerge as an idiosyncratic template of the structural and operational configurations of that community. It will not, as in the interdisciplinary approach, emerge as a predetermined template of the theoretical structure of the dominant discipline.

As a result, program designs constructed in this manner are deeply imbedded in the target community. They will develop as another ecological system among the many, thus greatly clarifying the context in which any program can be integrated into the life of the community as a whole. Furthermore, the delivery organization itself becomes viewed as a system with assigned tasks made up of subsystems performing subtasks including intraorganizational tasks. This allows for more clarity in the selection of staffing patterns, in the definitions of staff role functions, in the construction of communication systems and data collection (record-keeping) systems, and of the assignment of tasks within the organizational structure to staff members best equipped to handle them. Of special import to community programs is the fact that with the clarification of specific tasks to be performed comes the increased possibility of identifying those tasks that can be carried out by staff members or volunteers who need relatively little training.

At the *operational* level the strategies of evolution and change can be more clearly designed. More important, perhaps, use of the ecological systems approach allows for the development of a whole new technology in the production of change. Many techniques have, as a matter of fact, already appeared on the scene, largely within organized movements aimed at integration in its broadest sense, such as the Civil Rights Movement and the War On Poverty. Some community organization and community development programs, techniques using economic and political pressure, and techniques which change the rules of the game, such as the nonviolence movement, all represent a new technology and all have their relevance to the broadly defined health needs of socially isolated individuals, families, and groups.

In service programs working with individual people and families this new technology is also emerging, more slowly perhaps. Many new ways of coping with familiar situations are being developed. Techniques of treating families as systems, for example, represent one advance. In particular, an emphasis which stresses the organization of events in time and traces the movement of the developing infant-child-adolescent-adult-aged individual's degree of participation versus his isolation in relation to his family and to the flow of surrounding community life—such an emphasis makes it possible to determine with much more clarity in what life arenas the individual, the family, or a group of individuals needs assistance, and thus to combat more effectively the anomie and dehumanization characteristic of our age. The result is that the targets of therapeutic activity are much clearer and therapeutic work is more clearly focused on forces and situations that are truly etiological in a given problem situation. Techniques of producing therapeutic change can be brought to arenas much larger than the therapy room or even the home. I think that a single story will serve to illustrate more concretely what I mean.

In the story I wish to tell, two therapists, one a systems thinker, the other a member of an interdisciplinary team, became involved in the case of a runaway girl.

To give you some initial background, I should explain that I have been involved in designing and implementing a neighborhood health services system for provision of comprehensive biopsychosocial care to a so-called disadvantaged community. The main aim in setting up this unit was to find ways to avoid the fragmentation of service delivery which occurs when a person's problem is defined as belonging primarily to himself, and he is sent to a specialist who is trained to deal primarily with that type of problem. The specialist naturally sees the problem not only as an individual matter, but defines it still further according to the professional sector he inhabits. He is not accustomed to looking at the total set of systems surrounding the individual with the symptom or to noticing the ways in which the symptom, the person, his family, and his community interlock, and he is often in the position of a man desperately trying to replace a fuse when it is the entire community power line that has broken down. Furthermore, the specialist's efforts to solve the problem are apt to be confined to arbitrarily chosen segments of time called "appointments." And finally, there is that unfortunate invention, the written referral, a process of buck-passing that sends many a person in trouble from agency to agency till he finally gives up or breaks down. As a beginning we decided that we would have to pilot some cases in order to gain some experience with the different approach we felt was needed.

At this point, a case providentially dramatizing the points we had in mind fell into our hands. (We have since found that almost every case that falls into our hands providentially dramatizes these points.) One of our psychiatrists was wandering about the neighborhood one day in order to become better acquainted with it and to explore what sort of crises and problems our neighborhood program must be prepared to serve beyond those we already anticipated. I should say here that this psychiatrist,* by virtue of several years of pioneering work with families, including the experimental use of game theory and games in diagnosing and treating them, was particularly well qualified to handle the situation I will describe. His explorations that day had brought him to the local police station, and while he was talking to the desk sergeant, a Puerto Rican woman arrived to report that her 12-year-old daughter, Maria, had run away from home. This was apparently not the first time. She described the child to the police, who alerted their patrols to look for her and assigned two men to investigate the neighborhood. Our psychiatrist, whom I will refer to from now on as our "explorer," was intrigued and decided to follow up the situation himself.

He first identified himself to the mother as she left the police sta-

* Dr. Robert Ravich. I am indebted to Dr. Ravich for the case material reported.

tion and asked if she would be willing to allow him to help her with her current difficulty. She agreed. He learned that she lived a few blocks away with her now absent daughter and another daughter, aged 14. Her own parents lived nearby, and she had a paramour who also lived in the neighborhood. The father of her two children had long since deserted his family, and she was uncertain as to his whereabouts. The exploring psychiatrist learned also that the runaway girl had been seeing a psychotherapist at the mental health clinic of a local settlement house. In addition, he ascertained the location of her school.

He then decided that his behavior might appear unethical to the child's therapist, so he proceeded to the mental health clinic, a clinic which prided itself on the use of the interdisciplinary team approach. The original therapist turned out to be a social worker of considerable accomplishment and experience, who agreed to cooperate with him in his investigation after he explained what he was up to and that he had the mother's permission. He read the child's case record and discussed the girl with the therapist at some length. He learned that at a recent team case conference the diagnosis which was originally assigned to the girl, that of childhood schizophrenia, was confirmed. The team also decided that in the light of repeated episodes of running away from home, her behavior was creating sufficient danger to indicate that she be placed in a situation where that danger would be alleviated while her therapy continued. For a 12-year-old Puerto Rican girl in New York City, especially one carrying a label of schizophrenia, this almost always means hospitalization in the children's ward of a state hospital. Accordingly, the arrangement for her admission to the state hospital covering the district had been made and was due to be implemented within a few days.

The next stop for our explorer was the school, where Maria's teacher described her as a slow but steady learner, detached from most other children in the class, vague and strange, but somehow likable. The guidance counselor reported an incident in which she had been discovered masturbating an older boy under the school auditorium stairs. This behavior had led the school authorities to contemplate suspending her, but since they knew her to be in treatment they decided to hold off, temporarily at least.

The exploring psychiatrist also learned at the school that Maria was involved in an after-school group program at the settlement house. He returned there and got from the group worker a much more positive impression of the girl than he had previously encountered. She participated with seeming enthusiasm in the projects of the group and got along very well with the other children. The group worker, by way of providing evidence that Maria had much potential, showed the therapist a lovely and poignant poem she had contributed to a newspaper put out by the group. It was never ascertained whether the girl had written or copied the poem. She had, nevertheless, produced it, and there was general agreement that its theme of isolation was one which was expressive of her.

Back at Maria's home, our explorer talked to Maria's sister, who at first grudgingly, but then with some relish, admitted that she knew where the girl had gone during her previous runaway episodes. She was the sometime mascot of a group of teenage boys with whom she occasionally traveled for two or three days at a time. The sister did not know where she went or what she did during the junkets, but she suspected that sex was somehow involved. She also volunteered the information that neither she or her mother had ever found it easy to communicate with her sister, and that if the therapist really wanted to talk to someone who knew her, he should talk to her grandfather. So he went on to the grandparents' apartment.

The grandmother turned out to be a tight-lipped, highly religious Pentecostalist who was at first unwilling to say much at all about the girl.

The grandfather, however, was a different kettle of fish. Earthy, ebullient, jocular, and bright, though uneducated, his love for Maria was immediately apparent. He spoke of her warmly and bemoaned the lack of understanding that existed in her home. Remembering a passing reference in the case record at the mental health clinic to a suspicion that the grandfather may have engaged in seductive play with the girl, if not open sexual activity, our explorer raised the issue of the girl's emerging adolescent sexuality. This brought an outburst from the hitherto silent grandmother that confirmed the mutually seductive quality of the grandfather's relationship with the girl, followed by a return blast from the grandfather who revealed that his wife had refused to sleep with him for several years. He readily admitted his frustrated sexuality and the fact that he was at times aroused by his budding granddaughter.

I have presented only a sparse picture of the rich amount of information collected by our explorer up to this point. In a continuous five-hour effort, without seeing the absent Maria, he was able to construct a picture of her as a child who had grown up in relative isolation in a home where she received little support and guidance. Communication between herself and her mother had become more and more sparse over the years, most likely because of efforts of her older sister to maintain her favored position in the home. She had turned to her grandfather, who, feeling frustrated and himself isolated in his own marriage, brought his sexually tinged warmth willingly into a relationship of mutual affection with her. Furthermore, it seemed clear that with someone like the group worker who liked her and who, because the group was small, could spend time with her, Maria could respond with warmth and exhibit an intelligence that otherwise remained hidden. But, and this was, of course, speculative, the tools she perceived as useful in her search for a response from others would most likely be limited to infantile techniques of manipulation developed in early years prior to the need for verbal communication or, based on the relationship with the grandfather, some form of seduction where the currency of acceptance was sex. And, at the age of puberty, having been shut out of the female world of her mother and sister, she was using this currency full blast in the world of boys.

The next day our explorer talked again to the mother, who told him that the girl had been found by the police on the street and had been hospitalized at a large city hospital on the adolescent psychiatric ward. Before visiting her, he briefly questioned the mother about her paramour. It turned out that the subject of marriage had come up between the two of them, but because he earned a limited income, both he and the mother had decided against living together or getting married. Either action would result in loss of the support the mother was receiving from the Department of Welfare for herself and her two children.

All that had been predicted the day before was corroborated when our explorer visited the girl in the hospital. Her behavior with him, and, as it turned out, with the resident physician on the ward, alternated between childish manipulation and seductive behavior of a degree which appeared bizarre in a 12-year-old. But she was, at the same time, a lithely attractive girl with a lively wit which blossomed once she felt understood. She was ambivalent about the alternatives of going home or of going to a state hospital, mildly resisting both.

Our exploring psychiatrist then returned to the mental health clinic to discuss what he had observed with the child's therapist and the consulting psychiatrist. He suggested a plan of action as an alternative to hospitalization. By targeting on key issues in various systems surrounding this child, it seemed theoretically plausible that the conditions which held her fixed in a pattern of behavior that had been labeled as sick and crazy might be changed, thus freeing her to accept new coping patterns which she could be helped to learn. An effort to reestablish communication between the child and her mother, who had shown with her other daughter that she could raise a child with relative success, would be one step. It might not be feasible to work with the grandparents' unsatisfactory marriage, but an explanation to the grandfather, who had already tentatively understood his contribution to the girl's dilemma, might be useful. If the Department of Welfare were willing, and if the boyfriend's income could be enhanced by at least a little supplementary public assistance, the mother and her boyfriend might be induced to marry. Teacher and guidance counselor could be helped to understand the girl's behavior more fully and might cooperate on a plan for helping the girl learn new ways of relating in school. The group worker's investment in the girl could be used to a good effect in this joint effort to help her grow. And the original therapist, instead of concerning herself with defense systems and repressed conflict, could concentrate on helping the family provide the maximum of support and guidance possible, or, if she wished, could still work with the girl herself. With these suggestions, our exploring psychiatrist bowed out.

A month later, a follow-up visit to the mother revealed that the girl had been sent to state hospital on the recommendation of the resident on the adolescent ward who agreed with the diagnosis and felt that, since she was "a schizophrenic," she should be in a hospital. No one had made any countermove, and contact between all of the helping people except the state hospital doctor and the girl's family had been terminated. This outcome had occurred *despite the fact that the mother and her boyfriend had, after a conversation stimulated by our therapist-explorer, presented themselves at the mental health clinic and expressed their willingness to marry if it seemed wise, their wish to have Maria come home, and their hope that someone at the clinic would help them learn what they must now do for her as parents.*

I have, I realize, presented an unusual situation. Reasonable question could be raised, I suppose, as to how often this sequence could occur. And my own bias is obvious in the manner of my presentation. But I think the case illustrates the radical difference between the two approaches under discussion. The approach of the therapist from the interdisciplinary clinic and that of our exploring psychiatrist are not merely two points on a continuum of techniques. The ecological systems approach literally changed the name of the game. By focusing on the nature of the transactions taking place between Maria and the identifiable systems that influenced her growth it was possible for the systems psychiatrist to ascertain what strengths, lacks, and distortions existed at each interface. Two things happened when this was done.

The first was that Maria's behavior began to make sense as a healthy adaptation to a set of circumstances that did not allow her to develop more socially acceptable or better differentiated means of seeking a response to her needs as a developing child. Thus, the aura of pathology was immediately left behind.

The second was that the identification of lacks and distortions in the transactional arena of each interface automatically suggested what needed to be added or changed. Thus the tasks of the helping person were automatically defined. Rigidity of technique in accomplishing these tasks could not, under those circumstances, survive. Flexibility, ingenuity, and innovation were demanded.

The implications of what can happen if this approach is used universally are obvious. If proper data are kept, it seems inevitable that new clusters of data will occur to add to our knowledge, and a new technology of prevention and change develop.

The case of Maria has a certain uniqueness that separates it from most similar cases across our country. The uniqueness is not to be found in the interdisciplinary approach used, but rather in the quantity of skilled people who were trying to help her. Despite their dedicated efforts, all they managed to accomplish was Maria's removal from the only system that could be considered generic in terms of her growth and socialization—her family—and her removal from the school and community which should provide the additional experience she needed if she were to become a participant in the life of her society. In addition, they succeeded in stamping a label on the official records of her existence, a label which is a battleground of controversy among diagnosticians, but which means to the lay public simply that she is a nut.

By chance, Maria wound up in a mental hygiene clinic where her behavior was labeled as sick. She might just as easily have joined the many girls showing similar behavior who wind up in court and are labeled delinquent. Either label puts her in a category over which various members of interdisciplinary teams are in continued conflict. The needs of the girl, which are not clearly apparent in either arena, become hopelessly obscured. Decisions made by those charged with the task of helping her are likely to be made without cognizance of those needs, since they depend for their outcome too often on the institutionalized procedures and momentary exigencies in the caring organization or person.

As a final point, let me explore the nature of the communications breakdown that occurred between the two therapists.

In his explorations, our systems therapist collected a good deal of data that were not known to the interdisciplinary therapist and team in order to ensure that he understood the operations that had been going on at each interface in which he was interested. This additional data only supplemented the data previously collected and agreed with them in content. Thus the two therapists agreed substantially as long as they confined their communications to content and to inferred construction of the internal psychodynamics of the persons involved, Maria and the individual members of her family. And, as it happened, this was all they discussed until the exploring systems therapist returned for a final chat. At that time, having ordered his data in such a way as to clarify the transactions which had been taking place at the interfaces between Maria and the various systems contributing to her growth, his suggestions flowed from a plan designed to affect those interfaces. The interdisciplinary team, including the original therapist, had not ordered the data in this way. Since the dominant disciplinary framework used in their arena was psychiatric, they had ordered the data around a nosological scheme for labeling illness. The outcome of their plan of action, therefore, was to apply a label signifying the nature of Maria's illness, and to decide, reasonably enough within this framework, that since treatment of her illness on an outpatient basis had not been successful, the next step was hospitalization, a decision backed by the assumption that her episodes of running away were dangerous.

It was literally impossible, at the final meeting, for the suggestions of our systems therapist to have meaning to the interdisciplinary team. They fell on ears made deaf by a way of thinking which could not perceive them as meaningful. They came across as a dissonance which had to be screened out. Communication between the two approaches had broken down completely.

This instance of breakdown is characteristic of efforts of communication between people from the two arenas. Conversations I have had with a variety of persons who take the ecological systems view, backed by my own experience, seem to add up to the following.

There seems to be no serious problem of communication between the systems thinker who emphasizes structure, and the experimental behavioral scientist who does basic research in his laboratory or even the researcher who is attempting to deal with a wide range of natural data. Such researchers have selected and defined the structure of the theoretical framework in which they wish to work and are the first to admit that the outcome of their research carries the label of validity within that framework alone.

The clinical scientist, whose emphasis is more on the content of his data, is for the most part a different animal. Most clinical theorists, planners, and practitioners, regardless of discipline, seem caught in the highly specialized sequence of their own training and intradisciplinary experience, upon which they seem to depend for the very definition of their personal identity. Generally speaking, a situation seems to exist in which the integration of the cognitive apparatus of the clinician is such as to exclude as a piece of relevant data the notion that his intradisciplinary "truths," which he carries to the interdisciplinary arena, are relative. He most often will hear and understand the notion when it is expressed. But, again speaking generally, he treats it as unimportant to his operations, as peripheral to the body of knowledge he invests with meaning. Why should this be?

I think it is because the clinician is a product of the specialized fragmentation of today's world of science. To him, admission of this fact would mean that he would have to rearrange his cognitive style, his professional way of life, and, all too often, his total life style as well, if he were to maintain a sense of his own integrity. Not only would he have to renounce his idols, but he would have to go through a turbulent period of disintegration and reintegration. He would have to be willing and able to tolerate the fragmentation of identity boundaries such a transition entails. He would have to leave the safety of seeming truths, truths he has used to maintain his sense of being in the right, his self-esteem, his sense of values, and his status in the vertical hierarchies of his society. He would have to give up the games he plays to maintain his hard-won position in his discipline, games such as those which consist of labeling persons from other schools of thought as bright but limited, misguided, or insufficiently analyzed. More often than not, he would rather fight than switch.

I imply, of course, that he should switch. Thus the question must reasonably be asked: Why should he? Why should he attempt such a fundamental change? After all, he can point with pride to the many accomplishments and successes of his discipline and his own work within it.

But to rest on his laurels, in my opinion, is to abdicate responsibility. It is like crowing over the 70 percent or so of juvenile delinquents who become law-abiding citizens and ignoring the 30 percent who do not. The major responsibility of today's behavioral scientist is to those who don't or won't make it, not those who do, to Maria, not to Little Hans, whom he already knows how to help.

The least he can do is examine his labels and how he uses them. In the life-space of Maria's world, there is a serious question as to which system deserves the prefix *schizo-*.

21

The Individual and the Larger Contexts*

DON D. JACKSON, M.D.

WE ARE ON THE EDGE of a new era in psychiatry and the related disciplines of psychology, social work, anthropology, and sociology. In this new era we will come to look at human nature in a much more complex way than ever before. From this threshold the view is not of the individual in vitro but of the small or larger group within which any particular individual's behavior is adaptive. We will move from individual assessment to analysis of the contexts, or more precisely, the *system* from which individual conduct is inseparable.

Now this is obviously a very recondite area, one in which a beginning is just being made in family research, utilizing a patchwork of systems theory, cybernetics, and information theory, but I think there is a great promise that this group-oriented approach will tremendously enhance our knowledge of human behavior. Further, the conceptual problems we face in family study are shared by students of political, biological, and even artificial or inorganic systems, so there is a situation, rare and exciting in science, in which we can seek broad theoretical solutions of vital interest to incredibly diverse fields of study. At the moment, however, let us examine a few of the issues which arise when the family system of an individual is studied.

For over 10 years I have been studying family interaction to see whether and how such interaction relates to psychopathology or deviant behavior in one or more family members. For the past 5 years a group at the Mental Research Institute in Palo Alto has been studying the "normal" as well as the "disturbed" family in order to

* Presented at the 122nd annual meeting of the American Psychiatric Association, Atlantic City, N.J., May 9–13, 1966. Reprinted with permission from *Family Process*. Acknowledgment is made of the assistance of NIMH Grant MH-4916.

have some base line for the pathological. Our approach has been interaction-oriented because we believe that individual personality, character, and deviance are shaped by the individual's relations with his fellows. As Shibutani [4] has stated:

> . . . many of the things men do take a certain form not so much from instincts as from the necessity of adjusting to their fellows. What characterizes the interactionist approach is the contention that human nature and the social order are products of communication . . . the direction taken by a person's conduct is seen as something that is constructed in the reciprocal give and take of interdependent men who are adjusting to one another. Furthermore, a man's personality—those distinctive behavior patterns that characterize a given individual—is regarded as developing and being reaffirmed from day to day in his interaction with his associates. [Italics omitted]

We view symptoms, defenses, character structure, and personality as terms describing the individual's typical interactions which occur in response to a particular interpersonal context, rather than as intrapsychic entities. Since the family is the most influential learning context, surely a more detailed study of family process will yield valuable clues to the etiology of such typical modes of interaction. Whether one thinks in terms of "role," "tactics," or "behavior repertoire," it is obvious that the individual is shaped by, and in turn helps to shape, his family. This may not at first appear to be such a startlingly new approach but rather the most commonplace social psychology or, at best, merely a shift of emphasis, an accentuation of ideas which are implicit in many of the great theories of contemporary behavioral science which refer to "interaction," "relationships," and the like. But it has been our experience, which I want to share with you, that when one begins to approach or even gather the data, it makes all the difference in the world exactly where the primary emphasis lies. One finds oneself almost immediately faced with certain conceptual watersheds, certain discontinuities between interactional data and individual theories. I would like to discuss two of these critical problems.

The Individual Versus the System

The first is the question of the basic unit of study: is it the individual or the system which constitutes what L. K. Frank called an "organized complexity"? Operating from the interactionist view, our original approach to the family was, and somewhat unavoidably has continued to be, the search for common processes in families with a schizophrenic, a delinquent, an asthmatic, or other "abnormal" *individual* member. My dissatisfaction comes from the fact that, although this investigation of the influences of the family on the individual patient has yielded many new and useful concepts, hunches, and observations, it also contains inherent difficulties and potential fallacies. We must remain constantly alert to the dangers inherent in using an individual as a starting point from which to investigate family interaction data.

Especially if we use the symptom as our starting point, our problem is immediately that the psychiatric nosology or system for labeling deviance is almost totally individual-oriented, not at all well suited to considering the *interpersonal context* in which the patient's behavior takes place. The absence of psychiatric labeling, which is our only referent for normality, only seemingly avoids this inadequacy, for on consideration it is the same problem in converse. Further, the inadequacies of individual classification schemes can only lead to compounded confusion in family classification.

These pitfalls have led us to a few ground rules about interactional research, especially in regard to the concept of the individual. First, though our practical and clinical interests may be served by classifying families according to the presence or absence of individual pathology, we must avoid imposing the elements of individual theory onto the family model. That is, there is no evidence for the isomorphism of the two theoretical models. The shift from individual to interactional thinking may be a discontinuous one in psychiatry, and we must scrupulously examine the basic premises and methods of the former model before applying it to the latter. *It is likely that what we mean by the term "individual" when we take the family system into account may be quite different from what this term presently describes.* It is unlikely that a typology of families as systems will simply be able to use individual nosology. Specifically, we must not let our desire to understand, and to ameliorate, individual pathology carry us into family process with individual-oriented theories, lest we do disservice to both theory and therapy. This error has two forms: We might treat the family as only an additive compound of its individual members and neglect the transactions and the whole; or, out of habit, we might encase these members in hypothetical skin and apply to this unit the theoretical models of the individual.

Philosophy of Causality

In the second place, the tendency to these kinds of errors leads us to examine not only our theories of family structure but, more basically, our philosophy of causality. The behavioral sciences are only now coming to the transition made by many of the natural sciences in this century, i.e., from a mechanical to a systems theory. Specifically, our traditional model of causality does not encompass those feedback processes of a system which *achieve outcomes*. The problem of like causes which do not produce like results (or, conversely, identical results from unlike antecedents) has been analyzed in cybernetics in terms of positive and negative feedback mechanisms. A random event introduced into a system with deviation-amplifying tendencies, for instance, will produce a final result quite different from the same event in a system with deviation-counteracting processes. Thus the study of single elements or static "before and after" situations will not be too enlightening. Neither, then, is the study of essentially accidental historical events feasible. Adopting the premise of the family as a system requires us to *attend only to present* (*observable*) *process*, that is, to ecology rather than genesis.

The circular, or feedback, model of causality is a necessary corollary to our basic axioms of communication. It is impossible to think in terms of interpersonal systems with the ordinary cause-and-effect notions. The strictly individual point of view tends to minimize the two-way effect of persons on each other.

These assumptions have enormous influence on the daily practice of psychiatry, on parents' attitudes toward various child-rearing practices and on those most important of citizens, our educators. It is not uncommon to encounter a schizophrenic who has been denied a possible chance of recovery by the psychiatrist's attitude toward the label "schizophrenia" as a heredoconstitutional thought disorder. The label is framed by the parents' earnest willingness to supply various details that demonstrate that the disorder was present from birth. Our research group noted that in twenty-one of twenty-two first interviews with the parents of a group of young schizophrenics, head injury and the school systems were uniformly mentioned by the parents as important causative factors.

What seems like ordinary common sense can also lead us astray. If we remember John as a little boy, especially if he used to be a troublemaking little stinker, and we see John 20 years later in the clutches of the law, we are apt to be impressed with the fixity of the human character and the infant molding which has occurred. We may forget to be impressed with the fixity of the system within which John lived. The variation of responses permissible within this system may be small, and those responses which the social sieve quickly filters out might bring John to the attention of its agencies. Or we may forget that certain other kids were also stinkers at John's age but turned out to be decent, God-fearing adults. In fact, there have been a few studies of normal subjects who reported traumatic backgrounds essentially similar to those of the general run of analytic patients with neurotic disorders. If we exclude the clichés of genetic predisposition, the only explanation for such differences lies in family and social processes, in those vital relationships where the trauma is amplified or counteracted.

Maruyama, in a personal communication, points out that by a "deviation-amplifying mutual causal process" a relatively small kick can be enlarged by the system over time to sizable deviations. In this light, the difference between the environment of identical twins does not have to be large in order finally to produce sizable differences between them. He also points out that a small set of rules can generate a very complex pattern. This might be important to the family rules, because some will be more vulnerable to mutual deviation-amplifying effects than others. He states further that it requires much more information to go from the adult pattern back to the embryo than to study the rules of the embryo and understand how it becomes adult. This has relevance to the historical method in psychiatry, in which inferences and implications are made about the adult state based on assumptions about what the early state must have been.

Finally, in this question of the appropriate model of causality, it must be emphasized that the linear, cause-and-effect train goes by only once, and once past, is incapable of being retrieved. The accumulation of evidence is that self- or parent-reported histories are notoriously unreliable, filtering the past through the present as well as through the selective vagaries of human memory. Whatever an individual says about his past is also a comment on, or way of handling, the interviewer; that is, the "history" is a metaphor about the present relationship. Such methods, therefore, make impossible the distinction between cause and effect which they seek to clarify. The same charge, incidentally, must be leveled at an unfortunately large amount of present family research which, though purporting to study interaction, actually applies standard individual testing methods to individuals who happen to be related. The impossible question must logically be asked of such research: Are the family members such and such a way because one member is ill or is that member ill because the other family members are the way they are and, presumably, were when the patient was born? Only the study of the family as a contemporary, ongoing system with circular networks of interaction can avoid this pointless and irresolvable debate.

Systems Analysis

All this is not merely armchair philosophy. It is my contention that psychiatry must consider such systems analysis if it is going to fulfill its present obligations and open up new possibilities in the improvement of psychotherapeutic intervention. If, for example, a psychiatrist interviews a couple who complain of marital difficulty, he might describe the wife as hypochondriacal, ineffectual, and dependent, with hysterical tendencies, and the husband as cold, efficient, passive-aggressive, and so forth. Short of sending them immediately to an attorney, his recommendation is apt to be based on the notion that each of these individuals is disturbed and will require a good deal of therapy if he is to live with the other. However, there is another way of looking at this particular couple. They can be viewed as a mutual causative system, whose complementary communication reinforces the nature of their interaction. The therapist can look for rules that govern this system; therapy then consists of the therapist behaving in such a way that the rules must change. Rather than focusing on individual pathology, he might notice that this couple behaves in a remarkably consistent manner, the paradigm of which might be:

> The wife demands, in any one of dozens of ways, that the husband love her (paradoxical, since "love" has to be spontaneous).

The husband replies tangentially, or perfunctorily, that he does.
Wife is enraged and refuses his message, saying she doesn't need it anyway.
Husband is hurt and withdraws (paradoxical in the framework of marriage).
Wife responds by demanding he love her, and the game begins again.

If you will accept my description of this couple's interaction as no less accurate or more inferential than the usual psychiatric formulation, then let us consider some of the differences in emphasis from a more traditional view.

1. Neither is more wrong, or sick, than the other, and thus neither could be said to "start things." The wife's paradoxical command, "Love me," is of course impossible to respond to appropriately; but the husband's withdrawal within the context of marriage, which therefore amounts to a denial that he is *really* withdrawing, is equally impossible to decipher and deal with.
2. It matters little how this got started, since once under way it tends to be self-perpetuating and mutually causative. The responses each makes are about the only ones available to them, and the longer the process continues, the more rigidly such responses will be made, thus inevitably triggering the sequence.
3. Individual psychiatric symptoms can be seen as functional, even necessary, in such a system. Psychosomatic symptoms on the part of the wife would engender the husband's solicitude and involvement; or a third party, one of the children, might become such a problem that the parents are brought into a coalition in order to deal with him and are able to focus all their difficulties on him. The important point here is that the behavior which is usually seen as symptomatic in terms of the individual can be seen as adaptive, even appropriate, in terms of the vital system within which the individual operates.

It may be protested at this point that our present theories supply us with terms such as *sadomasochistic symbiosis* to describe the above, and perhaps any other, two-party relationship. However, I feel such a term is not useful because it reductionistically obscures the important elements of causality. Such a formulation implies that a sadist met a masochist and they lived happily ever after because they were "made for each other." On the contrary, we are constantly defining and *being defined by* the nature of our relationships. (Another objection to this particular description of a relationship is the connotation of force and morality attached to any of the "domination" words. The power of passivity has been demonstrated to us from Christ to Gandhi and the present success of sit-ins, and we must begin to consider the many levels of communication other than overt individual power-seeking which control the nature of relationships.)

In examining the role of the individual within a system, I have relied thus far on examples from the family, surely a vital, virtually universal, yet readily investigable interpersonal unit. Let us expand this kind of analysis to international systems with an example from the philosopher C. E. M. Joad [1]:

> . . . if, as they maintain, the best way to preserve peace is to prepare war, it is not altogether clear why all nations should regard the armaments of other nations as a menace to peace. However, they do so regard them, and are accordingly stimulated to increase their armaments to overtop the armaments by which they conceive themselves to be threatened . . . These increased arms being in their turn regarded as a menace by nation A whose allegedly defensive armaments have provoked them, are used by nation A as a pretext for accumulating yet greater armaments wherewith to defend itself against the menace. These yet greater armaments are in turn interpreted by neighboring nations as constituting a menace to themselves and so on . . .

This form of analysis makes it clear that the behavior of nation A stimulates the behavior of nation B, which in turn spurs on nation A, such escalation being abetted by one-sightedness on the parts of both nations which see only their individual *reactions* and not their mutual roles in an extremely dangerous system.

We know that the Wright brothers flew roughly 100 yards at Kitty Hawk in 1908, and today men travel in 1700-mile-per-hour jets. This is progress; it is sure and almost dull and commonplace. However, progress in psychiatry is never thought of in such terms, because we do not know in what direction to head or how to get there. That is, we cannot expand what we now have and consider it progress. Ten thousand psychiatrists doing what a thousand are now doing will result in increased coverage but little progress. To fulfill its promise, psychiatry must develop a frame of reference that is supraindividual while retaining its traditional model as well. Einstein did not nullify the work of Clerk-Maxwell, but his frame of reference represented a spectacular departure from his predecessors. We can keep our traditional ideas about individual motivation and personality, and learn also to examine contexts. This sort of tolerance has been necessary in other, more "scientific" fields. Several years ago, the Quaker mathematician Lewis Fry Richardson had this to say when he noticed that the outbreaks of war throughout history follow simply and precisely a well-known statistical formula, the Poisson distribution [3]:

> This explanation of the occurrence of wars is certainly far removed from such explanations as ordinarily appear in newspapers, including the protracted and critical negotiations, the inordinate ambition and the hideous perfidy of the opposing statesmen, and the suspect movements of their armed personnel. The two types of explanations are, however, not necessarily contradictory; they can be reconciled by saying that each can separately be true as far as it goes, but cannot be the whole truth. A similar diversity of explanation occurs in regard to marriage: on the one hand we have the impersonal and moderately constant marriage rate; on the other hand we have the intense and fluctuating personal emotions of a love-story; yet both types of description can be true.

As psychiatrists, we cannot view diverse theories in an "either-or" fashion, but must live with the idea that many discontinuous approaches should be investigated and given credence.

Thus, all this is not to deny individuality as a value or to belittle the importance of the subjective experience of ourselves as individuals. I mean rather to emphasize a new dimension in the study of human behavior, an interactional perspective which all who would involve themselves in the affairs of mankind have a responsibility to recognize. Attention to the individual in the extreme is artificial and cannot be the basis for realistic actions. Recently, the Nobel Prize–winning geneticist H. J. Muller [2] proposed a eugenic program which would:

> . . . aim at the ideals of *intelligence, creativity, cooperative temperament, joy of life, vigor,* and *perceptivity* that have allowed men to reach their present high position. [Italics added]

I feel that all the qualities which Dr. Muller cites as individual properties take form and meaning only in interaction and are in fact inseparable from the persons with whom the individual is involved. One who is joyful in isolation is very likely to find himself taken to a mental hospital. We have the enormous job of translating such intuitively obvious notions into precise scientific language.

References

1. Joad, C. E. M. *Why War?* Baltimore: Penguin, 1939. P. 69.
2. Muller, H. J. Perspectives for the life sciences. *Bull. Atom. Sci.* p. 3, Jan., 1964.
3. Richardson, L. F. Statistics of Deadly Quarrels. In J. R. Newman (Ed.), *The World of Mathematics.* New York: Simon and Schuster, 1956. Vol. 2, p. 1258.
4. Shibutani, T. *Society and Personality.* Englewood Cliffs, N.J.: Prentice-Hall, 1961. Pp. 20–23.

22

Intervention, Therapy, and Change*

FREDERICK J. DUHL, M.D.

THE PSYCHIATRIC CLINICIAN in the United States today is faced with an enormous responsibility, for society has literally legislated his tasks so as to exclude none of its members from his concern. Most often trained in an individual orientation, he is asked to provide care to such large numbers of people that individual psychotherapy seems no longer feasible.

At the same time increasingly detailed explanations of human behavior are being made, not only at the intrapsychic level but at cellular, neurophysiological, group, familial, social, and cultural levels of systems. New techniques of intervention at these many levels abound. Is there any reason the psychiatrist should not feel fragmented and compartmentalized as he finds himself shifting from his role as individual therapist to that of family therapist or clinic administrator, or consultant to the school system?

How can today's psychiatrist find a conceptual model that will permit him to function effectively as a psychiatrist at any level? How can he reconcile his understanding of psychoanalytic theory, small group processes, family processes, and community processes as he proceeds in his daily tasks? Too often these levels of explanation and intervention seem competitive, and his professional identity, formulated in residency, appears threatened or compromised by acceptance of other levels of theory.

It is the view of a number of psychiatrists that general systems

* Adapted from a paper presented at the 122nd annual meeting of the American Psychiatric Association, Atlantic City, N.J., May 9–13, 1966.

theory can provide a language and framework for their tasks without such a compromise. It is the proposal of this paper that there are recognizable parallel analogies in systems in which clinicians already function and that the presentation of such isomorphic analogies can offer a reasonable pathway for the understanding of the generalized systems theory.

The photographer who uses a zoom lens of variable focus understands that, while he is aiming his lens steadily at the same scene, a change in the focus of his lens from wide angle to telephoto changes the image viewed but at no point creates a different reality. Rather, it is the difference in focal point that provides the apparently different view of reality. So, too, the psychiatrist may funnel his attention down from the wide angle of the community to the narrower focus of the neighborhood, the family, the individual and, at its most "telephoto" point, the neurological system and the cellular or microbiological system. But he needs to recognize that it is the same reality he views despite his vantage point.

David Rapoport [10] divided the processes described in psychoanalytic theory by their topographical, structural, dynamic, genetic, economic, and adaptive vantage points. Clearly, all systems can be looked at in these ways. It is the similarities of relationships and processes that generalized theory must examine, without disallowing the uniqueness of each system. Anatomy, as a science, describes the appendix all human beings have; it does not demand that all appendixes be in the same unvarying location. In the same way, general systems theory indicates the processes to be searched for in each system; it does not demand that they be identical in time or space. Too often, the unique properties of each system as described by its devoted biographers or scientists have seduced clinicians acting within the system to believe they can understand no other system. Love is blind, but we cannot be lovers of Helen alone if we are to comfort all of Greece.

In this paper I will deal with the parallels in the system levels of individuals, families, and communities found in certain processes of change. Since causing or responding to change is our basic task, I would suggest this is not an arbitrary point from which to begin a journey into general systems theory.

I intend to cover three aspects: (1) the process of change; (2) the therapist, consultant or intervener; and (3) the process of intervention.

The Process of Change

All individuals, alone or in families or social systems, organize their experience or their perceptions of it into patterns that remain remarkably consistent and continuous. The inertial component of this consistency is personal and unique.

Experiential variations in locale, people, and time resemble new versions of an old movie (like *Auntie Mame* in book, movie, play, or musical form), recognizable though obviously different. Psychoanalytic theory talks of fantasies as generic patterns underlying repetitive behavior. Sociologists talk of norms or role-expectations as the cognitive patterns from which repetitive social or group behavior is derived. These cognitively organized patterns develop from a genetically determined base of behavior which is modified by the transactions with the environment. There is evidence that early rhythms remain in a recognizable state throughout life; obsessive children can be identified prior to the era of toilet training. It is a well-known fact that no therapy, including psychoanalysis, can completely change certain major styles of behavior or ways of thinking that underlie them, though it may enhance the socially useful and psychologically more effective aspects of it.

Organizations and communities, as we know, develop their own rules, mores, rituals, and patterns which have been described by anthropologists and sociologists alike. Families, too, have family myths, norms, or shared beliefs, as each school of family study or therapy has described. Despite the lack of a shared language and nomenclature, there is common agreement that the family is organized and maintained in dynamically stable ways through underlying cognitive patterns.

Such cognitive patterning is a property of the mind, which has as its major task the ordering of the massive amount of data that are constantly absorbed through all the senses. The mind's ability to focus, shut out, order, connect, select, and choose provides for the habituated responses that save time and energy in the multiple actions and tasks an individual must undertake. Messages, data, or information that do not fit already organized cognitive patterns are necessarily ignored as irrelevant or viewed as strange or distorted to fit the procrustean bed of the organized patterns.

A simple example comes to mind. Some persons who think only the insane see psychiatrists will hesitate to see one since it would label them "crazy" also. Psychiatrists, in turn, tend to think it "silly" of them to think this way since they do not see their role in this narrow a fashion. But any reasons given such people to undertake psychiatric help will be dismissed until their stereotype is changed.

These cognitive patterns, points of view, attitudes, fantasies, or norms are generally valued highly whether or not they are consciously perceived since they are related to the individual's sense of identity and feeling of self-esteem.

When introduced into awareness, information that is contradictory to these patterns, in spite of the defenses against it, may be taken as a threat to the integrity of the individual, family, or social organization. The degree of threat depends upon the relative importance of the viewpoint to the sense of identity. New data can be assimilated without stress if they do not threaten the basic cognitive organization. This is the way most facts, details, and specific small patterns of behavior are generally learned.

Many cognitive patterns are not easily available for self-observation, much less change, as any psychiatrist well knows. The air we breathe in our own home has no odor. Nor can we smell it the way our friends do until we have returned to it after being away. Breathing itself is taken for granted until it is difficult to breathe. That which is constantly experienced is neutral to awareness, being so immersed in the identity, so "egosyntonic," that it is rarely open to observation or challenge.

The basic *conservative* element in conscious and unconscious thinking must be respected and recognized. It is present when a person struggles with the effects of the loss of a loved one, waiting for his nonappearance at the usual dinnertime. It is present in adults who respond to authorities as they did to their fathers and in families whose second child is brought to the child guidance clinic with symptoms similar to those of the first child who had been successfully treated. It is present in the United States today, where de Tocqueville's [2] description of our country, written over 120 years ago, is still a basic guide to its attitudes and conflicts.

PROCESS OF CHANGE

Does this mean there is no such thing as new learning—that no adaptation or change is possible? Of course not. Individuals, families, and social organizations do change their myths and points of view; attitudes change; beliefs, rules, and norms shift, but never totally. The many cognitive patterns in each system are maintained in a value hierarchy. Some are linked more tightly than others to the ego-identity and its affect, self-esteem, so that some are less easy to abandon or modify than others. The value attached to each belief depends upon the culture (as Kluckhohn and Spiegel [6] have restated), the specific group, the specific family, and the unique individual.

The most basic concepts leading to self-esteem are those of (1) a sense of belonging, (2) a sense of participation, and (3) a sense of competence and mastery. Thus, any attempts to modify beliefs or norms or points of view in a system must take into account these aspects. Information which threatens the sense of belonging, participation, or competence will be tenaciously avoided. That which enhances them has a good chance for acceptance.

In addition, under varying conditions different patterns assume different relative importance. For example, it may be more necessary to maintain a particular construct in public than in one's family, as Miller and Westman [8] have stressed. Or, during extreme deprivation, adaptation may render some valued beliefs relatively meaningless in order to preserve life, as has been observed in concentration camps or in times of war. The fact that beliefs, patterns, and attitudes are more easily abandoned under stress or crisis is an important clue to understanding the thesis of the process of change.*

Erikson [3] points out, in considering individual development, that the steps of change are significantly bound up with crises—that is, a "radical change in perspective" accompanies each identity crisis. Lindemann [7] and Caplan's [1] elucidations of a theory of "situational crises" provide insight into the process of change precipitated by events not usually or automatically present in the normal developmental sequence of individuals.

Lindemann's original article on bereavement described the loss of

* Change itself may be valued, whether in the case of a Don Juan or of our American society. But the "depth" or degree of change is critical. "Plus ça change, plus c'est la même chose" is the response of most psychoanalysts to the Don Juan and most readers of de Tocqueville to his view of the United States.

a loved one leading to a typical sequence, that of (1) anxiety, (2) disorganization of erstwhile usual behavior, (3) a sense of loss or emptiness, (4) the search for ways to cope with the loss, (5) a loss of self-esteem, (6) an approach to others to aid in the process of coping with the loss, (7) internalization of the lost person with related body sensations, and (8) a return to usual behavior or symptomatic distress.

Stated in other terms, sudden new information or the sudden absence of characteristically expected data presents the individual with a dissonance between the usual patterns and the new data. This is a paradox which demands resolution, for the anxiety which accompanies it is distressing. The cognitive reorganization which occurs brings a decrease in the anxiety. The reorganization may be a return to an earlier point of view which leads to old ways of coping, as in regression (and symptom-formation in individuals), which limits the alternatives of behavior. More favorably, it may incorporate the new data and become more functional, which in turn would permit new alternatives of behavior. The instability of the crisis period is thus followed by the relatively dynamic stability of the resolution.

Families, viewed as social units, go through the change process in a parallel fashion. This is true not only in bereavement or in the loss of a home, as Fried [4] has noted in his study of the effects of urban renewal in Boston's West End, but in many other changes in the patterns which have provided a sense of continuity. Rhona Rapoport [11], following Erikson's model for individuals, has defined and begun a study of the normal steps in the family life cycle as crises whose healthy resolution leads to growth. Organizations and communities have similar crises, including loss or gain of an executive, new technologies, or new concerns for previously ignored issues, such as the plight of the poor or Negroes (or both).

Regardless of the unique aspects of conceptual models proposed by students of organizational change, such as those based upon group dynamics or structural aspects, one can recognize the process of major change as a crisis in which individuals play special roles in a total process that parallels the model of Lindemann and Caplan.

To repeat, the basic process appears to be one in which the dynamic patterns, norms, beliefs, or cognitively organized data are confronted with new concepts, new information, or a new goal or expectation that is dissonant. This leads to a period of crisis with attendant effects of shock, anxiety, and related behavior. Rejection of this apparent "threat" is attempted. If such rejection is successful, change does not occur. If change does take place, the sense of threat gives way to a belief that the new information can be of value and that change itself is useful or can be acceptable to highly valued self or group constructs which are part of a sense of identity and which encourage self-esteem.

PARADOXICAL CONFRONTATION

As the new data are allowed to have impact, they may be appreciated, as Haley [5] puts it, as a "paradoxical confrontation." The new information appears to be of positive and negative value at the same time. Solving this paradox is part of the resolution of the crisis. As the dissonance subsides, and if change does occur, the new patterns are available for integration with old, unthreatened, unchanged patterns that have been retained. If the basic, overall identity of the individual, family, or organization is to be recognizable and continuity maintained, the new alternative concepts, patterns, or viewpoints chosen are usually those which are seen as parallel or similar to, not threatening or enhancing of, the retained patterns.

The successful change process provides not only for a new dynamic equilibrium, but it may also provide the base for adaptation and future changes. This comes through the development of an awareness of the change process itself. This preparatory process has been recognized by a psychoanalyst who stated that "psychoanalysis is the best preparation for preventive intervention." In a parallel fashion, Sifneos [12], in his studies of short-term therapy, showed that the most common result was that patients felt they had learned a new way of viewing themselves which would help them solve their future as well as their present problems.

The Therapist, Consultant or Intervener

The role of therapist or consultant is attached to sets of beliefs or theories not unlike the systems which are treated. There are many theories at each system level explaining individual, familial, or social dynamics to which any one intervener may subscribe in his particular task. At this point I would suggest that, despite the uniqueness of any particular theory held, there are some basic components to be found in the attitudes of all successful therapists or agents of change.* These are:

1. The therapist believes change is possible and provides hope.
2. He has a theoretical model of the widest possible range to organize the information he receives from the system that is consulted—individual, family, or community—as well as those systems with which it is in contiguity or within which it functions.
3. He is focused at the interface. The *interface* is defined as the area of contact between two systems. Contact may take place through all senses in verbal and nonverbal ways. What is seen, heard, felt, and even smelled is of importance. For example, the therapist is concerned with how, when, where, to whom, and why a mother communicates what she does in a family session. How does the president of the community council use his eyes, voice, hands, and body to send what messages to whom? The psychiatrist may also focus on the language used by each representative of an organization at a meeting, thereby discovering the basic attitudes of the organization.
4. He is sensitive to all communications, taking them in empathically but with relative objectivity through the use of his models.
5. He can take in the disparate messages, then find the coherent underlying issues and communicate them back as a confrontation or confirmation.
6. He communicates to his consultees or patients in the "language of impact," that is, in the mode, style, and grammar—in body or vocal terms—that can be received by them.
7. He respects their frame of reference, their viewpoint, their ways of cognitively organizing their experiences, as important to them.
8. He can understand the process of change and communicate it in the language of impact.
9. He tolerates anxiety and other affects in others and himself as parts of this process.

* These capacities are not idealistic, but appear in my clinical experience to be the ones possessed in varying degrees by successful consultants or therapists. I would suggest that they should be taken as standards in psychiatric education, and that the training of future therapists should provide for their attainment more than that of belief in any one theoretical model.

10. He is aware of the time dimension so that he is sensitive to the timing of communications.
11. He is goal-directed in terms that are compatible with his clients' or patients' viewpoints and self-esteem, and is concerned with a process of problem solving.
12. He is open to change of his own viewpoints if new information presents itself.

The Process of Intervention

Therapists usually become involved with a system when it is referred to them because of anticipated expertise. Although individuals are easily referred, family systems are less so and community systems rarely so. With the referral there may be the opportunity to redefine the situation at a different systems level: for example, it can be a family issue when individuals come to our attention, or a school issue when a child is referred. The capacity to view the total field as though using a zoom lens allows the clinician to observe thoroughly and choose carefully which focus is needed to fulfill the task of coping with the distress.

The first step of intervention includes establishment of the equivalent to what has been called the "therapeutic alliance." This is the feeling of mutual respect that builds out of the helped-helper, patient-doctor, consultee-consultant roles. It is enhanced by the therapist's capacity to understand the viewpoint of the system and to respect it and respond accordingly. Using his capacity for empathy to feel out the basis of the consultee's self-esteem and identity, he arrives with the system at a task-oriented contract which fits the cognitive patterns of his patients or consultees. It is important that he indicate his commitment to the integrity of their identity despite his task as therapist or consultant. His image in the eyes of his patients is one of a person of value whose presence and skills will enhance most of the ongoing processes in the referred system at a reasonable cost to them of anxiety, time, money, and self-esteem. This period is a time of testing and clarification of easily modifiable stereotypes.

At the same time, since the individual or system is in crisis, it is open and searching for help. Thus it is receptive to those responses that indicate help will be available. In this first phase the capacity for change is explored through the past history of the system and its ability to deal with previous crises of development and unexpected situations. The capacity for accepting a new point of view or new data is tried by the therapist's communication of a sample of his own viewpoint—"So your wife is like your mother." "So this is not a new situation for your organization." A response without defensiveness is an indication of openness to change.

The acceptance of any possibility for change of any sort is a necessary part of this first phase. As long as it does not appear in any communicative form, there can be no contract between the therapist-consultant and patient-consultee systems. Change must be perceived as useful and not as a threat if a consultee is to remain available long enough for the therapist to intervene.

The first phase ends with the contract in which a task-oriented process is agreed upon between the intervener and the system. The field to be focused upon is chosen, and the therapist begins to zero in on the "point of leverage" which will allow for change to take place. In a family system the therapist may look for the one person who has the capacity and the power to initiate change or allow it to take place. He will explore which myths can be confronted or which patterns must give way to allow for change. In an organization he must find the key people whose change of attitude makes for the change in the organizational system. In an individual he searches out which attitudes or concepts are open to change without threatening the core identity.

In the second phase the therapist-intervener-consultant communicates his belief in the system's basic competence and raises the presence of paradoxical, dissonant data as a problem to be resolved. He helps control the amount of anxiety or tension in the system. Too much anxiety leads to either disorganizing regression or an attempt to extrude the therapist as the party responsible for the distress. Too little anxiety provides little motivation for change. This is another testing point of the therapeutic alliance.

If the system is already confronted with a task of adaptation by reality—for example a loss or a new law demanding racial integration—the therapist continues with his problem-solving approach. If there is no such task, one may be provided. Giving a father the duty of decision-making which the mother had previously performed, while asking the mother to withdraw from that role and to support her husband, is an example from family therapy as practiced by Minuchin and Auerswald [9]. This new task may produce a period of crisis with which the therapist deals as he would any other "natural" crisis.

The positive acceptance of new information leads to the possibility of new and adaptive ways of integrating it. New alternatives may come from the consultant or the system, but the development of them is a shared task. The ways in which the system trusts and shares with the consultant aids in this learning process.

Resolution takes place successfully with a new pattern of dynamic stability, a new cognitive organization of data, a decrease in anxiety, and feelings of increased competence and self-esteem. The process may be considered even more successful if an awareness of the process of change accompanies it and if the consultee believes that future crises can be coped with successfully.

The termination of the consultation or therapy does not preclude future contracts if it is ended to mutual satisfaction at a realistic level of achievement of change. There is no doubt that the therapist-consultant-intervener feels a sense of self-esteem if he believes that the beliefs, viewpoints, or cognitive patterns of the consultee are more inclusive and potentially more open to new information. But this success is proved real only if the therapist too has learned, changed, or incorporated new information to be used in developing further his own beliefs, attitudes, and patterns.

Conclusion

This paper has presented a process with which most clinicians are familiar; that is, the process of intervention, therapy, or change. But it has done so with an eye to a generalization of the process so that the individual therapist will see the analogues of his work with individuals in work with families or organizations and the community. The language used in this paper is not meant to be that of general systems theory, but the concepts of parallels or isomorphic qualities in each level of systems is part of the theory.

I have tried to show that what is already known can, through generalization, be extended to provide conceptual models for work with other systems. It is important to the psychiatrist's self-esteem that he not give up his old concepts totally to fulfill the new tasks asked of him. It is important, too, that he be competent in his new tasks and that he keep his identity. Above all, it is important not only to society but to the psychiatrist himself that he participate in the new tasks that his explosively expanding world needs to have done, that he provide the desperately needed skills and knowledge that few others have. Implicit in this paper is an awareness of the crisis in the identity of the psychiatrist. To meet this crisis with defensiveness is to protest and even fail out of fear. To meet it with adaptation is to grow and to change. This paper is meant as an intervention leading to success in the process.

References

1. Caplan, G. *An Approach to Community Mental Health.* New York: Grune & Stratton, 1961.
2. de Tocqueville, A. *Democracy in America.* New York: Oxford University Press, 1946.
3. Erikson, E. *Identity and the Life Cycle.* New York: International Universities Press, 1959.
4. Fried, M. Grieving for a Lost Home. In L. Duhl (Ed.), *The Urban Condition.* New York: Basic Books, 1963. Chapter 12.
5. Haley, J. *Strategies of Psychotherapy.* New York: Grune & Stratton, 1963. Pp. 179–191.
6. Kluckholm, F., and Spiegel, J. Integration and Conflict in Family Behavior. Report #27, Group for the Advancement of Psychiatry, Topeka, Kansas, 1954.
7. Lindemann, E. Symptomatology and management of acute grief. *Amer. J. Psychiat.* 101:141, 1944.
8. Miller, D., and Westman, J. Family teamwork and psychotherapy. *Family Proc.* 5:49, 1966.
9. Minuchin, S., Auerswald, E., King, C. H., and Rabinowitz, C. Study and treatment of families who produce multiple acting-out boys. *Amer. J. Orthopsychiat.* 34:125, 1964.
10. Rapoport, D. The structure of psychoanalytic theory. *Psychol. Issues* 2: No. 2, 1960.
11. Rapoport, R. Normal crises, family structure and mental health. *Family Proc.* 2:1, 1968.
12. Sifneos, P. E. Learning to Solve Emotional Problems: A Controlled Study of Short-Term Anxiety-Provoking Psychotherapy. In R. Porter (Ed.), *The Role of Learning in Psychotherapy,* a Ciba Foundation Symposium. Boston: Little, Brown, 1969.

23

General Systems Theory and Multiple Family Therapy*

H. PETER LAQUEUR, M.D.

IRA WILSON [28] quotes F. R. Kappel, head of American Telephone and Telegraph Company, the largest corporation of the world, as saying: "The successful union of business and science depends on systems thinking." With a little variant we might say that the successful understanding of mental and physical health of patients and the people around them depends upon systems thinking.

In the 1930's Charlie Chaplin showed in his film *Modern Times* the despair of man who, by the First Industrial Revolution, had been made a mere cog in the wheels of industrial machinery. Norbert Wiener, in 1954 [27], describes the Second Industrial Revolution, the introduction of self-controlling, self-reproducing, and self-teaching machines capable of replacing human beings, which perform judgments on a low level, and warns that they may ruin humanity unless they are used responsibly for the benefit of man. He speaks of the need not only of "know-how" but of " 'know-what,' by which we determine not only how to accomplish our purposes, but what our purposes are to be." And he emphasizes that we cannot transfer the choice of good and evil to a machine, whether it has "brains of brass and thews of iron" or is made of flesh and blood. "Whether we entrust our decisions to machines of metal, or to those machines of flesh and blood, which are bureaus and vast laboratories and armies and corporations, we shall never receive the right answers to our questions unless we ask the right questions."

* Presented at the 123rd annual meeting of the American Psychiatric Association, Detroit, Mich., May 8–12, 1967.

Since then the discussion of the dangers and promises of the new technologies has become increasingly emotional, so that today the alternation between fear and hope inspired by our tools for industry, economics, labor, and world politics as well as for education is truly a psychiatric problem for all.

There are those [2, 3, 5, 22] who advise us not to worry but to enjoy the tremendous increase of power and inventiveness brought about by the new technology, but there are also prominent voices [4, 6, 11] who see a potential danger that the new *technostructure*, a term coined by John K. Galbraith, will dehumanize and robotize us all.

In this context von Bertalanffy's work [25] of analyzing the difference between creativity and robotlike submission is of the greatest importance. Thanks to his work and to that of other investigators and thinkers in the field of general systems theory (W. Gray, F. J. Duhl, and others) we are able to take an objective view of the current debate.

My interest in systems thinking was aroused in my father's laboratory at the University of Amsterdam (1921–1939), where he and his co-workers were engaged in studies and discoveries in the field of endocrinology, which at that time was excitingly new. I learned about the slow but effective interplay between hypothalamus, pituitary, and the hormone-producing glands, which has all the characteristics of negative feedback for the purpose of controlling and stabilizing life processes, including the balancing of growth and metabolic functions. These control mechanisms work over relatively long periods of time—days and weeks.

There are other control mechanisms in the living organism which work within seconds and milliseconds. The central nervous system and the autonomous nervous system, our sensory and proprioceptive nerves, respond to changes in our external and internal environment with adaptive control and stabilization action within fractions of seconds. Modern technical apparatus (computerized factories, military defense systems, and the like) do their calculations and adaptations in even shorter units of time—microseconds and even nanoseconds (one thousandth to one millionth of a millisecond).

In contrast to the fast-working control and stabilization systems described above, the therapist who tries to restore balance and stability to a sick individual, family, or group of families must expect weeks, months, and sometimes years of work before unhealthy mechanisms can be changed into more useful and constructive behavior.

How control and stabilization processes occur in psychiatric therapy, especially in group work with several families (multiple family therapy), and how intrapersonal and interpersonal occurrences in the family system connect with the larger framework of society, are the subjects of this paper.

Program and Style

While a human being possesses material structures and processes that can be understood from physiology, cybernetics, and technology and that can be demonstrated on a robot model, concurrently with these run other processes best described as mental, spontaneous, creative processes that, although dependent on the material structure and function of the organism, are not identical with it. On the sensory side the selection and acceptance of information, in the associative field the spontaneity and originality of combination of ideas, and on the motor planning side the determination of the goal and purpose of life are of such nature that they cannot be described in the same way as the memory and pattern-recognition area, the program, and the motor preparation area of a robot system.

Each individual has a program which can be understood in part from statistical expectation and cybernetic models, but each individual also has a personal style about which no statistical predictions are possible. If two or more individuals interact, they do this not only with their statistically predictable structure and function—their program—but also with their personal selections, their spontaneous associations, and the creative goals and purposes each feels to be the most important things in his life; in other words, with their personal style.

The program for survival includes not only phylogenetically and ontogenetically acquired particulars but also much of what has been learned during the individual's lifespan; unconscious (repressed), preconscious, and conscious elements are mixed into the program, new things are learned daily, and the art of learning new ways to cope with the environment and create objects and communications of prolonged usefulness become part of the program. Such a program can develop either normally or pathologically. Faulty ways of dealing with stresses sometimes drift into the program as "bad habits," "addictions," and so forth, and must be corrected through new experiences such as therapy or catastrophic life crises.

The program becomes partially or completely disjointed in mental illness. The personal style of the individual may or may not be affected by illness, but his program for adaptation to the stresses of life is always affected. The program includes the border between original mental structure and robotlike physiological structure, and has functions in both fields.

Inorganic computers can be built that have many of the traits of the human brain. Pattern recognition, such as reading a variety of human pronunciations or handwritings, once thought to be too individual for any possibility of computer analysis, is now systematically performed by computers. Complex orders of great differentiation and refinement, including the command to reproduce its own organization in a new computer, to correct errors, and to bypass faulty channels while repair takes place, are now normal, everyday occurrences in computers.

However, in one area so far, namely, learning to create new ways of dealing with changes in internal and external environment, computers have not been successful. Von Bertalanffy, in his paper "General Systems Theory and Psychiatry" [26], speaks of the limits of the robot approach to creativity. These limits lie in the area of original handling of a new and unexpected situation, such as an overload of energy or information, a skillful readjustment of the internal control and feedback system if there are unusual internal malfunctions, good intuitive prediction of probable success in a critical situation by choosing pathways hitherto unused, and a sense of timing for scanning, probing, self-testing, and correction of probing techniques in order to obtain optimal data for prediction, anticipation, and output of product or information. All these functions are part of the personal style of an individual.

The concepts of program and style, of course, overlap in some areas, especially in the area of learning, but to make them clearer we might say this: *What* is learned (be it by a computer or a human being) is objectively and statistically measurable and is part of the program. *How* (by what method) it is learned is part of the original, creative, and spontaneous approach to learning which characterizes the personal style. A computer has a program but no personal style.

Placed before similar tasks, individuals can resolve problems by imitating others or by finding new ways of intelligently coping with the situation. Whether they choose the first or the second way determines whether they act as followers or leaders in complex situations, and it is their personal style (not the program built into all human organisms) that determines in which category they fall.

"Style is the stamp of individuality impressed upon our adaptive behavior," says Allport [1]. In a world where an ever greater attempt is made to manipulate our program, our adaptive behavior, to serve an industrial system whose only goals seem to be economic ones, as Galbraith shows in his book *The New Industrial State* [6], it may not be without interest to examine the influence of personal style on group interaction as it occurs in multiple family therapy.

Application of General Systems Theory to Psychotherapy

General systems theory is of particular interest to those engaged in therapy with individuals, peer groups, single families, multiple families, or larger groups of people because it helps them to a better understanding of how the individual (subsystem) fits into his environment, which consists of such larger systems as families, kinships, tribes, communities, states, nations, and the human species as a whole.

Whenever maladaptations or pathological conditions prevail in the function of an individual or a family, or in the family's relationship to the community, general systems theory permits us to analyze where the primary focus of disturbance lies and to devise methods for better adaptation and integration of disturbed individuals and families in the surrounding community.

Since general systems theory deals with the efficiency and effectiveness of systems created for the purpose of producing an output meaningful and useful to surrounding suprasystems, it is clearly relevant to psychotherapy systems such as individual therapy, group therapy, conjoint family therapy, and multiple family therapy. In all these cases therapist and patients constitute a system with input, transaction, and output. The input consists of sick people and relationships, the transaction is constituted by the therapeutic interaction taking place, and the output is healthier and better functioning of individuals and groups. A good therapeutic system will achieve these goals efficiently, that is, in a reasonable time, and effectively, that is, with a high incidence of good and lasting results.

The Multiple Family Therapy Group as a Special System

Multiple family therapy was described and its development from 1951 to the present traced in previous papers [15–17]. Therefore I will outline it here in only a few words as follows.

It has been amply demonstrated that even when patients have returned from a hospital in a fully functioning condition, a recurrence of the illness is probable as long as the environment produces stresses which may force the patient back into sickness. In general terms we could say that the patient and his environment have lost understanding for each other, and, unless this lack of understanding can be remedied, there can be no certainty or even much probability that the patient will remain well after his discharge from the hospital. Although the hospitalized patient is probably not always the sickest member of the family, this does not alter the concept that patient and family must be given a chance to develop a healthier mutual adaptation if optimal results are to be achieved by hospitalization.

The multiple family therapy group usually includes four to five hospitalized patients and their immediate families (parents, spouses, siblings) with a therapist, one or two cotherapists, and an observer, for the purpose of producing a better understanding between patients and families and changing their mutually harmful behavior patterns into more acceptable and mutually tolerant ones. Sessions take place once a week during the entire period of hospitalization of the primary patient.

Multiple family therapy can be described as a special therapeutic system consisting of several subsystems and being itself part of a larger suprasystem: (1) the therapist, his cotherapists, and observers form a subsystem; (2) each family is a subsystem of the first order within which the primary patient and the rest of the family constitute subsystems of second order; and (3) surrounding society (hospital, clinic, community, state, nation) is a suprasystem of higher order.

The efficiency of a multiple family therapy system can be measured by the time consumed in turning out better functioning families, and its effectiveness by the number and percentage, as well as the duration, of improvements. If predictions and expectations are clearly formulated, one can measure the proper functioning or malfunctioning of a multiple family therapy system in terms of results.

Why Multiple Family Therapy?

We reasoned that in a group of several families the therapist would be able to use some of them in specific "cotherapeutic" roles to achieve insight and changes in behavior quicker than he could do if he were to work with a single family.

There are two outstanding problems for the treatment of families containing a schizophrenic member: (1) the code every family seems to develop for internal verbal and nonverbal communication among its members and (2) the resistance of the family to change [12].

The therapist must "break" the code in order to understand what goes on within the family, but he must also be able to reach the family with his own messages. The therapist must learn to decode not only the normal intrafamilial language but the semantically distorted language of the sick person [7, 13]. We found that messages passing from the therapist to a family containing a sick member required translation into terms fitting with the internal family code in order to become understandable information for this family. The help of other families in this translation process is of priceless value. If four families are together, usually at least one can pick up the signal from the therapist and rephrase and amplify it to make it understandable to the other three families. When there are problems of patient-family communication, very often schizophrenics can translate for each other better than any therapist in the world could do.

If we want to break through the resistance to change of the family with a schizophrenic member, we must ask ourselves first: How do people integrate new experiences into their outlook on the world and their preparation for future behavior? Perhaps we can derive some clues from information theory, which postulates that those things have the highest information value which have the least probability of occurring and yet do occur. In other words, a new pattern, a new sequence of signals producing an excitation focus in the nervous system becomes significant information; or, as Norbert Wiener [27] expresses it: ". . . the more probable the message, the less information it gives. Clichés, for example, are less illuminating than great poems."

Translated into terms of multiple family therapy, a new, successful, more realistic type of behavior of one family, distinguished from its usual behavior, can act as a focus of excitation for the whole group if used skillfully by the therapist. Usually the patient and family who have succeeded in modifying their behavior toward each other become this focus of excitation, emitting suddenly a mass of new signals of significant impact on the other families.

Mechanisms in Multiple Family Therapy

In order to clarify further how a multiple family therapy group can be seen as a special system, we will review now the various concepts operating within the system. The concepts all have a bearing on the restoring of adequate adaptation of individuals and families to their environment through multiple family therapy.

COMPETITION AND COOPERATION

On one level competition, on another cooperation between systems leads to changes affecting the involved organizations. For instance, social changes occur much more rapidly when two systems (countries) compete sharply or even fight with one another than if reformers try to achieve such changes more or less peacefully from within. (Both Czarist Russia and Imperial Germany changed from autocratic states to socialist republics within two years, 1917–1918, although prior to the war anarchist socialists in Russia and democratic socialists in Germany had attempted during 40 years to change the governing systems.)

Similarly, two or more competing and clashing families in multiple family therapy change one another in their internal power distribution much more rapidly than an attack on a single family by a therapist could lead to change. A threat to the status of a family or an individual is felt as competition and results in a more rapid and deeper involvement and interaction of the participants, which means that they see messages from the therapist or other members of the group as more relevant to them.

At a later stage of the game cooperation between participants may take the place of competition. Jimmy may point out to Johnny how he and his family learned to handle a situation similar to the one that is disturbing or puzzling Johnny at the time.

The therapist makes full use of both aspects, competition as well as cooperation, in his handling of the group.

LEARNING BY ANALOGY

Multiple family therapy provides many opportunities to watch others respond to "analogue" conflict situations and to see how others play the game differently than oneself. A family in a conflict they cannot solve may observe another family behave in an analogue conflict more successfully and learn from this example new ways of negotiating. They may see that another family plays a "mixed-motive type of game," that is, it is not always necessary for somebody to win and somebody to lose, but that compromise may be a solution. The feeling of "having been there," in similar trouble as others, helps immeasurably in learning, and this feeling is more likely to occur in the multiple family therapy group, with its many examples of the most diverse types of conflicts, than in other forms of therapy.

LEARNING THROUGH IDENTIFICATION CONSTELLATION

In multiple family therapy young people watch each other in their dealings with the older generation. Likewise, fathers learn from other fathers, mothers from other mothers, and married couples show understanding for each other more easily after observing other marriages. In other words, the fellowship of experience helps in learning to cope with existential and situational problems.

LEARNING THROUGH TRIAL AND ERROR

Multiple family therapy affords its participants the opportunity for trying out new behavior and reinforcing it through the approval of the other members of the group. New insights may be achieved through role-playing. The therapist in discussing one family may, for instance, say to the son of another family: "What would you do if you were that parent?"; and the son, by taking up the challenge and acting "as if" he were that parent, will achieve for himself and even transmit to other children a greater understanding in general of parents in the given situation. Also, if one family is more successful on an interactional level, the more troubled families will feel challenged to emulate this successful handling of hurdles and obstacles they have not yet mastered themselves. Multiple family therapy provides many opportunities to practice such new behavior.

AMPLIFICATION AND MODULATION OF SIGNALS

The therapist emits signals to a sensitive patient who can then be used as an amplifier to sensitize his family. This first family, in turn, will promote further amplification and modulation of such signals to the other families who, without this amplification, might not yet have responded to signals from the therapist.

The specific "I and thou" relationship between therapist and first patient [9] can then set up reverberating communications and feelings leading ultimately to similar warm and identifying understanding between other members in the group, particularly those who feel that the therapist and the first patient and his family "have been there," that they have experienced the difficulties not yet resolved by the sicker families. This example can then produce confidence in the other families that such difficult situations can be resolved.

USE OF HEALTHIER ASPECTS OF CERTAIN FAMILIES AS MODEL

The therapist in multiple family therapy uses the healthier aspects of one family as a model and challenge for other families to change their own behavior. The persons with the most severe symptoms (the primary patients) usually profit first in a multiple family therapy group. The attitude of their relatives changes later, when more mature behavior in the primary patients becomes manifest.

It seems noteworthy in this context that the whole multiple family therapy group works better than a fragmented group from which many family members are absent, and certainly much better than the peer group, in which the so-called normal relatives are present only in the fantasy of the patient, but where no real perspective of mutuality and reciprocity in family life is obtained.

UNDERSTANDING THE TOTAL FIELD OF INTERACTION

Multiple family therapy attempts to achieve an understanding of the total field of interaction between subsystems (patient, family) and suprasystem (the total social environment). It teaches people that not just one individual becomes ill but that illness must be understood in the context of a field of interaction in which whatever one person does has a bearing on the actions, reactions, and the general behavior of all others in the immediate environment—that people react not only to their own family and circle of friends, employers, and fellow employees, but that each individual (the subsystems in the family) and surrounding society in its demands (suprasystem) all have a direct impact on the family and participate in the final result of treatment.

This awareness of the changing surrounding field and its importance for sickness and health [14, 23] is the major asset of multiple family therapy sessions. All this is, in principle, based upon Kurt Lewin's *Field Theory in Social Science* [18], which has made psychiatrists aware of this basically important interactional field.

USING THE FIELD OF INTERACTING FORCES

The therapist makes use of this field of interacting forces. Instead of exposing patients or relatives to single exchanges with himself, the therapist can make ample use of the multiple combinations occurring in a field in which several fathers, mothers, and children are present, and a variety of influences can be brought to bear on a person reacting in an abnormal or undesirable fashion. The therapist has, therefore, a much greater range and variety of therapeutic approaches to choose from, based upon mutual recognition of problems and needs, than he would in a one- or two-way communication with single individuals or single families.

THE FAMILY AND ITS COMPONENTS AS SUBSYSTEMS

Multiple family therapy considers the family and its components as subsystems of a larger field of interrelations whose interface problems can be studied and dealt with in the same way that a good troubleshooter understands and repairs component parts of a complex machine or missile system. If therapists see malfunction in a family as caused by specific malfunction of component parts or of feedback mechanism, they act as systems analysts (analyzing feedback, perception recognition, association, and planning for response). They can then propose to the sick family new and specific possibilities for repair and restitution that could lead to understanding and correction of the malfunction.

Style of the Therapist

The program of a multiple family therapy group is largely determined by the goals of the hospital or clinic in which it works. In other words, *what* is to be learned by the group is given, namely, a better mutual understanding of patients and their families for the purpose of assuring the patient's continued well-being after his return from the hospital. *How* this goal is approached is determined by the personal style of the therapist. All the qualities mentioned before as being part of the style of a person, namely, originality in the handling of new situations, skillful readjustment in case of unusual malfunctions, choice of hitherto unused pathways in critical situations, a sense of timing, and the like, are demanded of every therapist. But they are demanded to an especially high degree of the therapist of a multiple family therapy group, because such a group can show all the properties of a system threatened by failure through unexpected situations, overload of information, critical events, malfunction of a subgroup, or external developments, for instance, changes in policy and milieu of the surrounding system, that is, the ward and hospital.

The Therapist's Function as Systems Analyst

Following are some examples of situations in multiple family therapy that the therapist may have to cope with.

UNEXPECTED EXTERNAL EVENTS

A breakdown in the heating system, for example, forcing the group to meet in a room without heat or in another room than the one they are used to, may have a very disturbing influence on the group.

OVERLOAD

Several families at once begin to quarrel violently and demand of the multiple family therapy participants arbitration of their conflict, without paying any attention to the therapist's attempts to regulate traffic and assign priorities.

INSTABILITY OF SYSTEM

The therapist has a "bad day" and finds himself irritably overcorrecting or withdrawing instead of using his normal good sense of proportion and humor. Families become aware of his imbalance and reinforce his overcorrections by positive reverberation, throwing the group more and more out of balance, particularly by organizing "side shows" and private powwows among themselves.

BREAKDOWN OF TIMING AND FEEDBACK REINPUT (DELAY IN FEEDBACK LOOP)

Private disagreements in a family or between members of an identification constellation, when the therapist as yet does not understand their "private code" and does not know what is going on, lead to lack of information and temporary inability of the therapist to keep up with events.

INFLEXIBILITY

Instead of being flexible (more random), the therapist sees things according to a set of rules not fitting the group. His rigidity makes him "drive through warning signs," and he frightens and disturbs patients and families rather than helping them.

INSUFFICIENCY OF INTERNAL SENSORS

Lack of perception and sensitivity may cause some therapists to give academic, psychodynamic interpretations which, although they may be theoretically correct, will increase a patient's resistance. This must be resolved, or the multiple family therapy group will not succeed.

FAILURE OF EXTERNAL SENSORS

The therapist's failure to take into account changes in the administrative climate (in ward, hospital, or clinic) may bring members of the multiple family therapy group into conflict with the surrounding system and thus cause confusion detrimental to therapeutic progress.

BREAKDOWN OF COMPONENTS, REQUIRING BYPASS, OVERRIDING, OR REPLACEMENT

Cotherapists can make errors requiring quick face-saving correction or a changing of subject before greater damage is done. Sometimes another member of the group behaves in such an "impossible" way that the subject of discussion must be changed or the disturbing person may even have to be removed from the room. In rare instances it is necessary to eliminate a whole family from the group and replace it by another family.

Of course, a great deal of tolerance is expected and required from all participants, since the group is obviously existing with the enjoinder to promote freedom of expression in order to relieve existing tensions caused by insufficient communication between members of the family, generations, and the like. Therapists often find themselves between the devil and the deep blue sea: on one hand, they must "awaken sleeping dogs" that in a more sociable setting all participants would let lie; on the other hand, if too much shocking information is placed at the center of the stage, some families may feel extremely threatened. Extreme reactions can usually be averted by a skillful therapist, but bitterness and resentment in the threatened participants may delay the success of the therapeutic operation.

DELAY OF ACTIVITY THROUGH FRICTION BETWEEN PARTS (INTERFACE AND INTERPERSONAL PROBLEMS)

One patient could not give up his paranoid, withdrawing attitude because his wife, an intelligent and loving person, could not give up her unconscious protective, dominating attitude toward him. Her attitude changed only when, after many sessions, she became aware through deep emotions of hurt and frustration that her readiness to make decisions for and about her sick husband was not the product of loving, wifely concern for him but of an unconscious need to be superior to him and to his very quarrelsome and domineering parents. She then gave her husband more free choices and even took some risks by letting him come home when he was supposedly still unwise in some of his judgments. She saw him perform better than she had expected, and, as her control over him decreased, his self-confidence increased until finally both could enjoy his recovery and a mutually respecting and loving independence in decision making. This was tactfully and skillfully brought about by a wise and patient therapist.

PRODUCING A "HIGH VOLTAGE" FOCUS OF EXCITATION

A patient habitually talking in incoherent, schizophrenic, neologistic sentences, especially when his parents are around, is suddenly confronted by his therapist with the sarcastic statement: "Look here,

Irwin, we know that you are putting on this nonsense show to punish your parents. How about cutting out the nonsense?" No reply; the patient continues to ramble about helicopters and millions of dollars his father, a small shopkeeper, supposedly owns. Now the therapist suddenly lapses into a rapid, completely foolish, rambling, loud nonsense talk. Everyone watches him and Irwin. Suddenly, Irwin turns with an amazed expression to his parents: "What's the matter with him?" he asks, acting as if he had never been "out of it." This sudden turn into normalcy (flight into health?) had a powerful effect upon the other schizophrenic patients in the group, who also became less withdrawn and more to the point the excitement flowing off from the "improbable event" to the others who had not expected this.

A major success is obtained when a schizophrenic patient gives up artificial distortion of symbols, false syllogisms, creation of meaningless words, and incoherent talk and returns to normal language. This is a sign that he wishes to return to a healthier intrafamilial and interfamilial relationship. Our work with about 200 families shows countless examples of such occurrences that have prognostic value for the outcome of multiple family therapy.

The foregoing description of some of the typical situations the therapist in multiple family therapy is called upon to master gives an idea of the complexity of his task. The training of multiple family therapists therefore must include learning how to recognize and deal with the problems which are somewhat unique to this form of therapy. Once the therapist has learned this, it adds greatly to his capability in all forms of therapy.

Hierarchy and Stratification in Multiple Family Therapy

In the multiple family therapy system, as in all systems, hierarchies exist with regard to decision making.

1. The therapist is the center of the decision-making process, and while his choice of decisions displays his personal style, his program is for the greater part predetermined by the goals and needs of his hospital, clinic, or private practice.
2. The next level in the hierarchy is the professional subsystem, consisting of cotherapists, observers, and "observers of the observation process," sometimes also student therapists, who assist

the therapist in every scientifically recorded multiple family therapy session. They do not alter the style and program of the multiple family therapy group, and such a complex professional subsystem is not always present in ordinary working sessions.

3. On the last hierarchical level are the other (nonprofessional) subsystems; namely, the fixed determinants, that is, the families; and then the more fluid subsystems, that is, the so-called identification constellations, such as "fathers," "mothers," "primary patients," "husbands," "wives," and fellow members of a socioeconomic, religious, or ethnic group as they develop in a multiple family therapy group.

The weight or importance of messages emanating from the different hierarchical levels decreases from the therapist to cotherapists, to group members identifying with the therapist's goal, to neutral followers, and to opponents striving to do things "differently" who nevertheless contribute to the common goal by challenge and suggestions for changes of method.

Capability Requirements for Goal-Seeking, Open, Optimizing Systems

Multiple family therapy, like other such systems seeking to produce desirable outputs in an efficient and effective manner under constantly varying circumstances, requires the following characteristics to prevent system failures and breakdowns.

1. Ability to probe and scan the external environment continuously and repeatedly with adequate sensors.
2. Ability to probe, scan, and adapt its internal milieu and mechanisms and to do this faster than the changes in external milieu occur, as otherwise a lag in system functions would be produced.
3. Ability to change the timing, location, frequency, and direction of probing.
4. Capability for sufficient and qualitatively correct activity of motor-output effectors for reactive and productive behavior.
5. Ability to store and recall information (memory).
6. Ability to recognize and identify patterns (recall and checking by comparison).

7. Ability to feed information back (to know what was done the last time in analyzing a similar situation).
8. Capability for self-stabilization (to avoid excessive oscillations and overreactions which might throw the control system out of joint and cause it to lose balance). Example: Marbles in a flat dish lose their mutual positions easier than marbles in a deep dish.
9. Ability to be more random, that is, more flexible, than the environment. Example: Too infrequent probing while the environment changes from random input, such as a well-built, straight highway, to a patterned input, such as city streets with periodically changing traffic lights, can make a driver pass through a traffic light. If the driver is not flexible, meaning if he cannot change fast enough, he becomes dangerous on a street with fast-changing traffic lights.
10. Ability to adjust predictably and to change predictions with regard to changing probabilities.

Discussion of Systems Theory in Multiple Family Therapy

LIVING SYSTEMS

A biological system is characterized by its ability to regulate itself in an optimizing form for the purpose of survival. It has a regulated input, transaction, and output. The human organism is such a system. It is capable of regulating the quantity and quality of input, that is, material and information entering it. It transacts this input by recognizing, screening, selecting, and absorbing material or information, by associating it with memory traces of formerly received and transacted material. Checking this information with its general program, consisting of phylogenetically and ontogenetically received unconscious instinctive and learned instructions, and with its individual style, it will plan and prepare for purposeful action; and it will finally, acting under feedback controls, produce an output meant to contribute to its survival. This output consists of energy, matter (products and waste), and communication understandable and useful to the environment.

IMPORTANCE OF UNIFYING NERVOUS SYSTEM

Within the human organism there are subsystems, each of which follows similar principles. They are interconnected by the nervous tissues as the principal unifying and synthesizing network which helps the subregulators (thermostat, appestat, cardiovascular and respiratory, water and mineral controls) to interact efficiently. Equipped thus with a well-functioning material organism, man can intelligently correlate perceptive input (information picked up by the sense organs) with logical and purposeful output (communicating, moving to places where more adequate living conditions exist, using tools to change his external environment).

DYNAMIC DEFINITION OF HEALTH

Health can be defined as appropriate integration of the internal milieu of the organism with external activity in order to meet present and future demands and stresses arising from the internal and external environment efficiently and purposefully. Ill health (pathological responses) then can be defined as inappropriate and inefficient adaptation of internal milieu and external motor output to existing and future environmental demands and stresses, in the worst cases characterized by inability to function, derangement, withdrawal, and paralysis, or excessive, destructive agitation and purposeless overactivity (instability of controls). Both types of inappropriate and inefficient behavior may eventually lead to a total cessation of function—to death, whether suicide or homicide.

Karl Menninger [19] sees the organism under stress develop psychological adaptive reactions similar to the ones studied by Selye in physiology [21]. He describes a "hierarchy" of regulatory emergency devices in five orders of magnitude (from the simplest to the most complex and costly) which may be employed by the organism before homeostasis gives way and irreversibility of process and possibly disintegration and fragmentation of the psychological wholeness of the ego occur.

GROUPS AS SYSTEMS

In the same way that the human organism can be described as an integration of subsystems efficiently working together for the common goal of survival, dyadic, triadic, and larger groups of human organisms can be seen as systems that partly submerge individual needs in order to create jointly a new, larger system for ensuring survival in cooperation and competition with, or even in battle against, other similar systems.

DYADS

If two persons agree to form a partnership, we can call this the "inner circle" (*we*), while everyone else becomes the "out group" (*they*). Such a partnership, a dyad, requires a joint control system to prevent each individual (each subsystem) from defeating plans and activities necessary for the common purpose. The simplest dyad would be a young married couple or even any two persons deciding to do something together while jointly resisting attempts of the external milieu to separate them and thereby hinder their joint activity.

Such a dyad can function well as long as the joint goal is more important to each of the two individuals than a personal goal that would make each seek to use the other to his own advantage. In other words, the joint operation of the system must present to both something more desirable than the freedom to move as an independent system.

BEGINNING OF TRIAD

Such a young couple may wish to include a third person, a child, in their system as beneficial to the system as a whole. Evidently the child cannot have the same decision-making freedom as the original dyad but may be considered a future prospect for building an even stronger triad within the big external world. During the growing-up process information and guidance passing from the senior dyad to the junior member is corrective, sometimes even painfully limiting, but rarely destructive. The two senior partners wish junior to become strong, healthy, and capable of coping as subsystem with the demands of life. This is a normal form of symbiosis up to the point at which junior would be strong enough either to move out of the triad and start creating his own empire or to become a third partner with equal rights.

BEGINNING PATHOLOGY WITHIN TRIAD

Then we often see the two senior partners become uncomfortable, jittery, disagree with one another, and sometimes even compete for junior's love and loyalty, each hoping to set up a new dyad, more advantageous to himself, at the expense of the other senior. At this point pathological symbiotic reactions appear on the scene. Junior acts out his resentment of past limitations and plays the two seniors against one another in order to profit from their rivalry. The triad still holds together because all members find it in their best interest to present to the external world the image of a happy and healthy family. Only when tensions become excessive one of the three is sacrificed, becomes a victim or scapegoat, and then must be treated as a patient in a clinic or hospital. This act relieves the remaining two members of the triad, although they may feel guilty, but they can now as a dyad function better after having mourned over the elimination of the scapegoat.

PARTIAL RECOVERY WITHIN A TRIADIC FAMILY

When the sick person gets well, the triadic composition is restored and the potentials for tension and dissension are recreated. This recreation of the former disastrous situation can be avoided only if the two healthy partners are willing to surrender some of their homeostatically defended dominance and ascendency. Jackson [12] has shown how difficult it is to attain such surrender. When the former victim becomes stronger, it requires all the skill and energy of a first-class family therapist to bring about a more reasonable distribution of power within the triad without producing a new victim. Frequently this is nearly impossible and often, after release from the hospital, we see a patient make only a marginal recovery, remaining in a position of dependence in relation to the so-called healthy part of the family.

DEVELOPMENT OF MULTIPLE FAMILY THERAPY

Let us now examine what happens if two or three or more triadic systems, each with one so-called sick individual, meet with a therapist. We see immediately that now the victims or scapegoats (the patients) have an opportunity to unite to make their point directly or indirectly to the more powerful partners (the so-called "healthy" members of the family), who also may identify with each other. Since, however, the "healthy" partners are not all equally strong, cross-identification of the weaker partners may help to break the dominance of the least rigid among the strong partners. In other words, one or two fathers may ally themselves with a patient to point out to another patient's mother the faults of her relationship with her child. She could never accept such criticism from her own husband or her own child but may be able to see the point if it is made by someone outside her family. The victims can make their hitherto inaccessible partners painfully doubt themselves and learn that they must change their attitude if a happy family pattern is to be restored or created. The members of one system learn by observation from the members of another, rather than by head-on collision within their own family system.

Thus, two or more families, sometimes in competition, other times in cooperation, under the guidance of a skilled therapist, change each other in their internal power distribution much more rapidly than would be the case in interaction between a therapist and a single family.

This is in essence the principle behind multiple family therapy. Only such changes enable patients to return to society without having to fear that they will be pushed back into the victim position and thereby become chronic, deteriorating custodial cases.

Some doctors believe that schizophrenic patients must forever be separated from their families because a family will never permit the scapegoat to return as an equal on the social stage. I feel that such a viewpoint is unnecessarily pessimistic. Pathological family systems can be changed, as we have seen in our work with multiple family therapy, but tremendous forces are necessary to do so. Such forces can be provided by a single therapist in a single family only with prolonged efforts, but they come into play much more readily in the alliance between a therapist and several families.

OUTLOOK FOR THE FUTURE

To look at interrelations between human beings in this manner is descriptive of their movements and behavior within a small group. It does not account for the varieties in behavior caused by intrapsychic crosscurrents and conflicts. It remains for future research to formulate a comprehensive theory of human behavior which also explains the movements of the individual within systems of several triads with his special adaptation derived from his intrapsychic milieu as studied by psychoanalysis. Predictions of behavior depend upon a thorough knowledge of both interpersonal and intrapersonal dynamics and stresses.

The Language of Multiple Family Therapy and General Systems Theory

Terms such as *homeostasis, feedback, transmission of information, attenuation, delay,* and *memory-storage* are used by electrotechnology to denote concepts measurable in quantitative terms and capable of being treated mathematically. These concepts allow prediction of phenomena and permit construction of apparatus able to operate more efficiently than man when called upon to perform clearly programmed tasks.

Psychologists and psychiatrists borrow many terms from the physical sciences to describe rather than predict and measure phenomena and functions quantitatively. Older examples are such terms as *projection, sublimation,* and *tension- and stress-resistance.* I am afraid that we psychiatrists apply these terms mainly in a metaphorical sense while we do not seem to feel bound by the exact disciplines of the mathematician who likes to define each term in either numbers or units, or at least in number/time relationships, which can be expressed in meaningful equations, differentiations, or integrations. Even such complex matters as two messages traveling simultaneously over the same wire or using the same carrier wave can be analyzed in terms of exactly describable Fourrier curves, and the recipient can separate the two messages exactly without distortion, without loss of clarity, if he knows how to read such Fourrier curves. I fear that all our uses of symbols borrowed from physics, information theory, cybernetics, or statistical analysis are missing the quality of mathematically treatable statements, and therefore at best have only metaphorical value. This is true to some degree even of the originally physiological term *homeostasis* (Claude Bernard, Walter B. Cannon), now so often used to describe a concept in family life and family therapy.

General systems theory attempts to discover the common root of meaning in terms used by physicists and engineers on one hand and by physicians, psychologists, and biologists on the other. We once derived our terminology from the Latin or Greek scientific language used by medieval and Renaissance philosophers, for whom the physical sciences as well as the social sciences were only special cases which could be treated jointly in general statements pertaining to the theory of knowledge. In modern times the use of technical terms has gone so far that today the same term may be used for entirely different subjects, and in some cases may even denote almost contradictory concepts in physics and in social sciences.

It seems to me that we may be at the beginning of a new phase in science, the first symptom of which may have been the courageous attempt almost 30 years ago to write the *Encyclopedia of Unified Science* [20], in which common concepts were sought and defined and a true universality of knowledge, rather than the disparity existing in many specialties, was attempted.

I hesitate to use technological or physical terms in describing multiple family therapy but would rather do this, even if their use is metaphorical, than create neologisms understandable only to the few workers in our field. All my life I have striven for simplicity of expression, but in the treatment of such complex matters as multiple family therapy I am groping for words to describe, explain, perhaps even permit to measure, so that even those who may never have sat in a room with several sick families may become aware of what goes on in our work.

In a presentation of this material at the 1967 annual meeting of the American Psychiatric Association I allowed myself the use of the terms *control jet*, derived from the new fluid control systems used in industry, and *grid*, derived from characteristics of electronic tubes, to describe what happens when a therapist, with a few well-chosen words or gestures, transforms the internal situation in a patient, thereby causing him to redirect the full energy of his output toward other objectives than before. The patient and his family in turn were described as "control jets" or "charges on the grid" within the larger system constituted by several families in a multiple family therapy group. Thus, a few words from the therapist may reach the whole group of families in a twice-amplified manner and may become much more effective than if he had tried to preach didactically to twelve or fifteen individuals.

This seems to be the essential difference between multiple family therapy and an ordinary social gathering in which a professional scientist speaks to an assembly of students. The therapist changes the internal situation in the patient, subsequently in the patient's family, and finally in the whole assembly of families, through a method opposite to one used by a teacher or preacher. He does not try to synchronize and uniformly occupy the attention of all, but, by "awakening" sleeping dogs" in one individual, produces an atmosphere which arouses more "sleeping dogs" and which finally massively attacks the defenses normally built against such an incursion of feeling and new insight. There is some justification for the use of the terms *control jet* or *grid* because there are analogies in multiple family therapy with the comparison of the amount of energy used primarily and the discharges of larger amounts of energy following changes in the control organs. However, the control jet represents small hydraulic forces, the grid deals with electrostatic charges, while the therapist uses the often pooh-poohed Freudian libidinous energies to produce his effects. Hydraulic forces can be measured in psi's, grid charges can be measured in esu's, but nobody has yet been able to define and measure numerically units of libidinous energy transmitted through a field between people. In electrical engineering such terms as *capacity*, *induction*, and *resistance* have real and measurable meanings leading to exact calculations of quantities of energy flowing through an electromagnetic field. In his field theory the social psychologist Kurt Lewin [18] made a first attempt to speak of forces operating between individuals and groups which could be expressed in statistical terms, such as frequency of responses occurring in one part of the field as compared to other parts; but verbal responses are not comparable to electronic displacements, currents, and magnetic forces. This can best be shown by observing interaction between people [10]. Not only the number or speed of words contained in an answer but also the special energy expressed in emphasis, gesture, and sometimes self-defeating countergesture is of importance for the understanding of the response, and such rituals, operating in the way in which a response is given, do not lend themselves to quantitative analysis of displaced energy.

It is therefore with great caution that I would like my statements to be received. I will be the first to accept better terminology if social science should make it available.

Summary

Therapeutic measures in psychiatry are designed to help the individual to maintain his ego (his psychological wholeness) intact. While psychoanalytical and other uncovering and history-taking approaches in individual therapy strive to explore events and situations longitudinally, along the life axis, the modern forms of group therapy seek to explore horizontally the transactional, interpersonal plane of interactions between the patient and his environment in the present ("here and now") and in the past.

In multiple family therapy not only the patient's reaction to his familial environment but also the needs and responses of the members of his family are explored. General systems theory helps us to achieve a better theoretical understanding of the mechanisms and conditions affecting the results of multiple family therapy. It gives us a better preparation for the observation of the therapeutic processes and results of multiple family therapy, and provides us with better methods of learning to learn. Thanks to general systems theory a fruitful beginning was made in analyzing the interaction of the therapist's and his co-workers' general program and individual style with the programs and styles of each family and individual in the multiple family therapy group. We also see a possibility for analyzing and preventing system failures and breakdowns by learning from living and nonliving systems [24].

Although we are only beginning to understand the full impact of this new theoretical approach, we are already profiting from it to the highest degree because, as Kurt Lewin once said: "There is nothing so practical as a good theory."

References

1. Allport, G. W. *Becoming. Basic Considerations for a Psychology of Personality.* New Haven: Yale University Press, 1955.
2. Berkeley, E. C. *Giant Brains or Machines That Think.* New York: Wiley, 1949.
3. Burck, G., and the Editors of *Fortune The Computer Age and Its Potential for Management.* New York: Harper & Row, 1965.
4. Dechert, C. R. (Ed.) *The Social Impact of Cybernetics.* Notre Dame, Ind.: University of Notre Dame Press, 1966.
5. Diebold, J. *Automation. The Advent of the Automatic Factory.* New York: van Nostrand, 1952.
6. Galbraith, J. K. *The New Industrial State.* Boston: Houghton Mifflin, 1967.
7. Goldstein, K. Methodological Approach to the Study of Schizophrenic Thought Disorder. In J. S. Kasanin (Ed.), *Language and Thought in Schizophrenia.* New York: Norton, 1944.
8. Goldstein, K. *Language and Language Disturbances.* New York: Grune & Stratton, 1948.
9. Goldstein, K. The Organismic Approach. In S. Arieti (Ed.),

American Handbook of Psychiatry, Vol. II. New York: Basic Books, 1959.

10. Haley, J. Family experiments: A new type of experimentation. *Family Process* 1:2, 265, 1962.
11. Hilton, A. M. *Logic, Computing Machines, and Automation.* New York: World (Meridian Books), 1963.
12. Jackson, D. D., and Weakland, J. H. Conjoint family therapy. *Psychiatry* 24:2, 30, 1961.
13. Kasanin, J. S. The Disturbance of Conceptual Thinking in Schizophrenia. In J. S. Kasanin (Ed.), *Language and Thought in Schizophrenia.* New York: Norton, 1944.
14. Laing, R. D. Family Relationships Crucial in Treatment of Schizophrenia. Lecture sponsored by William Alanson White Institute, New York, 1967.
15. Laqueur, H. P., and LaBurt, H. A. The therapeutic community of a modern insulin ward. *J. Neuropsychiat.* 3:3, 139, 1962.
16. Laqueur, H. P., LaBurt, H. A., and Morong, E. Multiple Family Therapy. In J. H. Masserman (Ed.), *Current Psychiatric Therapies.* New York: Grune & Stratton, 1964. Vol. IV, pp. 150–154.
17. Laqueur, H. P., LaBurt, H. A., and Morong, E. Multiple family therapy—further developments. *Int. J. Soc. Psychiat.* Congress Issue, Section K, pp. 69–80, 1964.
18. Lewin, K. *Field Theory in Social Science.* New York: Harper & Brothers, 1951.
19. Menninger, K. A. Psychological aspects of the organism under stress. *Gen. Syst.* 2:142, 1957.
20. Neurath, O., Carnap, R., and Morris, C. (Eds.) *International Encyclopedia of Unified Science.* Chicago: University of Chicago Press, 1955.
21. Selye, H. *The Physiology and Pathology of Exposure to Stress.* Montreal: Acta, 1950.
22. Silberman, C. E., and the Editors of *Fortune* *The Myths of Automation.* New York: Harper & Row, 1966.
23. Strodtbeck, F. L. Personal communication, 1964.
24. Thomas, E. L. Abnormal Behavior in Non-living Systems. Presented at the Tenth Annual Psychiatric Research Meeting of Group Without a Name, Cincinnati, March, 1967.
25. von Bertalanffy, L. General system theory. *Gen. Syst.* 1:1, 1956.
26. von Bertalanffy, L. General Systems Theory and Psychiatry. In S. Arieti (Ed.), *American Handbook of Psychiatry,* Vol. III. New York: Basic Books, 1966.
27. Wiener, N. *The Human Use of Human Beings: Cybernetics and Society.* Garden City, N.Y.: Doubleday Anchor Books, 1954.
28. Wilson, I. G., and Wilson, M. *Information, Computers, and System Design.* New York: Wiley, 1965.

24

A General Systems Approach to Human Maturation and Family Therapy*

NORMAN L. PAUL, M.D.

THE NEED to develop more effective treatment programs is one of the major challenges to psychiatry. Psychiatrists are continually searching for ways to guide patients toward the achievement of a sense of mastery in coping with inner fantasies and feelings and in acquiring more satisfying relationships with the people in their lives. We hope that the gains achieved during treatment will continue after termination of treatment. Many of us also attempt to formulate ways in which accumulated clinical knowledge can be translated into meaningful educational programs and experiences which will enable the general public and ourselves to become better acquainted with the dynamics of the human maturation process.

One intriguing approach to these goals is to set about understanding the human life cycle in the light of what recent work in general systems has taught us. In such an approach all of an individual's experience from conception to death would form a system. It would then be useful to classify various experiences into subsystems of sequential emotional experiences that the human being must pass through in order to master the empathic relational skills he needs for coping with the different demands of different life stages.

* Presented in part at the 123rd annual meeting of the American Psychiatric Association, Detroit, Mich., May 8–12, 1967.

Conceptual Pioneers

Birth, and copulation, and death.
That's all the facts when you come to brass tacks:
Birth, and copulation, and death.

T. S. Eliot, *Sweeney Agonistes**

Skilled novelists, poets, and playwrights have generally been attuned to many aspects of the living process that elude scientists. Stanley Cobb [3] said, "Poets are always years, if not centuries, ahead of scientists. They see and understand what goes on; the scientist labors along in the rear, trying to explain why it goes on." Dewey and Freud in their respective ways contributed to bridging this gap between the poet's perception and the scientist's labors.

John Dewey saw experience as the essential mode of human life. Events, the external world, have no reality to the individual except as they are perceived through experience: "I begin with experience as the manifestation of interactions of organism and environment. . . . Experience is not a veil that shuts man off from nature; it is a means of penetrating continually further into the heart of nature. . . . Things interacting in certain ways *are* experience" [4].

Dewey also believed that life experiences have an essential continuity in their effect upon the individual, so that the outcome of one experience can affect the way in which an individual enters into a new experience. For instance, he stated that "Any experience is miseducative that has . . . the effect of arresting or distorting the growth of further experience" [5]. This, of course, has relevance for psychiatry and emotional growth since at all times growth-promoting experiences exist alongside growth-arresting ones. Also of therapeutic relevance is his conviction that the participation and sharing which occur in the communication of experience are key elements in the growth process; this is akin to my own concept of the meaningful dialogue.

Dewey's growing concern for the continuous and total qualities of experience led him to adopt *transaction* as a key word in his last work [6]. This word replaced *interaction* and more adequately focused upon continuity and process within the whole rather than among the parts. Dewey's writings, however, include no references to anxiety, frustration, anguish, and loneliness—some of the subcutaneous experiences of life.

* From T. S. Eliot, *Collected Poems* 1909–1962. New York: Harcourt, Brace & World, 1962.

Sigmund Freud's additions to the Dewey frame of reference deal with these subcutaneous experiences. Through his researches, particularly on the nature of fantasies as revealed by the free association technique, Freud broadened the notion of continuity and history to include the development of man's emotional and fantasy life. His presentation of the fantasy antecedents of dreams highlights this major contribution to our understanding of man.

Other early contributors to the matrix of a general systems approach were Sullivan and von Bertalanffy. Sullivan relied heavily on Dewey's point of view in the development of his "interpersonal theory of psychiatry" and made Dewey's insights more readily accessible to modern psychiatry. He believed that the individual is not a separate entity but part of a field, his behavior and mental state affected at every instance by whatever transpires in that field. He regarded anxiety in mental disorders as largely resulting from experiences of inadequate communication [8]. Von Bertalanffy, a biologist, proposed a general systems theory as a means for coping with the multiplicity of variables influencing an organism. He legitimated inquiry into the "field," where there is an ongoing dynamic equilibrium between parts of an organism and its total functioning [9].

These four diverse investigators all emphasized the continuous and fieldlike qualities of human experience, qualities that make a general systems approach a useful hypothesis. Such an approach can make possible substantial contributions to our understanding of how an individual is formed by his environment.

Systems and Experience

One common definition of a system is that of an entity which acts through the interaction of its components toward some end or goal. We are surrounded by such systems. A useful example would be an airplane with its pilot. For achieving its goal of moving through the air from A to B, this system requires constant interaction between the whole man and the whole machine; the linkages, though, are really more minute and complicated. The pilot's thought that he wants to achieve altitude causes him to make certain muscular motions to operate instruments which will transmit the energy necessary for lowering the ailerons on the plane's wings. This lowering causes the whole machine to rise, a motion which reacts on the plane's altimeter, which feeds the information in visible form back to the pilot. In any such system there are thousands of possibilities for interaction. One can think of an individual person in his environment as a system, a system which comes into being with his conception and terminates with his death.

The components for interaction in such a system are single experiences that build up over time in the consciousness and stored memories of the individual. At first glance it may seem that an experience is an unnecessarily abstract component to be conceptualized as the basic unit of environment, that something more concrete, like persons, would be more suitable. But an individual organizes his life not in terms of people and objects, which are wholly external, but in terms of experiences, which simultaneously include the internal with the external. A person recounting an episode, even one in which he may have been merely an observer, typically begins, "I had a funny experience today." Asked to describe someone he has just met, the describer can speak only of what was perceived while interacting with this person and inferences drawn from his perceptions; it is not the person himself but the describer's experience of that person that is meaningful for him.

Only some experiences, the more striking ones, seem to be clear-cut components, but actually an individual is always living through a succession of experiences with beginnings and endings, whether or not he is conscious of them. The moment a particular experience begins or ends cannot be determined objectively; the termini depend upon the participant's point of view or the point of view imputed to him by the observer. For a child the experience of a school day may begin as he leaves his house and starts to think of the friends and classes that await him at school or about the situation he left unresolved at home in order not to be late for school. For his teacher, on the other hand, this child's school day begins as he enters the classroom.

An individual's experiences are, in one sense, successive; the child just mentioned ended his breakfast-at-home experience as he left the house and then began his experience of the school day. Some experiences are linked over time by a kind of subject continuity. For example, the experience of a relationship with another person may be seen as lasting for a moment, for a few days, or until severed by death or other permanent separation. Actually, the experience has all these durations because, in certain ways, the relationship changes and in other ways it remains the same. Many of these longer experiences are referred to in words that express their unity over time—"that movie," "that journey," "that friendship." This unity connects but in no sense eliminates any of the component parts.

Time

One of the most important dimensions in the system of an individual in his environment is time. Experientially, time and space are woven together. Consider again the model system of the pilot in his plane. The pilot's judgment, vision, action, and reaction for any given situation are all influenced by past experiences in similar, related, sometimes seemingly unrelated situations. The plane's functioning is influenced by its past design, construction, operation, and maintenance. The pilot's past experiences with that particular plane may influence present thought and action; if he once had trouble with the landing gear, he may tend to lower it a few seconds early in order to be sure of its present operation. Similarly, an individual within his environment physically and emotionally carries with him the cumulative influences of the past. Unlike a machine, he is influenced in the present by what his memory and recollection tell him of the past, stored in his mind not with the rigid orderliness of clock time but in a kind of emotional time; certain past experiences may suddenly spring forward or, indeed, may never recede into the past at all.

We can all cite instances when the past has intruded into our present. The reader of this page may find his attention turning away from the text toward something in his past experience, even something apparently unrelated, that my words recall. This kind of reverie is the simplest form of the immanence of the past. There are more subtle forms, some of them closed to conscious appraisal. For example, A may respond to another person, B, not only in terms of his past experience with B, but actually as though B were another person, C, encountered in the past. This is the situation of the passive-aggressive college student who perceives the college authorities as though they were his parents, demanding and expecting that he will conform to their values for scholastic excellence. Another example is preoccupation with an unresolved past experience to the point that the person cannot devote his attention to the present; this phenomenon is often loosely labeled "lack of concentration," a label which fails to recognize that the individual is indeed concentrating, but not on the present. Then again, the wish to deny the past may throw the individual into a replication of it. The chronic alcoholic who has repeatedly lost jobs because of drunkenness vows to himself and others that there will be no further recurrence of drinking, thus denying the extent of his alcoholism and his compulsive need to drink, as though he could will his alcoholism away.

The immanence of the past also includes a situation frequently observed in family therapy, a complex situation in that it requires individuals to assume new roles in a reenactment of the past. A mother who in her own childhood was a very "good" child, actually nourished rebellious fantasies that she could never express. She now has a rebellious child. Nobody is consciously aware of the historical influences, but it becomes obvious that the mother has unwittingly taught her child to relive her own past as she wishes it had been. That the past is being suppressed becomes obvious in the mother's continuing denial of her own rebellious fantasies and feelings. If she can be brought out of this denial through a reliving of these fantasies and feelings and if she does this in her child's presence, enough empathy may be generated in the child for the mother as she once was to neutralize the acting out. The mother, having purged herself of this unresolved inner experience, would then be free to modify her behavior toward her child.

Persons

The various experiences which form an individual's environment can, for convenience, be divided into three subsystems—the individual in his transactions with himself, with his permanent objects (usually, in our society, members of his family), and with strangers. These subsystems are not absolutely distinct but, in some ways, form a continuum, particularly those classified as transactions with permanent objects and with strangers. Yet most Americans experience relationships with intimate family members as qualitatively different from other relationships. A girl may, for example, live for 3 or 4 years in good harmony with her college or apartment roommate, then, on getting married, find her living accommodations suddenly complicated by a tendency to replicate patterns observed in her family of orientation.

It is important to realize that the subsystems are interlocking in their effects on the individual, that experiences not only influence subsequent experiences within the same subsystem, but spill over to influence behavior in both the other subsystems as well. Every relationship in each of the subsystems generates feelings and fantasies that frequently have an enduring existence within a person and mingle with external events in his memories. Such feelings and fantasies form an important part of the past and contribute substantially to modes of present behavior.

The subsystem of transactions with family members or permanent objects represents the soil out of which the child's sense of self first emerges. The manner in which his mother feeds, bathes, and handles a child communicates to him whether or not he is loved, to what extent he is regarded as part of the mother. In these relationships, as in others, people are not always the best judges of their own behavior. The mother may have the fantasy of loving her child, but, if a physically normal child responds with evidences of distress, such as feeding or toilet-training problems, there is indication that the mother is transmitting some emotional dissonance to the child that reverberates between them.

To take another example, a parent may subject an older child to repeated verbal expressions of rejection or degradation. The parent may consider such expressions unimportant or easily counteracted, but they usually generate fantasies of a similar ego-alien nature in the child; in his transactions with himself he will adopt such fantasies and feelings as one parameter of his oscillating sense of self-esteem. If the child becomes sufficiently inhibited from expressing his sense of being hurt, he may begin to respond to the world around him with constricted awareness and interest. The diminished self-esteem, sense of hurt, and loss of interest generate feelings of loneliness in the child that increasingly isolate him as he negotiates life in and out of the home. These are all negative examples, reflecting my own clinical concerns. But a comparable pattern holds true for supportive parents who contribute to the positive self-image and self-esteem of their child.

Whatever self-concept the child begins to form in his early life with his family he will take with him as he moves out into transactions with strangers, that is, with friends, classmates, and teachers. Generally he will tend to solicit responses from those strangers that reenforce his idea of himself. This is not inevitably so; sometimes experiences with strangers can neutralize experiences in the home. (It might be noted that an individual meets his psychiatrist as a stranger.)

Marriage is a critical point in the lives of most persons. At this point, two individuals move their relationship out of the stranger subsystem into the subsystem of permanent objects. Although most people are aware of the seriousness of this venture, few fully appreciate the multiple ramifications of the change.

During courtship, love is associated with fantasies of oneness and idealization of the future marriage bond. Recollections of unhappy parental marriage patterns are brushed aside with the optimistic belief that replication can be avoided by act of will. Once the new marriage is a reality, however, repetition of behaviors learned in the family of origin may be difficult to avoid. The difficulty is enhanced by the fact that an individual is often not aware that he is imitating his parents. A child may readily observe that his parents have arguments which degenerate into scenes of physical violence, and he may later resolve not to let this happen in his own marriage. At the same time he is probably not aware of the sequence of more subtle actions that precipitate the visible violence; it is these more subtle antecedents that he may unwittingly replicate, with the result that he finds himself inexplicably involved in a physically violent scene with his own spouse.

It is easier to be aware of patterns of behavior in someone else than in oneself. Ideally, each spouse could act as a watchdog for the other, pointing out or neutralizing the other's unconscious actions. For this, it is necessary that each accept the fact that he does not realize how he appears to the other. If he recognizes that he can and does make mistakes in transactions with the other, feedback presented in good will can be accepted. Actually, it is difficult for most couples to deal with problems in this way; they find it hard to accept the existence of error in their own transactions with each other or to review enough of their past together to enable each to perceive the historical antecedents of current behavior. Ideally, courtship should be a time when both partners review the history of their relations with permanent objects and strangers so that they will be better prepared for whatever difficulties emerge in their own marriage.

Maturity

Any human being can be viewed at any point in time as the sum of his experiences since conception. Or he can be considered to be composed of a number of traits, with each trait regarded as the result of a variety of experiences and having its own history. Maturity, then, can be considered a constellation of positively valued traits.

In 1930 Adolf Meyer described maturity, which he considered the goal of the living process, in some detail [7]:

> Expressions of mature living are the balance of expectation against reality, and the capacity to fit into groups: in business; in home life, with its nonsexual affections as well as with its visions of sexualization; in our allegiances as well as in our emancipation. It implies the capacity to accept illness, disappointments, bereavements, even death, and all that which is largely beyond our own control and influence; to accept our own make-up, the perfections and imperfections of self and others, success and failure, sportsmanship, and the social comparisons which we call advice, criticism, and authority. Finally, maturity assumes a philosophy of objectivity about the past and a vision of creative opportunity for the present and the future.

Meyer's description provides a point of departure for reflecting further on the experiential aspects of maturity, whether achieved naturally or through treatment. What should a mature adult be expected and able to do in relation to the people in his life? He should be able to perceive, think, and feel empathically about himself and others so that he is motivated to differentiate the external from the internal world and to accept and assess feedback from other people. He is then able to meet the changing needs both of others and of himself. He should be able to effect an interdependence with family members and close friends in which he and they communicate appropriately about their wishes, joys, pains, disappointments, and other meaningful life issues without retaliatory or annihilative anxiety. He should also be able to relate comfortably to peers and authority figures in both occupational and recreational settings. He should be able to accept being gradually extruded from his own nuclear family, concomitant with developing relationships outside his family—often to select a mate of his own and to establish a new nuclear family. If he can do these things, he can pursue his activities with a sense of achievement and mastery. The system, one might say, would be working so as to meet its goal.

Another critical feature of maturity is the ability to cope with a succession of losses, ranging from a child's concern about his mother's leaving for the afternoon to his reaction when she dies. Mature behavior in such situations includes the ability to admit the existence of one's inevitable fears, guilt, and anguish and then to share such feelings and associated fantasies with intimates.

A precondition for such sharing is the receptivity of others. Maturity, like everything else about a person, cannot be understood in isolation. Maturity is a function of past and present experiences, and experiences are not shaped by one individual alone. A mature individual is part of a mature environmental system; a mature environmental system contains many mature individuals.

Treatment

These observations have several implications for psychiatry. First, psychiatric intervention must be seen as a series of experiences which can correct or neutralize the effects of previous experiences so as to alter undesirable behavior and traits. Second, psychiatry must take due note of time, recognize the significance of the past, and make use of the vagaries of emotional time to bring the past into the present where it can be dealt with. Third, psychiatry must take as the unit of treatment not the individual (who probably cannot be effectively treated in isolation), but the whole system, that is, the individual in his environment. The total system can be broken down into the three subsystems already delineated, and each subsystem can be treated by bringing the appropriate group of people together with the therapist. At all times one must keep a transgenerational perspective and realize that time is an important dimension, not only for the individual, but for the entire system. Bowlby's conception of defense or symptom, grief or mourning, and separation anxiety as parts of a single process in an individual represents a useful point of view for understanding maladaptive transactional patterns converging on the labeled patient [2].

The permanent object subsystem is dealt with through individual family group therapy, a setting in which the two or more persons present, other than the therapist, are close relations. A married couple, a mother and son, or two siblings are possibilities for individual family group therapy sessions. This setting provides each family member, as well as the entire unit, with a controlled family communicative experience—controlled in that physical violence is prohibited and the therapist is present to direct attention to material that is relevant to the problem. The goal is to unravel the dead ends of previously unrecognized and unresolved conflicts, generally rooted in both displacements and projections. At times grandparents, domestics, or close friends are invited to provide historical data and recollections relevant to current conflicts. Reality testing of the historical sources of inner hurts, depression, anxieties, guilt, and hostility is impossible unless family members are seen together so that fantasies can be sorted out from reality in a shared enterprise.

In such a setting, inquiry into and exposure of reciprocal affective responses that have developed in the course of family life can often help to unfreeze long-established faulty transactional patterns that have reenforced undesirable character traits in the labeled patient. The therapist has the opportunity to help not only the patient but also other family members to grow and live comfortably by encouraging them to become empathically acquainted with each other. Each family member can be helped to recognize that under the other's narcissistic shell lies the helplessness, uncertainty, sadness, and loneliness common to all people. The poignancy of what is mutually experienced here confirms Robert Ardrey's statement [1], "There is nothing more moving—not even acts of love or hate—than the discovery that one is not alone." Such reciprocated sharing of one's history of inner hurts and other distressing feelings is a goal of therapy for this subsystem. The process initiated is termed a meaningful dialogue or a shared being-experience.

Self-confrontation through either audiorecorded or videotaped playback of such experiences provides each member with the feedback he needs to empathize with himself. The feedback is essentially an opportunity to see himself as he is exposed to others and leads to increasing intimacy with oneself, a prerequisite to developing intimacy with other people. Long-range goals of this process are increased self-esteem, spontaneity, and effectiveness for each person involved.

One particular variety of individual family group therapy deserves separate discussion, namely, conjoint marital therapy. Establishing the marital pair as a distinct treatment unit aids in developing or reenforcing generational boundaries between parents and children. This makes it possible for the family to accept comfortably the inevitability of its own dissolution when, in the course of time, the children are extruded and the parents die. The focus in this setting is on the delineation and exploration of those experiences cumulatively leading to marital incompatibility. A historical review of each partner's knowledge, fears, and expectations about marriage and parenthood is useful. Many couples have characteristic inappropriate reactions to one another that adversely affect their child's development. I have yet to encounter a disturbed child whose parents have a stable marital relationship. Accumulated data suggest not that marital instability is caused by living with a disturbed child, but that the entire situation is primarily rooted in unresolved and unrecognized legacies and internal hurts from the parents' own childhoods, displaced and projected onto their children.

Individual psychotherapy for the labeled patient and other family members as indicated is designed to help the individual achieve a more comfortable relationship with himself, both in the present and in terms of the history of his fantasies and feelings. The patient is encouraged to begin discriminating internal from external realities, a process that can be continued in family therapy. Hostile fantasies with their associated internal hurts and guilt can be reviewed in this less threatening setting to neutralize some of the unrealistic retaliation fantasies (e.g., "she'll kill me") prior to sharing them in either individual or multiple family group therapy sessions or in stranger group therapy.

The sequence of therapeutic experiences for purging shame-ridden historical data will illustrate the process. The material is first reviewed in an individual therapy session, then shared with the spouse (if we are dealing with a married person) in conjoint marital therapy, and finally shared with the rest of the family in individual family group therapy. It is vital that the patient learn not only to discriminate between internal and external realities in himself, but also that he learn that other family members share his kinds of defects in reality testing.

In stranger group therapy or multiple family group therapy, one goal is to enable the patients to see the similarities and differences between themselves and their family and the others. The observation of conflicts, patterns of projection, displacements, and distortions in others is a step on the way to self-awareness. The historical relevance of the stranger subsystem for the individual or family can also be explored in this setting.

Since the therapist is and must remain a stranger rather than a permanent object to the patient-family, the fantasy of eventual termination has to be reenforced by the therapist. Otherwise, the possibility that he may permanently join the family becomes very real to patients. The loss of the therapist is not only necessary for the therapist himself, but can provide a corrective emotional experience for family members. Learning to deal with one loss in a mature fashion will facilitate coping with other losses.

In some families the techniques used in therapy, such as empathic sharing and reality testing, can occur spontaneously and without intervention. In these families it seems that parents learned to take part in meaningful dialogues within their own families of orientation. It would be useful to carry out transgenerational research into those forces that encourage and perpetuate such dialogues in family life. Most of the family members we encounter in therapy, however, have not been able to achieve meaningful dialogues among themselves and, consequently, demonstrate various limitations in their capacities for empathy and reality testing. A goal of therapy is to overcome these limitations. I suggest that this goal can be approached with more confidence and understanding if the individual in his environment is seen as a system. To illustrate the application of this concept, I cite the case of Herbert Y. and present extracts from three therapy sessions included in the history of that case.

Case Study of Herbert Y.

Herbert Y., the labeled patient, was 21 years old, single, a stock clerk, a schizophrenic. When he was 6 years old, his father had a prefrontal lobotomy for a schizophrenic reaction. The father was continually hospitalized from that time except for brief trial visits with his family immediately after the lobotomy and again when Herbert was 17. The first excerpt, from a Y. family meeting, illustrates the schizophrenic process in action. Herbert always felt rejected by his mother, whereas she saw herself as good and loving. Here, rage and guilt are expressed and projected. In the second excerpt, from a family meeting 10 months later, the mother finally shares with her son a shame-ridden secret. The final excerpt from an individual meeting with Herbert 1 week after this revelation includes Herbert's description of his reactions to his new knowledge.

FAMILY MEETING NO. 5.
(TEN MINUTES INTO MEETING)

DR. PAUL: How has he [Herbert] been since last Tuesday?

MOTHER: He's been pretty good, and you know I'm almost—

DR. PAUL (*To Jonathan, 20 years old, the younger son*): Would you agree that he's been better?

JONATHAN: I haven't been home in more than a week. But I feel that today there were some things that were pretty normal, you know. I'll take my mother's word for it, you know. My mother's usually pretty graphic.

DR. PAUL: I wanted to find out what you people thought about our last meeting, what you discussed with one another about our last meeting—

HERBERT: I can't discuss anything with her.

DR. PAUL: —if anything, how you felt about it.

MOTHER: Well, I can't discuss things with him, what is it, Herbert? Can I say anything? Why are you looking at me with such hatred? Why do you grit your teeth when I—

HERBERT: Because I'm mad.

MOTHER: Well, why are you mad right now?

HERBERT (*in a rage*): Because you took advantage of me.

MOTHER: Well, just what do you want me to do now? Just what do you want?

DR. PAUL: Well, you just have to let him work this out.

HERBERT: No, she looks at me when I go and listen to the radio. She looks at me like this. I was a flunky. Everything I did was wrong. I was an idiot. She made fun of me. She laughed at me. She always speaks for me. She looks at me. She can't stand me. Because I look like my father and I made fun of my father. She hates me; she does. She can't stand me because I never got good marks. She hates my guts. She does. She said it.

DR. PAUL: When did she say it?

HERBERT: She looks at me and torments me. She never smiles.

What are you doing? Where are you going? She hates me. She always did. I was stupid. I was a dumb kid. I was stupid. I was sad about my father. She can't stand me this way. She can't.

DR. PAUL: This way, meaning what?

HERBERT: The way I am now. Anytime I go home, she can't stand it. She hates me.

DR. PAUL: Can you stand him this way?

HERBERT: No, you can't.

DR. PAUL: Why not?

MOTHER: No, I can't stand him this way. When he carries on at home this way. It's awful. I can't cope with it.

DR. PAUL: Why not?

MOTHER: Because it's sickening to see a boy of 21 talking this way.

HERBERT: I've always been this way. It's sickening at 12 doing this. It's sickening at 11 doing this.

DR. PAUL: But what is it that's so difficult in seeing him this way? (*Both talk at once.*)

MOTHER: This is an unusual sight. This is an unusual thing. This is not normal. This is not right.

HERBERT: Why isn't it normal?

DR. PAUL: It's natural for someone who's lived through what he has.

HERBERT: I made fun of my father. I made fun of my father. He fainted.* "Maw, come here, Daddy's bothering me." He fainted. "Go in the other room. Look what you did. You made him faint." You blame me for him. You blame me. I made fun of him. I called him a loon. Now—

DR. PAUL: Just a second—what do you mean she blames you?

HERBERT: She blamed me.

MOTHER: I never did.

HERBERT: She blamed me. Yes she did. My father came home. I

* Herbert refers here to an incident that occurred when he was 6 years old. Herbert's father was staying home after convalescing on a farm in Vermont. Herbert and his father were wrangling over which was to have his favorite television program tuned in, a conflict that the mother repeatedly tried to resolve, but without success. Suddenly the father fainted—actually he had a postlobotomy seizure—which terrified Herbert and his mother. The myth later developed that the father became crazier after this seizure for which Herbert was responsible. Shortly after this episode the father went to a state hospital.

was 17. I couldn't stand it.* A week later I got into an argument and she said "I got rid of your father on account of you." He fainted. He got worse after he fainted. "You made him faint. Don't go in there, you'll make him faint again. He got worse." She blamed me. She blamed me. She blamed me for everything that went wrong.

MOTHER (*defensively*): That's not true.

FAMILY MEETING NO. 44.
(FINAL PORTION OF MEETING)

MOTHER: Supposing I told you you were, to a point were illegitimate. Supposing I told you?

HERBERT: I'd feel better. At least I'd know how I stand.

MOTHER: How do you know? Because the doctor told you you'd feel better?

HERBERT: No, no. Because I'd know where I stood.

MOTHER: What do you mean, you'd know where you stood. What would you gain from it?

HERBERT: I'd know where I stand and why I stand and why I do act this way toward you.

MOTHER: That doesn't mean if one is born like that that I hate you.

HERBERT: No, I want to know why, I don't understand why I'm this way and why I act crazy.

MOTHER: Now, do you know why . . . Supposing I say this is the truth? I'm only supposing. I'm not telling you.

HERBERT: You told me this wasn't the truth. You said no.

MOTHER: Well, now I'm telling, supposing I change my mind.

HERBERT: Then I'd know where I stand.

MOTHER: All right, supposing Daddy and I got married and I had you too soon afterward. What would you think of that?

* Herbert visited the father about once a year from the time he went to the state hospital when Herbert was 6. When Herbert was 17, the hospital staff decided to send the father home and neglected to prepare the Y. family adequately for this change. Herbert was, at that time, attending a local psychiatric clinic. He discussed with his psychiatrist the fear of having his father come home and revealed his perplexity over the return. He had felt that his father was dead; when he saw his father, he wondered whether the dead return to life. The psychiatrist spoke with Herbert's mother about this and suggested that she return the father to the hospital; she followed this advice. Two weeks thereafter, Herbert awoke one morning in terror, convinced that he was immediately to become bald like that "bald-headed bastard," his father. He also felt that his body was shrinking in size so that he might disappear altogether.

HERBERT: I'd know where I stand, but that isn't it.

MOTHER: Now this is it, now what, now how do you feel? Do you feel better? Do you feel—

HERBERT: No, but—

DR. PAUL: What do you mean, this is it?

HERBERT: I don't believe that's the truth.

MOTHER: What do you mean? Supposing I said you were born 6 months after I was married instead of—

HERBERT: It wasn't because you said you, you, uh, you were married in 1939.

MOTHER: Supposing I told you I lied.

HERBERT: No, because I found out that was the truth.

MOTHER: How did you find out?

HERBERT: I asked your sister.

MOTHER: Well, my sister would tell you—

HERBERT: In March of '39.

MOTHER: All right, supposing I said we were all lying.

HERBERT: You're not all lying.

JONATHAN: Why not?

MOTHER: Why not? My sister would tell a little lie to help me.

HERBERT: Because I do look like my father's family.

MOTHER: You are your father's child.

HERBERT: Yeah.

MOTHER: Without a question you are. By all means. No one ever doubted that you were Daddy's son, ever, ever. I didn't say that. But I said supposing you were born too soon. Supposing.

HERBERT: Yeah, what about it?

MOTHER: Well, how would you feel?

HERBERT: I'd understand a lot more.

MOTHER: What do you understand?

HERBERT: Why you feel this way toward me, why, why I'm dirt in your eyes.

MOTHER: I didn't say you were dirt in my eyes.

HERBERT: But I feel it.

MOTHER: I would say supposing I didn't want to marry Daddy and I had to because you were on the way? I resent you because I had to marry him.

HERBERT: Yes.

MOTHER: Because of you. How would you feel?

HERBERT: I'd understand a lot more.

MOTHER: Well, of course, when Jonathan was born maybe I didn't have that attitude because I was safely married. I might not have wanted him but I was respectably married and he could have been born at any time.

HERBERT: I'd understand a lot more.

MOTHER: Does that make you like me?

HERBERT: No.

MOTHER: All right, then, what difference does it make in your life if I tell you this? (*Pause*) I'm asking you a question.

HERBERT: I don't know. I'd have to think about it.

DR. PAUL: Well, how do you, is this what's supposing or what? You said supposing he—

MOTHER: Well, I'm just saying—

HERBERT: Is this it? Is this the truth? I don't know. You said you got married in 1939.

MOTHER: Well, supposing that was a little lie. Or a big lie.

JONATHAN: Well, supposing the sisters lied to cover her, which I'm absolutely sure they would.

HERBERT: I don't know. I'd feel, uh . . . I'd be kind of upset about it.

MOTHER: I'm sure you would.

DR. PAUL: Upset in what sense?

HERBERT: That, uh, that my mother and father would have to get married because my mother was pregnant.

. . .

MOTHER: I can also say, and it's very true, that my resentment against my husband, against my marriage, and my whole bitterness has come out on this boy [Herbert]. I know it and I don't know how to stop it.

DR. PAUL: Okay, this is how you're stopping it right now. And it's taken a lot of guts on your part to be able to do this. But I think this can be the beginning of the end of a lot of family misery.

MOTHER: You just have to make him understand it. I'm going to try. I've talked to Dr. X about this. I can talk to you about this. But he's got to be made to understand.

DR. PAUL: Look, he's been living with this over a period of years. He has been living in a quagmire where he has been feeling certain things and not being able to find out what this is all about.

INDIVIDUAL MEETING WITH HERBERT Y.
(BEGINNING OF MEETING)

HERBERT: Well, I'll tell you about my reaction—I felt better. I can't explain it; when I left here I couldn't even look at my mother. I just took a walk, I took a walk all the way down Brattle Street.

DR. PAUL: Yeah.

HERBERT: Uhm, I j–, I went into the Hotel Commander and I bought cigarettes, for me to sit down, and I bought myself a candy bar.

DR. PAUL: You bought what? A pack of cigarettes?

HERBERT: Yeah.

DR. PAUL: What kind?

HERBERT: I don't even remember what kind.

DR. PAUL: Hmm.

HERBERT: Uh, and I sat down; I just couldn't believe it. I just sat there in amazement; and I sat there and sat there.

DR. PAUL: Smoking cigarettes?

HERBERT: Well, not constantly, but I was just sitting there and sitting there and sitting there . . . and sitting there. And then, it didn't hit me, and then I thought about it; it happened.

. . .

HERBERT: . . . I couldn't sleep, I, I mean, I don't know, I, I just kept thinking of my mother and father intercoursing, you know?

DR. PAUL: Uhum.

HERBERT: I kept thinking of it, and I just couldn't believe it. . . . And I just kept thinking of it; so then I went to work the next day, and I asked this woman, this woman that I like, I said, "What would you think if you knew a girl that you respected and, uh, you found out that she had to get married because she was pregnant?" She says, "It all depends on the girl, what kind of a girl she was." I asked somebody else, "What would you think of a girl you knew . . ." So then I said, "What happens if they don't love each other, what happens, how does it affect the kid?" She says, "It hurts, it must hurt the fellow, the boy or girl they have, uh, if they don't love each other." So what happens to the woman and the man if they don't love each other? They, the woman usually goes out with other men or the man goes out with other women. So she says, "Yeah, that's right." I asked somebody else, "What do you think of a woman, the girl, I knew a girl, I respected her—"

DR. PAUL: Uhum.

HERBERT: "—and you find out that she's gone, she's got, she's pregnant." Uhm, well, "It depends on the kind of girl." This is what I asked, so, one of the girls says to me, "Gee, Herb, you must like this girl. You're worried about her." I said, "Yeah, yeah." I, so, I got up and I went to work, you know.

. . .

HERBERT: Why did I feel so horrible? Why did I feel so horrible? Why did I do such things? I began to fit the picture. I was thought of as horrible. I had my mother's mixed feelings. I had my mother's feelings. Everything seems to be coming back. And then I thought of my mother's family. And she said the older ones, uh, and I always wondered why, maybe I'm only projecting or something, but, wondered why they looked at me certain ways . . . and then I thought back about my mother, and, uh, if it wasn't for this kid, if it wasn't for this kid. (*Laughs*) I wondered why I was frightened, and then it made sense. My mother used to want me to keep clean; I never kept clean. (*Laughs*) She liked me to socialize; I never socialized. She wanted me to get good marks; I never got good marks. When she wanted to go out and enjoy herself, I was just the opposite. It seemed that I was doing the opposite—it made sense. She was trying to like me, and unconsciously I knew she didn't. Is that it? (*Laughs*) I've been trying to figure out why I'd do the opposite of her.

DR. PAUL: You see—

HERBERT: All these years . . . she says, "I had the baby. My father loved this baby, accepted this baby." But she didn't say she did, she was looking at her family's op–, way of looking at it. "My father loved this baby; my sisters loved this baby." You know?

DR. PAUL: Uhum.

HERBERT: And that puzzled me, why did she want her father (*laughs*) to like this baby?

In subsequent sessions Herbert commented that, after the revelation of this information about his birth, he felt that he became a separate person for the first time. Even these brief transcripts illustrate clearly some of the points raised earlier in this analysis. They show, for example, how Herbert's transactions with permanent objects seriously affected his ability to live with himself and others.

They further emphasize the ingrained effects of the past, especially the hidden past, upon the present, and indicate that knowing the past makes one better able to deal with the present. Finally, they demonstrate that the true nature of the past can be established only when several concerned family members come together for a review of secret family history. The therapy sessions helped to provide members of the Y. family with new experiences, or components, to neutralize and correct the effects of other components, so as to increase their chances for achieving the goals of the mature adult, the expectant mastery of emergent situations.

References

1. Ardrey, R. *The Territorial Imperative*. New York: Atheneum, 1966. P. viii.
2. Bowlby, J. Separation anxiety. *Int. J. Psychoanal.* 41:89, 1960.
3. Cobb, S. *Borderlands of Psychiatry*. Cambridge, Mass.: Harvard University Press, 1948. P. 76.
4. Dewey J. *Experience and Nature*. Chicago: Open Court, 1929. Pp. 19–21.
5. Dewey, J. *Experience and Education*. New York: Collier, 1963. P. 25.
6. Dewey, J., and Bentley, A. *Knowing and the Known*. Boston: Beacon, 1949.
7. Meyer, A. The Meaning of Maturity. In E. Winters (Ed.), *The Collected Papers of Adolf Meyer*. Baltimore: Johns Hopkins Press, 1952. Vol. 4, pp. 425–434.
8. Sullivan, H. S. *The Interpersonal Theory of Psychiatry*. New York: Norton, 1953.
9. von Bertalanffy, L. An outline of general system theory. *Brit. J. Phil. Sci.* 1:134, 1950. Reprinted as General system theory. *Gen. Syst.* 1:1, 1956.

25

Steady State and Change in Family Interviews*

EDWARD J. CARROLL, M.D.

PSYCHIATRISTS ENGAGED in research problems are often baffled by the complexities of the variables they deal with and the vast amount of data they confront. General systems theory, as the repository of common features of the behavior of complex systems, offers many generalities that are attractive to the researcher and the theorist. It is a difficult matter, however, to apply specific concepts from general systems theory to specific problems.

This paper is the report of the application of the concept of steady state to the clinical material of one interview.

Steady State

Steady state, or dynamic equilibrium, is a condition of rest, or equilibrium, in a dynamic system caused by the interaction of forces which balance each other out. In a steady state the variables under consideration remain within defined limits. With change they move outside these limits.

Cannon made a major contribution to this field in his study of the ways in which vital processes constantly fluctuate, yet remain the same, and described this as homeostasis [2]. W. Ross Ashby, a psychiatrist and one of the pioneers in general systems theory, has for several years studied change as a cybernetic process [1]. Ashby's

* Presented at the 122nd annual meeting of the American Psychiatric Association, Atlantic City, N.J., May 9–13, 1966. Acknowledgment is made of the assistance of NIMH Grant MH-05433.

model of a system is an organism and its environment being impinged upon by a disturbance to which the organism responds by regulatory functions which counterbalance the disturbance and maintain the equilibrium. If the regulators do not contain the disturbance, then an output results which changes the system. The stable system in equilibrium with its environment, if undisturbed, remains the same from one selected unit of time to the next. An unstable system under the same conditions changes.

The Research Project

In a research project in the area of family therapy we set out to evaluate the usefulness of interviews with different members of the family by means of tape-recorded interviews. We gradually focused upon the central question of whether any significant change took place in the interview. In an interview changes of all sorts are always going on, and the problem is that of distinguishing changes that are merely transients from changes which are enduring and therefore really make a difference.

In order to study change within the therapeutic situation, we had to use quantitative methods and we had to measure specific things which presumably had some relationship to clinical criteria. There are no established scales for assessing a clinical interview with two or more family members. Rather than developing new scales, Dr. Thomas Mellett of Pittsburgh adapted four scales that have been devised for measuring some aspect of interaction and developed an additional one. The use of these scales is described in the research report [3] and will not be detailed here. Suffice it to say that if we select and quantify various elements of the interview and find that the values of the variables show related patterns over time, then we can assume that some element is affecting the variables and that they all have some relationship to each other.

In the collection of data for the research there were six interviews with each family: three individual interviews with mother, father, and referred child and three group interviews with the sibling group, the parents, and the whole family. The particular interview we chose for intensive study was with the parents and was the sixth and last of the research interviews with this family.

The Research Family

The parents are middle class, educated Negroes who came to the clinic because of underachievement in school by their 14-year-old son. The interviewer for all six interviews was Dr. Morton Johan, a psychoanalyst. The well-established and stable pattern brought out in the first five interviews was the mother's dominance and control of the family and her chronic dissatisfaction with all aspects of the 14-year-old's behavior, and the discrepancy between the high expectations of the mother in regard to all family members and the more relaxed attitude of the father.

The interview will be described with reference to the Chapple Interaction Chronograph [4]. This device is a computerized stopwatch which we used to record who was talking at each instant. It has no reference to what was said. We divided the interview into eighteen 3-minute segments and plotted the percentage of time that each interviewee talked in each segment. This is shown in Figure 25-1, which may be referred to in order to follow the action in the

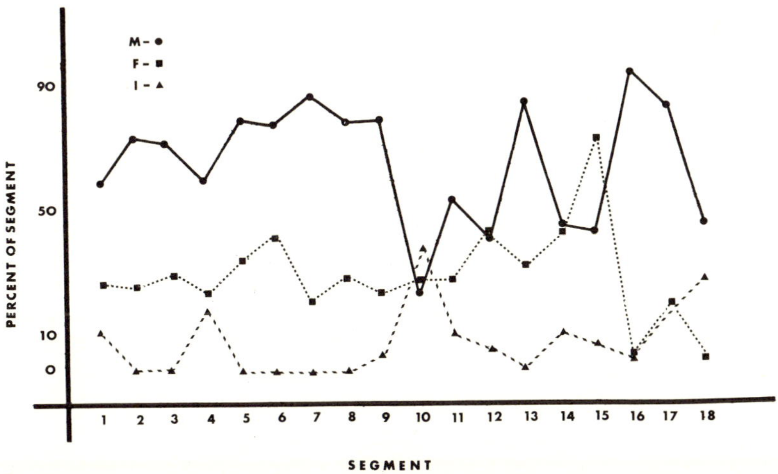

FIGURE 25-1. INTERACTION CHRONOGRAPH: DURATION OF SPEECH BY EACH PARTICIPANT.

interview. The content of the first half of this interview resembled that of the previous five interviews. A very condensed account of the interview follows. In the left margin the segment numbers are given.

The Interview

1. As the parents enter the office, they are discussing an obliga-
tion owed them by a friend. The interviewer joins in this dis-
2. cussion to elicit further information and asks how they gen-
erally handle such situations. They do not follow this lead but
3. get on to the topic of Jack, his poor work in school, and his
difficulties with a particular teacher. This continues, and fi-
4. nally the interviewer rather sharply questions the mother as to
just what she had said to the teacher. He fails to get any spe-
5. cific answer and asks a more general question as to their feel-
ings about corporal punishment. This gets nowhere and the
6. parents discuss Jack's behavior back and forth, with numerous
7. examples of mother's high expectations and father's low ex-
8. pectations, and the interviewer is completely shut out for 17
minutes.
9. Finally, in the ninth segment, the interviewer takes advantage
of a momentary silence to say that he feels he is listening to
the prosecuting attorney and the defense attorney. Mother
responds to this by saying she knew that was what he was go-
ing to say, then launches an attack on nonconformists, indi-
cating her husband and the two sons as being in need of
guidance. The father seems to be thrown off balance and
10. comes in with an unintelligible statement. Mother attempts
to get the focus back to Jack, but the interviewer continues to
press as to whether her concern about nonconformists is not
really a fear. He directly asks her, "What are you afraid of?"
The parents respond by focusing upon the words *alarmed* and
11. *concerned*, and then return again to Jack. The interviewer
continues to press, again asking the mother, "Why are you so
worried?" and indicates that she fears underlying delinquent
trends. The mother laughs this off, and the interviewer turns
12. to the father for support, but the father defends the mother's
high expectations. The father and mother then continue

13. an exchange with each other about their expectations, and
the interviewer is shut out. The interviewer finally takes
14. advantage of a silence to change the subject by asking, "How
do you feel about this talk today?" They parry this by saying
15. it is the same as their usual talk. They get back to the boys
again, and the father talks about their short, short trousers,
with much laughter until mother shushes him for talking too
16. loud. They go on, but the interviewer interrupts them to point
out that she corrects him the same as she does the children,
and she always needs to keep her eye on the children. The
mother agrees and then goes on to introduce the problem of
her children and the attitudes of white teachers toward Negro
children.
17. As the mother continues to document the attitude of white
teachers, the father tries to get back to the matter of pants.
The interviewer tries to get at their feelings about discrimina-
tion, and finally confronts the mother with her anger, saying
that she is not only angry at whites, but angry at herself for
her need to comply. She confirms this vigorously with, "It's a
whole vicious, almost futile process, but you keep going
through it."
18. She continues to talk of the demands of the white world and
of her unremitting efforts to do the right thing. The inter-
viewer points out how exhausting this is, to which she agrees,
and the father adds, "Exhausting for the children, too." With
this, the interview ends.

Analysis

We consider the interview as a system composed of the three participants. As will be seen from Figure 25-1, the first half of the interview is stable, with mother dominating the proceedings. Father accepts mother's direction and guidance, but noisily and good naturedly protests about the details. The interviewer introduces a minor disturbance in segment 1 and again in segment 4, which simply result in increased regulatory activity on the part of father and mother.

Beginning with segment 9 the interviewer introduces more disturbances which are not neutralized and which upset the equilibrium, as indicated in Figure 25-1.

One of the other scales we used was Lorr's Dimensions of Interaction in Group Therapy [5]. This was scored on seven factors which characterize various overt behavior characteristics of therapeutic

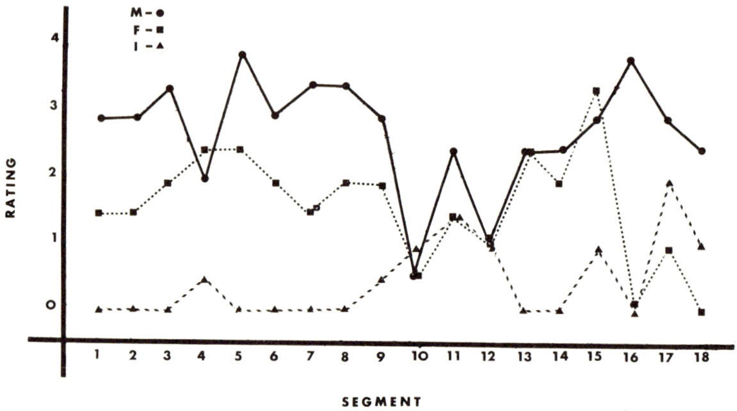

FIGURE 25-2. LORR'S GROUP SCALE: RATINGS OF ATTENTION-SEEKING CONTROL.

groups. Attention-seeking activities are shown in Figure 25-2, and this pattern is very similar to the Chapple Interaction Chronograph of Figure 25-1.

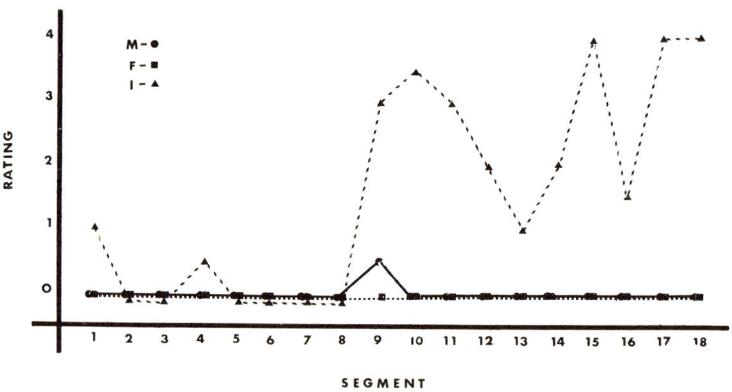

FIGURE 25-3. LORR'S GROUP SCALE: RATINGS OF THERAPIST ROLE.

Another Lorr rating is Therapist Role, which scores items such as interpretations, pointing out feelings, and restating feelings. This scale is shown in Figure 25-3, which indicates that much of the interviewer's activity reflected in Figures 25-1 and 25-2 falls within the category of shifting the level under consideration from facts to the significance of the facts.

The Riskin Family Interaction Scale [6] was another rating method we used. Among the items scored on this is Topic Change, shown in Figure 25-4. We see that a large percentage of the interviewer's ac-

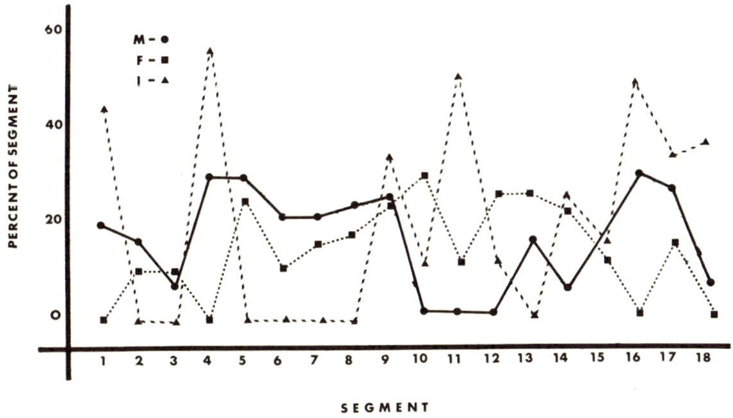

FIGURE 25-4. RISKIN FAMILY INTERACTION SCALE: PERCENT OF SPEECHES SCORED AS TOPIC CHANGES.

tivities are interventions, namely, that he tries to influence the course of the interviews by changing the subject to what he wants to talk about. In segments 15, 17, and 18 the topic changes fall off compared to Therapist Role Activity.

Both the Lorr and Riskin scales support our notion that the interviewer is a disturbing factor in the interview.

We will now attempt to indicate some evidence of reactions to disturbances. The Chapple Interaction Chronograph gives us the duration of silence when no one is speaking and the duration of interruptions when more than one person is talking. This is shown in Figure 25-5. Interruptions represent competition for dominance and an attempt to silence the person speaking, and these interruptions were mainly by mother and father. The peaks of interruptions in segment 6 seem to be related to disturbances in segment 4, and the peaks in 13 and 15 may be related to disturbances in segments 9, 10, and 11. The falling off in segments 16, 17, and 18 may indicate

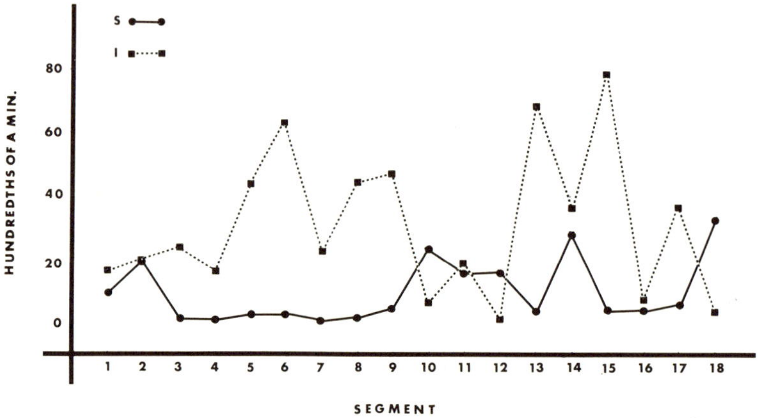

FIGURE 25-5. INTERACTION CHRONOGRAPH: DURATION OF INTERRUPTIONS AND SILENCES EXPRESSED IN HUNDREDTHS OF A MINUTE.

that the parents are no longer fighting off the interviewer but are now going along with him.

We think silence indicates tension, suspense, and uncertainty, and also may reflect attention, consideration, and deliberation which allow for integration and organization to take place. We suspect that the silence in segments 1 and 2 reflects the introductory uncertainty as to where the interview should go, while the later silences indicate openness and reduced oppositional activity.

As we look for change in the interview, in the sense of outcome

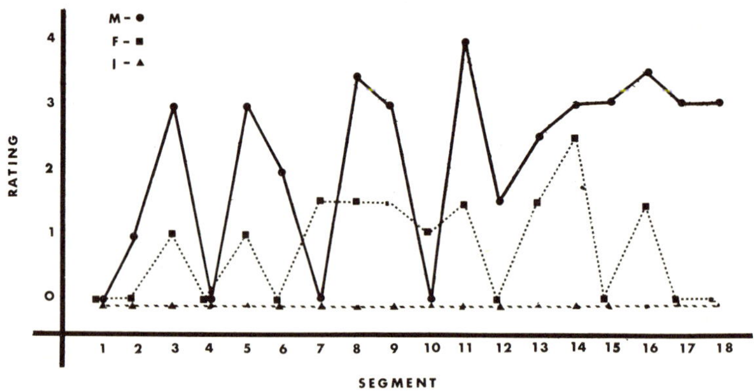

FIGURE 25-6. LORR'S GROUP SCALE: INTENSITY OF HOSTILITY.

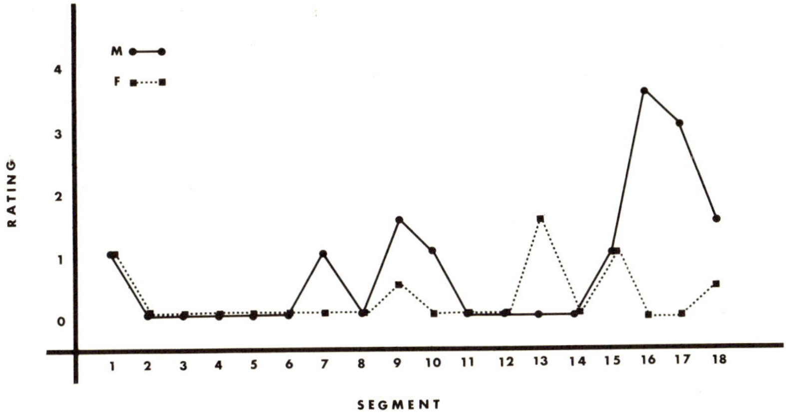

FIGURE 25-7. LORR'S GROUP SCALE: RATINGS OF SUCCORANCE NEED.

rather than variability, we note two other results from the Lorr scales, the Hostility Scale (Fig. 25-6) and Succorance Seeking (Fig. 25-7), both of which fluctuate then become sustained toward the latter part of the interview. We suspect that this reflects greater openness of expression toward the end of the interview.

Another scale we used was the Process Scale, developed by Carl Rogers [7] and his associates, to measure progress in individual psychotherapy. This is done in terms of the patient's *relationship* to his therapist, his *experiencing* of inner processes, the way he *construes* his significant experiences, and the way he *expresses* his problem.

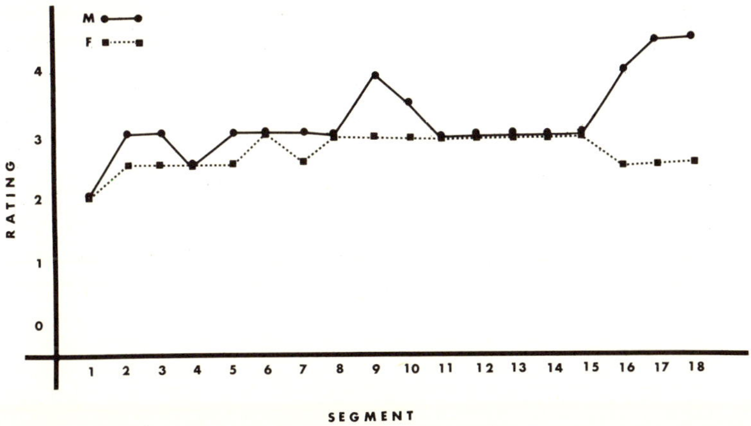

FIGURE 25-8. ROGER'S PROCESS SCALE: GLOBAL RATINGS OF PROGRESS IN THERAPY.

This scale taps a different time dimension and relates to the more durable items of a person's attitudes rather than to a person's behavior moment by moment. The results given in Figure 25-8 show little change for the father but quite a significant change for the mother.

From the systems viewpoint we can say that an initial steady state was maintained by reaction against introduced disturbances, and there was gradual reduction of the regulatory activity so that further disturbances brought about a change in the system.

Information, Relationship, and Decision

When we consider the various graphs and the variations in them that we have interpreted as disturbance, regulator activity, and change, and compare these with the protocol, we believe the interview content can be divided into information, relationship, and decision.

In the first half of the interview the exchange was largely at the information level, with the interviewer attempting to introduce relationship by asking about typical relationships from the details the parents were giving, but they only gave him more and more information. He again returns to the meaning of this information, and how the parents relate to each other and to their feelings about the interview.

The mother finally moves into her relationships with white school teachers and their relationships with her children, and then into the opinions of white sociologists, to infer the relationship between them and the interviewer. He indicates that this is not an impersonal conceptual matter for him but a matter of personal feelings, with the inference that he shares their feelings.

On the basis of this relationship, then, the parents can look with him at their attitudes and the ways in which their conflict about their attitudes is a burden to them and to their children. This constitutes a change in them and a decision in terms of their problem.

Conclusion

We believe that the concept of steady state and change offers a significant tool for the study of psychotherapy and a contribution toward a theory of therapeutic change. If further work verifies the usefulness of this concept as a divining rod to indicate the presence of change, the time of change, and the direction of change, then we can judge the effectiveness of therapy of various sorts and clarify the nature of the therapeutic process.

References

1. Ashby, W. R. The set theory of mechanism and homeostasis. *Gen. Syst.* 9:83, 1964.
2. Cannon, W. B. *The Wisdom of the Body.* New York: Norton, 1932.
3. Carroll, E. J. *A Study of Interpersonal Relations Within Families.* Pittsburgh: Craig House for Children, 1966.
4. Chapple, E. D. The interaction chronograph: Its evolution and present application. *Personnel* 25:295, 1949.
5. Lorr, M. Dimensions of interaction in group therapy. *Multiv. Behav. Res.* 1:67, 1966.
6. Riskin, J. Family interaction scales. *Arch. Gen. Psychiat.* 2:484, 1964.
7. Rogers, C. R. A Tentative Scale for the Measurement of Process in Psychotherapy. In E. A. Rubenstein and M. B. Parloff (Eds.), *Research in Psychotherapy.* Washington, D.C.: American Psychological Association, 1959. Pp. 96–107.

Index